ENVIRONMENTAL CHANGE AND HUMAN ADAPTATION IN THE ANCIENT AMERICAN SOUTHWEST

Environmental Change and Human Adaptation in the Ancient American Southwest

Edited by

David E. Doyel and Jeffrey S. Dean

Contributors

Julio L. Betancourt
Jeffrey S. Dean
Fred Espenak
Suzanne K. Fish
Donald A. Graybill
George J. Gumerman
Fred L. Nials
J. Jefferson Reid
Ann Clair Seiferle-Valencia
Alan P. Sullivan III
Darrell Creel
David E. Doyel
Paul R. Fish
Gary S. Funkhouser
David A. Gregory
W. Bruce Masse
James A. Parks
Anthony H. Ruter
Alan H. Simmons
Michael R. Waters

The University of Utah Press
Salt Lake City

The Defiance House Man colophon is a registered trademark of the University of Utah Press. It is based upon a four-foot-tall, Ancient Puebloan pictograph (late PIII) near Glen Canyon, Utah.

10 09 08 07 06 5 4 3 2 1

Index prepared by Andrew L. Christenson

LIBRARY OF CONGRESS CATALOGING-IN-PUBLICATION DATA

Environmental change and human adaptation in the ancient American Southwest / edited by David E. Doyel and Jeffrey S. Dean ; contributors, Julio L. Betancourt... [et al.].
p. cm.
Chiefly revisions of papers presented at a symposium held at the 1996 Society for American Archaeology Annual Meeting in New Orleans.
Includes bibliographical references and index.
ISBN-13: 978-0-87480-853-7 (cloth : alk. paper)
ISBN-10: 0-87480-853-7 (cloth : alk. paper)
1. Indians of North America—Southwest, New—Antiquities. 2. Paleoindians—Southwest, New. 3. Human ecology—Southwest, New—History. 4. Paleoecology—Southwest, New. 5. Southwest, New—Antiquities. I. Doyel, David E. (David Elmond), 1946– II. Dean, Jeffrey S., 1939– III. Betancourt, Julio L. IV. Society for American Archaeology. Meeting (61st : 1996 : New Orleans, La.)
E78.S7E68 2006
979'.01—dc22 2006007672

This volume is dedicated to
our colleague and friend
Donald Allan Graybill (1942–1993)

Contents

Figures

Tables

Preface

THIS BOOK IS DEDICATED TO OUR COLLEAGUE AND FRIEND DONALD Alan Graybill (1942–1993). His love of archaeology led him to field work in Africa, Israel, and the United States. For his Ph.D. research at the University of Arizona (1973), he conducted an archaeological survey in the Mimbres area in southwestern New Mexico. After several years at the University of Georgia in Athens, Don, wife Kate, and baby daughter Jessica Kathryn returned to Tucson, where he was employed at the Laboratory of Tree-Ring Research until his death from cancer at age fifty-one. For two decades, he was fully engaged in the intellectual foment and synergy permeating the Tree-Ring Lab, the Department of Anthropology, and the Arizona State Museum, where many of the contributors to this book were faculty or staff (see Reid and Dean 1996 for biographical information).

The idea for the present volume came to Doyel in 1994, and shortly thereafter he cajoled Jeffrey Dean, a colleague of Graybill's at the Tree-Ring Lab, to add his unparalleled experience and expertise in Southwest archaeology and paleoenvironmental research to the project. A symposium that was held at the 1996 Society for American Archaeology annual meeting in New Orleans, "Culture and Environment in the Ancient Southwest," became the basis for this book. The chapters by Creel, Doyel, Graybill, Reid, Simmons, and Gumerman are revised versions of work presented there, and the other chapters were invited subsequently.

Although Don devoted much of his academic life to dendroclimatology, his Ph.D. in anthropology and his archaeological experiences gave him an active and astute interest in human-environment interactions. At the Tree-Ring Lab he was able to combine his interests in archaeology, paleoenvironment, and computer applications. His skills included the ability to develop regional models of complex processes, an example of which is his reconstruction of prehistoric streamflow for the major rivers in central and southern Arizona (Graybill et al., this volume), a contribution that revolutionized research in the area. The integration of environmental and human behavioral components endowed his approach to the study of the past with unusually high levels of scientific commitment, methodological rigor, and standards for evaluating evidence. This outlook was reflected in his dismay at the scientific naiveté of many archaeologists and his frequent assertion that if archaeology could not be done within a strict scientific framework, it was not worth doing.

Overwhelming demands and responsibilities, a changing list of contributors, and other factors led to delays in completing this book. Little did we realize that our communal effort for a fallen colleague would take a decade to accomplish. Don would have found humor in this situation, as meeting deadlines was never top priority for him.

We sincerely acknowledge the unfailing support of Jeff Grathwohl and the University of Utah Press during this lengthy process. We thank those authors who waited patiently for other colleagues to complete their contributions. We acknowledge and thank two anonymous reviewers for providing constructive suggestions to improve the final volume. Owing to the long preparation process, the authors had an opportunity to update their chapters after the external reviews were completed. We thank Jessica Graybill, University of Washington, Seattle, for finding and permitting the use of the photograph of Don Graybill shown on the dedication page. It was taken in August 1988 in a forest somewhere in Siberia.

This book offers a sample of institutional contexts and approaches focusing on the study of cultural and environmental interaction in the ancient southwestern United States. It also underscores the diversity of archaeological research in which Graybill was involved, research for which he often contributed the analytical expertise and computer methods to realize the outcome of a particular project or line of thinking. Although wide-ranging and dissimilar in some ways, the chapters share the rigorous, multidisciplinary spirit in which Graybill approached research. The volume contains many of the elements that constituted his career—archaeology, cultural and human behavior, paleoenvironmental reconstruction, refined chronological control, and spatial variability—topics that define much of the research being conducted in the Southwest today. We also like to think the book contains a level of scientific rigor that might have impressed even Don Graybill.

David E. Doyel
Jeffrey S. Dean

1

Culture, Environment, and Adaptation: Perspectives from the Ancient Southwest

Jeffrey S. Dean and David E. Doyel

SCHOLARLY OPINIONS ON THE RELATIONSHIPS BETWEEN HUMAN behavior and the natural world have run the gamut from doctrinaire environmental determinism to nearly total cultural control of human-environment dynamics (for discussion, see Diamond 1997; Huntington 1914; Redman 1999; Weiss and Bradley 2001). The truth undoubtedly lies somewhere between these extremes: the degree to which one component dominates the other is determined by mutable situational factors unique to individual contexts. Therefore, both components must be understood as completely as possible to investigate and explain any particular human adaptation and to comprehend human-environment interaction in general. Just how much influence environmental factors actually have on culture, and vice versa, must be ascertained empirically. Case studies and the growth of knowledge regarding specific historical instances and patterns illuminate the general processes that drive human behavioral responses to environmental stability, variation, and change.

The approach adopted in this book to the complex dynamics between human behavior on the one hand and the natural environment on the other is articulated in detail elsewhere (Dean 1988b, 1996a, 2000; Dean et al. 1985) and is briefly summarized here. The conceptual model used is a product of a twenty-year effort by an interdisciplinary team working on prehistoric cultural and environmental interaction among the Kayenta Anasazi in northern Arizona.

Human behavioral adaptation to environmental variability and change is governed by many factors, including interactions among environmental, demographic, and cultural variables, each of which alters and is in turn altered by the others in specific situations. Important among these factors are the attributes and temporal and spatial structure of environmental variability and the cultural mechanisms by which human societies adapt to variations and change in their physical and social environments. Because these natural and cultural processes are structured by long-term variation in the relationships among the natural, demographic, and cultural factors involved, they must be studied over long spans of time. Archaeology, therefore, is especially suited for investigating human-environment interaction as long as the relevant variables are adequately controlled.

As is frequently pointed out, the American Southwest has proved to be an exceptional natural laboratory for investigating interrelationships among the environment, human demography, and culture. The archaeological record in this region is uncommonly vast and detailed and spans thousands of years. Profuse historical and ethnographic data on modern Native American descendants of the prehistoric populations are a rich source of testable hypotheses about past human behavior in general and sociocultural mechanisms for adapting to environmental fluctuations in particular. Furthermore, a variety of paleoenvironmental analyses provides detailed reconstructions of many environmental variables across a wide spectrum of temporal and spatial scales. The quantity and quality of these data provide numerous opportunities to explore human behavioral adaptation to and impact on the physical environment (Redman 1999; Redman et al. 2004).

ENVIRONMENTAL VARIABILITY

The natural environments of the Southwest encompass a wide array of elements, ranging from relatively "inert" components (rocks, soils, topography, etc.) that require only one basic adaptation over long intervals to much more "active" factors (alluvial geomorphology, climate, flora, fauna, etc.) that require continual or periodic behavioral adjustments from humans. Furthermore, individual environmental factors vary from millimeters to continents and from minutes to millennia. Because of these wide spatial and temporal spreads, investigation of human-environment interactions must concentrate on adaptively significant units of time and space. For example, daily variations in climate and annual fluctuations in alluvial groundwater levels likely had little or no effect on the occupants of the region. Temporal variability ranging from seasons to millennia and spatial variability ranging from localities to the region, however, probably had important adaptive consequences.

Because environmental conditions vary in gradational fashion through space and time, the concepts of environmental stability, variability, and change must be precisely defined and consistently applied (Dean 2000). *Stability* denotes a lack of change in environmental factors over time and space ranges defined by the geographic and temporal boundaries of the study involved. *Variability* refers to quantitative variations around temporarily "stable" temporal and spatial means and types. *Change* involves "permanent" transformations in the natural processes that regulate environmental variability. Temporal change creates qualitatively different environmental regimes characterized by new long-term means and variances, while spatial change involves the transition from one defined environmental type to another.

Environmental attributes generally can be grouped into amplitude, temporal, and spatial classes. *Amplitude* denotes the magnitude of variation and includes aggregated measures of amplitude such as means, medians, modes, ranges, variances, and extremes. *Temporal variability* is characterized by a range of time frequency cycles, patterning in time and space, direction and duration of trends, recurrence intervals, and other parameters. *Spatial variability* is characterized by synchronic measures of magnitude, central tendency, variance, and extremes. Human actions commonly impinge on many of these attributes enough to modify the environment.

Scientific investigation of human relationships with the natural environment depends on our ability to discern the amplitude, spatial, and temporal attributes of environmental stability, variation, and change. Few documentary environmental records extend more than a century into the past. Beyond that point, a number of paleoenvironmental reconstruction techniques provide information on past environmental conditions. Each of these methods records different ranges of temporal and spatial variation in a limited number of environmental factors.

Given the problems inherent in relating human behavior to the broad and diverse range of environmental variation, it has proved useful to partition the temporal and spatial aspects of the latter into specific categories. Based on the periodic structure of the environmental elements involved and on the time resolution of relevant paleoenvironmental techniques, temporal variability has been divided into four classes (Dean 1988b, 1996a, 2000; Dean et al. 1985). *Stable factors* do not vary appreciably over the time period of the study, although they can change over longer periods. Some stable factors of the last 5,000 years in the Southwest are bedrock geology, large-scale topography, climate type, elevational zonation and geographic distribution of plant communities, and the division of the region into Basin and Range, Mogollon Highland, Colorado Plateau, and Rio Grande Rift physiographic provinces. If the study includes Paleoindian groups, some of the above factors, such as climate type and the distribution of vegetation zones, cannot be treated as stable. Because stable elements do not vary during the study period, modern conditions accurately reflect past conditions, and they do not have to be reconstructed. Environmental factors that vary during the study interval are divided into two types based on the length of one human generation, 25 years. Factors whose variation is regulated by natural processes with periodicities equal to or greater than 25 years are called *low-frequency process (LFP)* variables. LFP variation characterizes factors such as the rise and fall of alluvial groundwater tables, the deposition and erosion of floodplain sediments, changes in the elevational distribution and composition of vegetation communities, and some aspects of climate. *High-frequency process (HFP)* elements are controlled by natural processes that fluctuate with periods less than 25 years. Most observable HFP variability involves seasonal to multiannual fluctuations in climate. Because present conditions do not reliably represent past conditions, both LFP and HFP variations must be reconstructed by various paleoenvironmental techniques, the former by methods such as alluvial geomorphology, palynology, packrat midden analysis, and archaeobotany and archaeozoology and the latter by dendroclimatology and certain high-resolution palynological studies. *Episodic factors* have no underlying periodicities, or their cyclic structure is unknown, and they are manifested as more-or-less random, unpredictable events. Among the myriad episodic possibilities are volcanic eruptions, floods, freezes, hailstorms, insect infestations, and plant disease outbreaks.

Far less archaeological attention has been given to describing spatial environmental variation than has been devoted to characterizing temporal variability. Spatial patterning can be homogeneous or heterogeneous and exhibits many configurations within these broad categories, including mosaic, gradational, coarse-grained, fine-grained, and even (Plog et al. 1978). Geographic variation is high when change occurs across relatively short distances and low when uniformity prevails over long distances.

CULTURAL AND ENVIRONMENTAL INTERACTION

Human behavioral adaptation to environmental stability, variation, and change involves the interaction of environmental, demographic, and behavioral variables (Dean 1988b, 1996a, 2000; Dean et al. 1985; Kirch 1980) that are conjoined in the concept of *carrying capacity* (Hassan 1978), that is, the number of humans a particular subsistence system can sustain in a particular environment. Ordinarily, an oscillating but effective balance among the factors is maintained. When the limits of such situations are exceeded, adaptive responses are required to alleviate the resulting stresses. Stress most commonly develops when population growth exceeds an environmental threshold or when environmental deterioration lowers the threshold below the population level (Dean 1996a:Figure 1). In such situations, environmental factors usually behave independently, and population and behavior usually function as dependent variables.

Successful demographic responses to stress involve population reductions to maintain the number of people below the carrying capacity. Initial behavioral responses to systemic stress are likely to involve technological change and subsistence system adjustments that include using different tools, superior crop varieties, improved food processing and storage, more appropriate field locations, better production techniques, and more productive labor practices (intensification). Social changes in the procurement, accumulation, storage, and redistribution of food and the allotment of access to critical resources such as farmland, water, wild plants, game, and raw materials also may be entrained to alleviate systemic stress. Finally, intensified interaction with other groups through trade, tribute, raiding, and warfare is a common response to stress (Doyel 1991b; Haas and Creamer 1993; LeBlanc 1999). These potential responses comprise a stockpile of behaviors that can be mobilized as conditions require. Changing emphases on successful responses extracted from this behavioral reserve allow societies to accommodate "normal" variability and adapt to permanent systemic changes that create qualitatively different adaptive configurations (Dean 1996a, 2000).

CASE STUDIES

The case studies presented here reveal the dynamic nature of contemporary research regarding cultural and environment interaction in the American Southwest. The chapters integrate well-defined chronology, time depth, data quality, and refined methods to record, measure, and evaluate this interaction, addressing 4,000 years of human history beginning in the preceramic period and continuing into the historic period and using study areas mostly in Arizona and New Mexico (Figure 1.1). The presentation sequence is structured in part by time period and in part by geography.

The contributors were asked to consider the general model (known as the Anasazi Behavioral Adaptation Model, or ABAM) relative to their experiences from their research. Elements of the cultural-environmental-ecological approach suggested in the model, including low-frequency processes, high-frequency processes, scale, and amplitude, are illuminated by specific examples and discussions. The contributors bring a variety of backgrounds and approaches to the topic, some critical.

In Chapter 2, Alan Simmons reviews the origins of agriculture in the Southwest, concluding that cultigens were introduced into the region by 2000 B.C. He suggests that the introduction of cultigens across the region had differential impacts on local adaptive strategies. Agriculture was adopted rapidly in the southern Southwest and more slowly in the north, but recent research points to a concordant increase in complexity. He also notes the unlikelihood that the high-frequency resolution and precision of paleoenvironmental data available for later periods will emerge for the Archaic period in the near future.

In Chapter 3, Michael Waters reveals his interest in how landscapes affected cultural behavior and how people affected their landscapes. To fully understand culture change, he believes, we must reconstruct the environment and landscapes within which people lived, including the biological environment (floral and faunal communities), the physical setting (landscape), and the prevailing climate at the time of occupation. Because biological forms, landscapes, and climate change through time, the reconstructions should show how environments changed through time. When coupled with archaeological

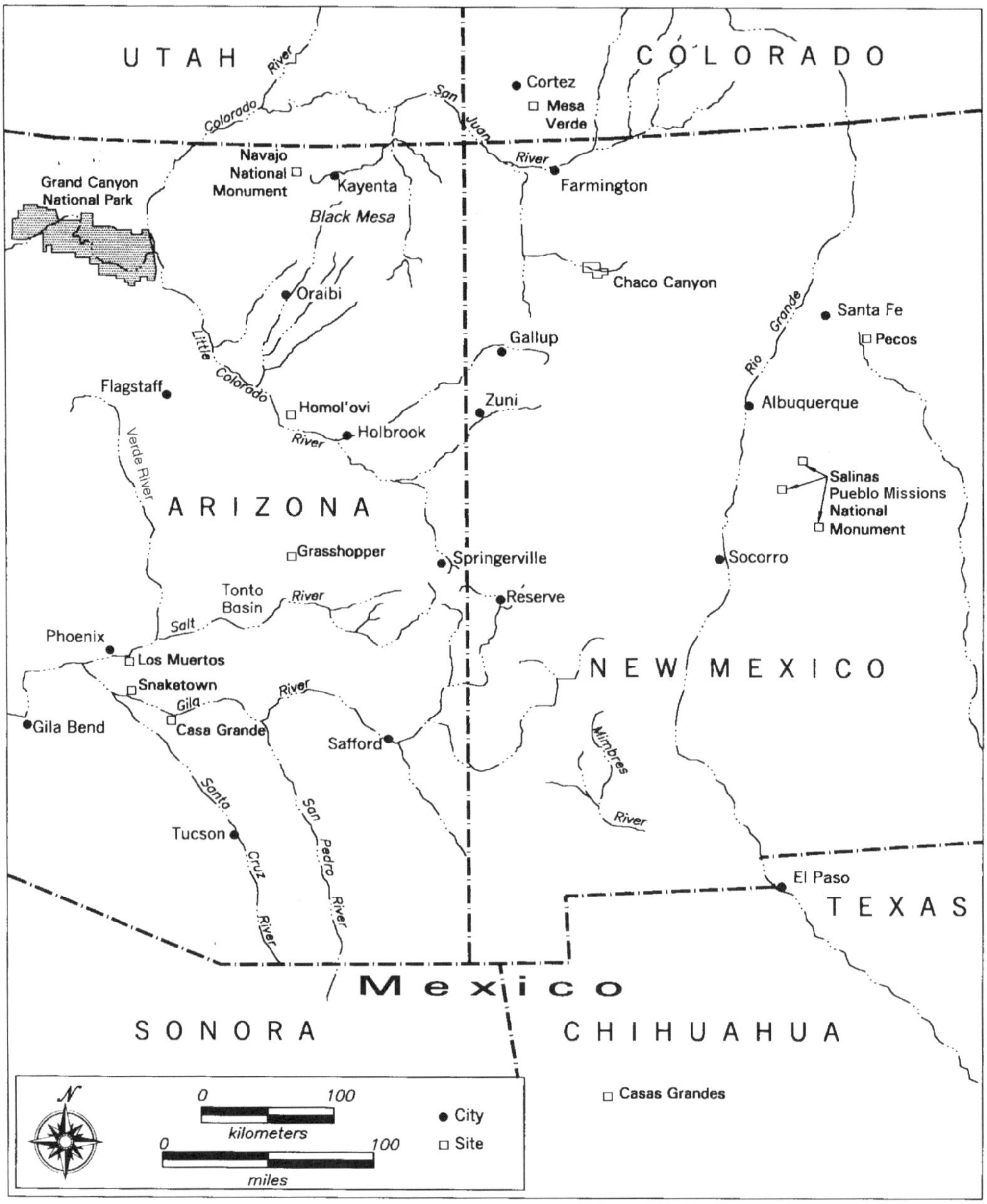

FIGURE 1.1. The Southwest: locations of study areas in this volume.

data, environmental reconstructions provide a powerful explanatory tool to address questions regarding why people chose certain areas for occupation, why some localities were abandoned and others remained occupied, why certain adaptive strategies were chosen, and why they changed. To this end Waters presents examples of high- and low-frequency processes operative in the Sonoran Desert of central and southern Arizona.

In Chapter 4, Suzanne Fish and Paul Fish confront a significant problem facing other researchers in the Southwest. Prehistoric cultivators occupied multiple areas that were not inhabited historically, such as the desert areas of southeastern Arizona and southwestern New Mexico and the forested regions at Point of Pines to the north. Ethnographic accounts of indigenous cultivation are lacking for such areas, and records of cultivators located elsewhere do not adequately describe the residential or land-use patterns that would account for the ancient remains. A similar situation exists for the Lower Verde Valley of central Arizona, where the development of appropriate models of agriculture for the past is problematic. An alternative and innovative approach to understanding indigenous cultivation is to consult traditional native farmers, whose experience has contributed to an understanding of the agricultural potential and specific adaptive practices of many areas. The evaluations of the study area by Zuni, Apache, Pima, and Tohono O'odham consultants are compared and found to converge in instances, and the consultants' reactions bring into focus the challenges that would have faced new groups entering the Lower Verde in the past.

Chapter 5 presents a multidisciplinary approach to a significant research issue in the Southwest, the behavior of the rivers and streams that past societies depended on for their livelihoods. Variation in streamflow and its effects on river channel morphology are significant environmental variables faced by irrigation farmers. At the time of his death, Donald Graybill was developing a streamflow reconstruction for the Gila River to complement an existing model he was instrumental in developing for the Salt River (Graybill 1989; see Gregory 1991). In Chapter 5, Graybill's colleagues David Gregory, Gary Funkhouser, and Fred Nials continue this work by presenting long-term reconstructions of Salt River and Gila River streamflow. They review the effects of variation in discharge on channel morphology and explore the implications of both high- and low-frequency variation for aboriginal irrigation agriculture along these rivers.

In Chapter 6, J. Jefferson Reid, Donald Graybill, and Ann Seiferle-Valencia present a tree-ring-based climatic reconstruction for the Grasshopper area of east-central Arizona to provide the context for explaining rapid cultural change in the fourteenth century. Canyon Creek Pueblo, a cliff dwelling partially excavated by Emil Haury, is used to identify a subsistence-settlement model based in traditional techniques of agricultural management under conditions of recurrent climate fluctuations. The success of agriculture in contrast to wild plant and game management provides insights into the rapid development and long-term stability of Pueblo village farmers as well as a more balanced understanding of human ecological relations in prehistory.

In Chapter 7, David Doyel summarizes the limited data on prehistoric agricultural diversity, settlement patterns, and population for an understudied portion of the Upper Little Colorado River Valley in east-central Arizona. A series of small sites in a small, nonriverine valley are used to model former land-use practices, resulting in a description of a specific type of agriculture, sand-dune farming, that was used extensively for a portion of the cultural sequence. The results of this analysis are placed in a broader regional perspective in further discussion on potentially significant cultural variation and possible influences of climate on agricultural practices and settlement patterns in the region.

In Chapter 8, Jeffrey Dean focuses on food storage as one component of agricultural subsistence systems in the northern Southwest. Human behavioral adaptations to subsistence stress are acknowledged as major stimuli of sociocultural change and as important factors in the evolution of cultural systems. Storage should vary directly with the degree of variability in the productive system as a viable means of compensating for shortfalls resulting from reduced yields of leaner years. Such studies must characterize variability in the independent and dependent factors involved, including crop yields and storage behavior, and establish a plausible relationship between the two variables. The thirteenth-century site of Kiet Siel in northeastern Arizona is used for a locality-level examination of storage responses to subsistence stress. The excellent preservation of this cliff dwelling provides unusually detailed data on room area and volume, as well as detailed architectural data and large numbers of tree-ring dates that document development of the village through time. These data permit both the documentation of changes in storage behavior and control of population variability through time.

In Chapter 9, Alan Sullivan and Anthony Ruter study the extent to which prehistoric societies in the Upper Basin of northern Arizona, near Grand Canyon, were agriculturally dependent. They explore paleoeconomic data from archaeological sites to evaluate how variation in food-provisioning systems may have been affected by environmental fluctuations. They also summarize past modes of inquiry and standards of evidence used to support inferences about ancient Grand Canyon livelihood. Modern weather data and tree-ring-based paleoclimate data provide insights into planning and scheduling problems posed by the variable temperature and precipitation patterns. The authors suggest that economies in the Upper Basin focused on use of native plant resources and not domesticated plants. In their evaluation of the predictions and assumptions of ABAM developed by Dean, Euler, and colleagues, they conclude that because of the economic focus on wild plants, prehistoric livelihoods in the Upper Basin were generally immune from environmental fluctuations that affected agricultural economies between A.D. 1050 and 1200.

In Chapter 10, Darrell Creel addresses how environmental variability may have imposed constraints on prehistoric adaptation in the Mimbres River valley in southwestern New Mexico. Tree-ring records document precipitation patterns (high-frequency processes), but records of other events, such as stream entrenchment and alluvial deposition (low-frequency processes), are lacking. Divergent views exist regarding the nature of events occurring after the A.D. 1130s, when drought conditions seem to have lowered carrying capacity, but there is agreement that changes in settlement patterns, architecture, and ceramics occurred. Yet the Mimbres valley seems to have been as good a place as any during drought conditions. Creel identifies precipitation and streamflow characteristics that may have affected Mimbres populations in the A.D. 1130s drought, and he addresses cultural changes in the twelfth century.

In Chapter 11, James Parks, Jeffrey Dean, and Julio Betancourt provide a tree-ring chronology that reconstructs precipitation for Socorro, New Mexico, back to 1598. This Sevilleta chronology bridges a gap between chronologies from modern pinyon in Tajique Canyon and archaeological wood from the Chupadera Mesa area. The revised Chupadera Mesa chronology reconstructs the annual precipitation of Mountainair, N.M., back to 1294. These reconstructions are used to study the pueblo abandonment of south-central New Mexico in the 1670s. Quantification supports documentary evidence of drought but reveals it to have been insufficient alone to cause abandonment. Political, social, and cultural instability wrought by the Spanish colonists forced the collapse of pueblo societies that otherwise would likely have survived the drought.

In Chapter 12, Bruce Masse and Fred Espenak address an important factor that figured prominently among traditional cultures, and one too often overlooked by archaeologists, including those in Southwest: namely, the sky as a part of the effective environment. The sky is as much an element of our environment as topography, climate, vegetation, and water availability, but models of the physical environment have ignored this critical variable in the development of human cognition and the evolution of complex societies. Masse and Espenak integrate archaeological and paleoenvironmental records with oral history and ethnographic data to focus on the role of episodic or transient celestial events, in this case solar eclipses, among the Hohokam of southern Arizona. Some fascinating coincidences, if not correlations, are revealed in cultural changes in the region, including settlement patterns. The authors demonstrate that the Hohokam, and by extension their neighbors, responded to a variety of celestial phenomena and episodic events, and they urge that greater efforts be made toward understanding the role(s) of the celestial realm in prehistory.

In the final chapter, George Gumerman, a contributor to the model explored in this volume and editor of *The Anasazi in a Changing Environment* (1988), connects the past with the future in terms of cultural ecological studies in the American Southwest. Changing theoretical perspectives have produced numerous "ecologies"—for example, historical, political, and symbolic—to frame research problems, depending on the nature of the data and the questions being addressed. Gumerman notes that many of the studies in this book represent refinements of traditional cultural ecological perspectives but also reflect the dynamic nature of the new ecologies and are more informed than earlier work about the complex dynamics of nonlinear interactions. Speaking to the future of ecological studies, he suggests that one productive direction is computer modeling, using large databases and rigorous quantification to generate a better understanding of the past.

2

Early People, Early Maize, and Late Archaic Ecology in the Southwest

Alan H. Simmons

OUR UNDERSTANDING OF THE ORIGINS OF AGRICULTURAL ECONOMIES in the Late Archaic in the Southwest has benefited substantially from new research. Chronological and paleoenvironmental precision, however, still suffer. It is unlikely that the high-frequency resolution enjoyed in the study of later, Anasazi, remains will emerge for the Archaic in the near future. Despite such problems, a complex picture is emerging indicating that the impact of cultigen introduction exhibits considerable regional diversity. In the northern Southwest, agriculture appears to have been adopted relatively slowly, whereas in the southern Southwest, research points to more rapid reliance on cultigens, with an increase in site complexity. Furthermore, it is now apparent that cultigens were introduced into the Southwest by or before 2000 B.C. This chapter summarizes these developments and places them within a broader context for interpreting the first farmers of the Southwest.

Don Graybill is best remembered professionally for his paleoenvironmental research, much of which focused on Formative southwestern cultures. He did not devote a great deal of time to earlier periods, such as the Archaic. This did not mean he was unconcerned with these earlier groups—indeed, quite the contrary. In particular, agricultural origins (both in North America and in the Near East) were of special interest to Don. He realized, however, that for the Southwest, far better environmental data were available for the later periods. Nevertheless, Don often displayed his displeasure with the lack of an explicitly scientific and ecological approach to much of the archaeology of the Southwest. Here I examine some contemporary views of the Late Archaic and the introduction of domesticates. I have attempted to structure this discussion from a Graybillian perspective, which is no easy task.

By way of explanation, I should explain my involvement in this issue. I no longer am active in field research in the Southwest, focusing instead on early food-producing societies in the Near East. Some years ago, however, I worked extensively with southwestern Archaic materials and had the good fortune of directing excavations at Late Archaic sites that contained evidence for some of the earliest use of cultigens in North America (Simmons 1982, 1986). Although I have switched my research orientation to another part of the world, I continue to be interested in developments on early agriculturalists in the Southwest. My goal here is not to provide a comprehensive summary or site-specific commentary, which I leave to far better qualified scholars actively working in the area. Rather, this contribution is a somewhat personal assessment of this intriguing issue by someone with a long-standing interest in it.

CONTEMPORARY RESEARCH CONTEXT

Much of Don Graybill's research was framed within an explicit and precise reconstruction of environmental conditions that affected human adaptations. The unifying theme of this volume follows the model of cultural-environmental interaction elegantly developed by Dean and colleagues (Dean 1988a, b; Dean et al. 1985). By using dendrochronological data with high-frequency resolution (literally to the year) in conjunction with lower-frequency paleoenvironmental techniques such as palynology and alluvial geomorphology, Dean detailed paleoclimatic variability on the Colorado Plateaus. This work provided a foundation for the use of paleoenvironmental reconstructions to evaluate Anasazi cultural responses to environmental stability and change (Dean 1988a:166). Dean's research was, of course, the type of study that Graybill embraced.

What needs to be made explicit for the following discussion, however, is that we are nowhere near this type of resolution with the Archaic in the Southwest. Most researchers working with Archaic materials are delighted if they can discuss chronological periods within several hundreds, if not thousands, of years. With dendrochronological data, a compromise often is to *decrease* resolution to periods of 10- or 20-year intervals (Dean 1988a:135). So, at the outset, it is apparent that for the Archaic we will never, at least not in the near future, obtain the high-frequency chronological or environmental resolution evident during the Formative period. Much more realistic is achieving what Binford (1980) has referred to as a "coarse-grained" resolution. With this caveat, what is the current status of research on the Late Archaic and its relationship to the introduction of cultigens in the Southwest? In short, discoveries at numerous sites over the last two decades have revolutionized our understanding of this process.

In the dark days of Archaic research, there was little innovative work. The focus was primarily on establishing "base-line data" (see Judge 1982; Simmons 1981, 1989), which was necessary, and on the detailed description of chipped stone tools, which may have been less necessary. The Archaic was considered a sort of poor stepchild, sandwiched between its glamorous siblings, the Paleoindian and Formative periods. There seemed little to recommend systematic study of the generally unimpressive and small Archaic lithic scatters.

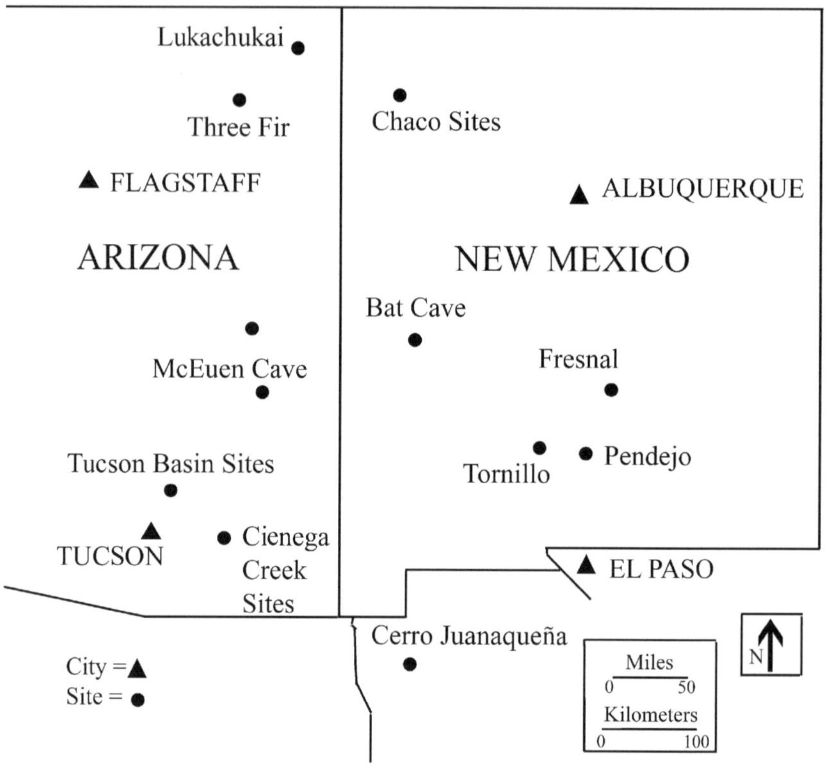

FIGURE 2.1. The Southwest: locations of sites referred to in the text. Map by Ramzy Ladah.

Fortunately, this changed with the realization that such sites can, in fact, contain important information (see Sullivan 1998). This is particularly true with regard to the Late Archaic and the introduction of domesticates.

There is, at the beginning of a new millennium, reason to be optimistic about Late Archaic studies. Can the same be said for research on earlier Archaic phases? Despite distinct improvements (e.g., Vierra 1994), the situation here is less than satisfactory. An overriding emphasis on artifact description remains a dominant theme. We could take lessons from our colleagues in the Great Basin, who are doing innovative research on similar sites. Their emphasis on landscape archaeology and geomorphic variability of site locations offers considerable promise (e.g., Beck and Jones 1992; Madsen and O'Connell 1982).

But the focus here is on the introduction of cultigens into the Southwest in the Late Archaic. Recent research has provided exciting discussions relating to this issue, based on both old data and new excavations (Figure 2.1) (e.g., Berry and Berry 1986; Berry 1982; Fish and Fish 1994; P. Fish et al. 1986; S. Fish et al. 1992a, 1992b; Ford 1984; Hard and Roney 1998; Huckell 1990, 1995; Mabry 2000b; Mabry, ed. 1997a, b; Mabry et al. 1997; Matson 1991; Minnis 1985a, 1992; Roth 1988, 1989, 1992, 1993; Simmons 1986, 1987; Upham et al. 1987, 1988; Wills 1985, 1988a, 1988b, 1992, 1995). Great progress has been made, although it would be a mistake to conclude that we have all the answers. Indeed, the situation is far more complex than initially imagined; this scenario would have delighted Don Graybill.

The study of agricultural origins has tended, until recently, to be an odd sort of competition to find sites containing the "oldest" cultigens. Fortunately, we are now beyond the search for the "first" cultigen, an endeavor that Flannery (1973:308) realized was hopeless many years ago. The pioneering research of Cynthia Irwin-Williams (1967, 1973, 1979) in the Arroyo Cuervo of New Mexico interjected a degree of theoretical discussion into the issue. She suggested that cultigens were introduced into the Southwest from Mexico by about 4000 B.P. This was an early date, but it was later than the previously accepted dates of up to around 6000 B.P. for Bat Cave (Dick 1965). Thus Irwin-Williams's dates seemed rather conservative at the time. By the 1970s, however, it was apparent that the Bat Cave sequence was grievously flawed (Wills 1985; Wills et al. 1984) and that the oldest maize from the site was actually around 3500–4000 B.P. (Wills 1999). Indeed, during the 1980s, a conservative trend hit the study of southwestern agricultural origins. Some researchers (e.g., Berry and Berry 1986; Berry 1982) convincingly argued that there were few defensible cases of early agriculture and that most claims were based on poor data sets and unverified assumptions. This argument was also applied to the poorly published Arroyo Cuervo information and garnered additional support from the reexcavation of Bat Cave. Some of the neoconservatives believed that cultigens were relative late-comers to the Southwest, perhaps no older than about 2500 B.P. (e.g., Berry 1982).

EARLY CHACO CANYON EVIDENCE

This situation was challenged by research near Chaco Canyon, where I was fortunate to direct excavations suggesting that cultigens had indeed been in the northern Southwest since nearly 4000 B.P. (Simmons 1982, 1984, 1986). Evidence of maize was recovered at three Late Archaic open-air sites: LA 17337, LA 18103, and LA 18091. Eight radiocarbon determinations ranged from 3985 to 2675 B.P. (Simmons 1986:79). The occurrence of maize was unexpected and highly unusual, because of both the early dates and the fact that these sites did not contain the accoutrements one normally associated with agricultural communities. Instead, they were similar to "typical" Late Archaic camp sites, with the addition of some cultigens.

Our data were not universally accepted. Part of the difficulty was that the two earlier sites, LA 17337 and LA 18103, dating to around 3600–4000 B.P., contained only maize pollen, which could not be directly dated and raised the issue of contamination. Furthermore, there were questions regarding the radiocarbon dates, since one from LA 17337 was considerably later (225 B.P.) than the others (Simmons 1986:79–80). Subsequent research at both LA 18091 and a rockshelter, Sheep Camp Shelter, however, recovered macrobotanical remains that could be directly dated. At LA 18091, carbonized maize was dated to 2720 B.P.; at Sheep Camp shelter, two determinations of 2290 and 2170 B.P. were obtained on maize, and two squash samples yielded determinations of 2900 and 2220 B.P. (Simmons 1984, 1986:79, 1987). This provided unequivocal evidence for the use of two of the major southwestern cultigens throughout the first millennium B.C. In our view, it also bolstered the argument for the earlier dates at LA 17337 and LA 18103, even though no macrobotanical remains were recovered. These data indicated a pattern of limited cultigen use in the Chaco region of some considerable antiquity.

Satisfied that the association of cultigens with the sites and the dates were indeed valid, we proposed a scenario in which domesticates represented an additional, and initially supplemental, resource to help Late Archaic populations through sparse seasons. We did not feel that cultigens had a significant impact for some time, and we believed that their initial introduction had limited effect on Late Archaic mobile lifestyles. There was no evidence to support any large degree of sedentism, even though cultigens were a component of the economy. This interpretation is in accord with other perspectives that favor gradual incorporation of cultigens in the Southwest (e.g., Ford 1984; Haury 1962; Irwin-Williams 1973; Minnis 1985a, 1992:121; Nelson 1990). Such arguments often are made within the framework of an evolutionary selectionist model proposing that "hunter-gatherers adopting agricultural strategies were concerned with the success of hunting and gathering strategies, not with developing a dependence on cultivars. In this view, domesticates were first adopted not for food production itself...but rather for their contribution to the success of existing foraging strategies. Stated more simply, foragers adopt domesticated plants not to become farmers but to remain effective foragers" (Wills 1988b:36).

Other researchers, however, preferred a rapid adoption of agriculture, often invoking population migrations from Mesoamerica (e.g., Berry 1982; Matson 1991, 1999) as opposed to indigenous hunter-gatherer adoption of agriculture (see Whittlesey and Ciolek-Torrello 1996; Wills 1988a, 1988b). Refining such arguments, Berry (1982:2–33) has suggested a "punctuated equilibrium" model. Regardless of which perspective one takes, however, it is important to realize that although the spread of cultigens to the Southwest may have been relatively rapid, based on available radiocarbon determinations (see Smiley 1994), its impact was quite variable and region-specific.

By the 1990s, the situation was far more complex than previously anticipated. New sites contributed to a growing database, and bipolar models existed for what might be termed, in the context of this volume, a "high-frequency" introduction of cultigens (as in the rapid introduction by outsiders) or a "low-frequency" incorporation into existing Archaic subsistence strategies. Biological/climatic scenarios independent of indigenous or migration models (Matson 1991) also were proposed.

By and large, most of the pertinent data for these models came from small, limited-activity sites. This has changed dramatically with new research from southern Arizona and northern Mexico (see below), but it is important to recognize the nature of much of the preexisting database. Though we now recognize that ephemeral sites often contain more information than previously believed (see Simmons 1998), they still have severe limitations. One simply cannot obtain the most robust data from small sites. Thus the Chaco localities initially were met with skepticism, and the early dates are by no means universally accepted (e.g., Berry 1982:27; Matson 1991:252–258). Indeed, getting our data published outside the gray literature of cultural resource management (CRM) was not an easy task. Even today, I admit to some surprise when I see the Chaco data ignored in many discussions of early agriculture in the Southwest. Criticisms of the Chaco sites have not seriously taken into consideration the ephemeral nature of nonsedentary archaeological adaptations. Even though Don Graybill was not involved directly, he had some colorful descriptions regarding the conservative nature of archaeologists and their refusal, or at least reluctance, to accept evidence that was contrary to conventional wisdom. Make no mistake, though: Don demanded the utmost scientific rigor, which he felt we had demonstrated with the Chaco data.

I will confess to a sense of vindication when cultigens started occurring at other Late Archaic sites throughout the Southwest that were roughly contemporary with the Chaco sites. From the central and southern portion of the Southwest, these included sites from the Tucson Basin (e.g., Fish et al. 1986; Mabry 1990, 2000b; Mabry, ed. 1997a, 1997b; Mabry et al. 1997), southeastern Arizona (Huckell 1995), central Arizona (Huckell and Huckell 1999), extreme southern New Mexico (Upham et al. 1987, 1988), and northern Mexico (Hard and Roney 1998, 1999). From the northern Southwest were sites from the Chinle (Gilpin 1994) and Black Mesa regions of northern Arizona (Kohler 1993:273–276; Smiley 1994) and the Zuni area of New Mexico (Damp et al. 2002). Radiocarbon determinations from these sites tended to cluster around the early or middle parts of the third millennium B.P., although a date of 3610 B.P. was obtained from Three Fir Shelter on Black Mesa (Smiley 1994:174).

It now seems apparent that cultigens have indeed been around the Southwest for some time. It is equally clear that simplistic explanatory models were inadequate. Many contemporary researchers are addressing this issue with compelling explanatory scenarios. It is useful to examine a few examples by way of characterizing the tempo of current research into the Late Archaic and the introduction of cultigens.

Wills (1995, 1999) has thoroughly summarized the issues. He shows that in the north, agriculture initially had a relatively low impact and that "after more than 2000 years of involvement with maize cultivation, populations in the northern Southwest continued to occupy small settlements for relatively brief intervals...[and] agriculture...did not promote large sedentary communities" (Wills 1999:13). In the south, however, the incorporation of agriculture was much different. Although the chronology for the earliest appearance of cultigens is not yet completely resolved, by ca. 800 B.C. southern Late Archaic sites with cultigens often were large and substantial and contained structures. In fact, researchers believe that some of these sites contain more cultigens than subsequent Hohokam villages (Fish et al. 1992a, 1992b; Huckell 1995; Roth 1992, 1993)! This is in sharp contrast to the north, where we see little evidence for much sedentism until Basketmaker times. This could be a perceptual problem, however, an issue I address at the conclusion of this chapter.

Wills (1995, 1999) and others seem to be on the right track, although details of their interpretative scenario are open to question. For example, Wills describes an emergent food production system for the northern Southwest comprised of two temporal divisions. The first began with the initial introduction of domesticates at 3500+ B.P., which roughly conforms with the Chaco sites. So far, so good. During the first use of cultigens in the northern Southwest, there was essentially no profound impact on settlement or subsistence. Rather, cultigens essentially served to enhance foraging efficiency. This likely was due to the "difficulties of coupling a highly mobile lifestyle with effective cultivation" (Kohler 1993:275).

Where Wills needs to refine his model, I think, is in his belief that sites in this early phase consist primarily of rockshelters, except for a few open-air sites (Wills 1995:231–234). This contradicts the Chaco data, where the early sites containing cultigens were, in fact, open air, while the later sites were rockshelters. Conversely, during Wills's second phase, from 2200 to 1600 B.P., he proposes an increase in the frequency of open-air habitation sites containing maize, although rockshelters continued to be used. During this second phase, he suggests a system of shifting, short-term residential camps around permanent cultivation localities. These areas were chosen to minimize the costs of cultivation while allowing access to key wild resources. Wills (1995:233–234) still envisions seasonal mobility but probably over smaller geographic areas. This, in turn, suggests greater dependence on annual resource cycles, in which higher maize productivity was probably required.

Smiley (1994) proposes a similar scenario in which rockshelters on the Colorado Plateau were used for primary camps and storage until ca. A.D. 1, and Plateau populations were more mobile than their counterparts in the southern Southwest (Smiley 1994:186). He concedes that open-air sites could have been used earlier, with the introduction of cultigens in the fifth millennium B.P., thereby accommodating the Chaco data (Smiley 1994:181).

Although these scholars present plausible arguments, I remain unconvinced that most of the early sites were restricted to rockshelters. Data from Chaco Canyon (Simmons 1982) and the Chinle region of northeastern Arizona (Gilpin 1994), and possibly the upper Rio Grande in New Mexico (Seymour et al. 1997), contradict this. My disagreement with the lack of open-air sites in the early phase may seem trivial, but it is not. Wills argues that it was during the second phase (that is, after 2200 B.P.) that local episodes of intensive occupation occurred. I think this pattern, in fact, was established earlier, with the initial introduction of cultigens. This does not mean that these people were sedentary but indicates only that their mobility may have been more restricted than it was before the incorporation of cultigens as a part of the subsistence strategy.

Clearly, the use of rockshelters versus open-air sites and the levels of occupational intensity have implications for issues of sedentism and the processes associated with incorporating cultigens into Archaic diets. This issue can only be resolved with additional research at early sites in the northern Southwest. The problem of both low visibility and poor preservation at open-air sites has perhaps pushed researchers toward investigating rockshelters and cave sites, despite the fact that significant data often can be retrieved from open-air occurrences (Simmons 1998). Given the state of our current knowledge, deciphering the configuration of early sites is a tricky proposition, since "the structure of the early agricultural data base is that the agricultural spread may be archaeologically *invisible*" (Smiley 1994:184), at least in the north. Thus far, I see little compelling evidence arguing against the use of both shelters and open-air sites in a contemporary, complementary fashion.

EVENTS IN THE SOUTHERN SOUTHWEST

In the southern Southwest, we have a very different Late Archaic strategy. Wills (1995:234–238) and others argue for a quicker sedentism, and this model is bolstered by a recent flurry of research (e.g., P. Fish et al. 1986; S. Fish et al. 1992a, 1992b; Hard and Roney 1998; Huckell 1990, 1995; Mabry 2000b; Mabry, ed. 1997a, 1997b; Mabry et al. 1997; Roth 1989, 1992, 1993), much of it stimulated by CRM archaeology. Certainly something quite different was occurring in the southern Southwest as opposed to the north.

Wills (1995:238–242) feels that the emergence of Late Archaic settlements in the south created a corresponding trend toward specialized hunting tactics that in turn caused economic crises for northern populations. This competition may have "forced" northern groups to rely more heavily on cultigens. Although this is an intriguing scenario, there may be more likely alternate explanations here—in a sense, this sounds like the old "population movement" arguments. While not denying that groups of people moved within the Southwest, I believe that more compelling augments must be generated for more independent intensification of agriculture in both the north and south.

I am unconvinced by arguments that early agricultural developments in the south were stimulated by actual population migrations from Mexico, which seems a favored scenario for some researchers (e.g., Berry 1982:31–33; Huckell 1990, 1995) and certainly enjoys wide currency in contemporary studies of later Formative southwestern peoples. Conclusions such as "No longer can the Anasazi tradition be thought of as being descended from indigenous hunter-gatherers who became dependent on agriculture only after the BM II stage. The origin of maize cultivation on the Colorado Plateau now appears to have much more direct links to Mexico" (Matson 1999:11) remain unverified.

There must be a *reason* for such arguments, rather than simply "proximity," which may be more apparent than real. After all, we need to remember that between an agricultural Mesoamerican point of origin and the Southwest proper there is the huge, and largely archaeologically unexplored, northern Mexican "periphery" (see Wills 1988b:150–155). Only recently have intriguing developments emerged from this region (see below). There is no doubt, however, that developments in the southern Southwest were distinct from those in the northern Southwest, and this pattern demands explanation.

Several southern areas have yielded intriguing claims for early agriculture, and these are worth summarizing. Much information comes from southeastern Arizona and the Tularosa Basin of south-central New Mexico (Wills 1995:234), as well as northern Mexico (Hard and Roney 1998, 1999). There also may be evidence for agriculture at New Mexico localities such as Fresnal Shelter, Tornillo Shelter, and Pendejo Cave (Tagg 1999; Upham et al. 1987, 1988), although this is not as compelling. Interestingly, the New Mexico data come from rockshelters whereas the stronger evidence comes from open-air sites. Furthermore, the rockshelter information does not appear to break from a "typical" Archaic settlement mode whereas much of the open-air data represent substantial settlements.

It is from these substantial open-air sites, primarily two areas in southeastern Arizona and one in Chihuahua, Mexico, where early agriculture has been well documented and is quite distinct from events occurring in the northern Southwest. The Arizona localities are the Cienega Creek region and the Tucson Basin and consist of sites dating to the San Pedro (ca. 1200–800 B.C.) and Cienega Creek (ca. 800 B.C.–A.D. 200) phases of the Late Archaic. The Chihuahuan example is from Cerro Juanaqueña and dates to ca. 1000 B.C.

Cienega Creek had been investigated early on by Emil Haury (see Huckell 1995:1–16), and Huckell's (1995) excavations at two sites in 1983 provided conclusive evidence that a mixed foraging/farming economy existed by ca. 800 B.C. These studies have provided an expanded perspective on Late Archaic adaptations in southern Arizona. Huckell notes that "the emerging picture of late preceramic life in this region is one of sedentary village communities along the permanent streams of southeastern Arizona, supported by a well-balanced farming-foraging subsistence economy.... Agriculture was a crucial element in their lives, underwriting the creation of these villages and helping to sustain them" (1995:140).

Similar agricultural developments were occurring at roughly the same time in the nearby Tucson Basin (Bernard-Shaw 1988; Fish et al. 1992b:13–19; Freeman 1998; Gregory 2001; Huckell 1995; Mabry 1990; Mabry, ed. 1997a, 1997b; Mabry et al. 1997; Roth 1988, 1989; Roth and Wellman 2001). In this area, Fish et al. (1992b:13) have argued for "extended residence" as a likely correlate of food production at many Late Archaic sites. "Extended residence" implies sedentism, an immensely complex issue that can be difficult to document, as Fish and Fish (1994:93–95) readily acknowledge. Nonetheless, evidence from the Tucson Basin suggests that cultigens were important dietary components as early as the beginning centuries of the third millennium B.P. Furthermore, in contrast to the northern Southwest, cultigens at these sites are abundant, even at small, ephemeral localities (Fish et al. 1992b:17). Fish et al. (1992b:13) suggest that Late Archaic groups had many similarities with later Hohokam agricultural peoples.

In the same region, even more compelling information has emerged in recent years at several sites in the Middle Santa Cruz valley, where Mabry and colleagues have uncovered an amazing array of Late Archaic accomplishments (Mabry 2000b; Mabry, ed. 1997a, 1997b; Mabry et al. 1997; Muro 1998a). Not only has maize agriculture dating to ca. 800 B.C.–A.D. 200 (or the Cienega phase of the Late Archaic) been documented, but the use of tobacco and possibly cotton, communal or ceremonial structures, rectilinear architecture, and ceramics also appear to have occurred within a Late Archaic context. Maize is remarkably abundant, unlike at many of the other "early" sites, and it seems clear that intensive floodwater farming and even irrigation (see Mabry 2000b) sustained the occupants of these apparently sedentary sites. It is, however, important to note that these more elaborate developments do not appear to occur immediately after the first evidence for cultigens in the region, and thus one must be cautious in arguing for too rapid of a settlement/subsistence transition from mobile foragers. Nonetheless, this major cultural shift appears to have been faster than in regions farther north.

Although it may be difficult to identify concepts such as "sedentism" and "villages," it is likely that a degree of "extended residency" (see Fish et al. 1992b:13) existed in the Santa Cruz region. For example, the large Santa Cruz Bend site, of which only ca. 15 percent was excavated, contains 176 pit structures among 730 features. Although it is unlikely that these are all contemporaneous, Santa Cruz Bend still represents the largest settlement site of this date found in the Southwest (Mabry 1997:1). Furthermore, chronology is well controlled with a series of 34 radiocarbon determinations from four sites (Mabry 1997:3). There is little doubt that the data from these sites have exploded earlier views of the origins of agricultural settlements in the Southwest (Muro 1998a).

So substantial is the cultural dichotomy between these early villagers and "traditional" Late Archaic groups that Huckell proposes to designate the terminal 1,500 to 2,000 years of the preceramic period the "Early Agricultural" period (1995:140). Although Huckell's conclusions are well reasoned, I caution against the casual acceptance of this suggested name change, which already seems to be enjoying some limited success (e.g., Hard and Roney 1999; Mabry 1997:6). The reason for this caveat is that "Early Agricultural" period may imply a near total reliance on cultigens or could suggest that all Late Archaic peoples participated in this economic strategy. A more realistic concept is that of "in between" transitional territories and landscapes (Smith 1998:1651). Thus at this point, it may be more prudent to retain the more commonly used Late Archaic San Pedro (ca. 1200–800 B.C.) and Cienega (ca. 800 B.C.–A.D. 200) phase designations. What is important here is that during the earlier phase, settlements are relatively modest, whereas during the Cienega phase, they are much more substantial.

Finally, another dramatic discovery has fueled the complexity of Late Archaic adaptive strategies. This comes from northern Mexico, at the site of Cerro Juanaqueña in Chihuahua (Hard and Roney 1998, 1999). What is remarkable about this site is not only its early maize (ca. 1000 B.C.) but its large size (over 10 ha) and the use of over 8 km of complex terrace systems (*trincheras*) for apparent residential purposes. The labor investment in constructing such features implies a degree of organization not typically associated with this time period (Hard and Roney 1999), and Cerro Juanaqueña seems to be yet another example of a sedentary, or nearly so, Late Archaic settlement that used cultigens to considerable extent. Equally significant is the occurrence of this large settlement in the presumed "periphery" between the Southwest and a central Mesoamerican "core." How similar this development is to those in the Tucson Basin is yet to be determined.

Given the exciting new discoveries just summarized, it is clear that what was happening in the southern Southwest was distinct from the situation in the north and in many ways more complex. An adaptive dichotomy exists, although the chronological framework in the north and south are roughly comparable. In the southern Southwest, cultigens apparently had a more pronounced influence than they did in the northern Southwest, where current evidence indicates that cultigens were gradually incorporated into existing subsistence strategies.

There are reasons why the situation may be more complex in the southern Southwest, and certainly the proximity of the southern region to the ultimate origin of cultigens may be a factor. This area, after all, is what Fish et al. (1992a:1) believe is the Mesoamerican fringe. Core-periphery relationships, both between Mesoamerica and the Southwest and between riverine and nonriverine regions within the southern Southwest (see Fish et al. 1992a:3–5), may have played a role in the spread of cultigens. I do not believe, however, that this necessarily involved waves of Mesoamericans physically relocating to the Southwest. Documenting actual migrations in prehistory is tricky business, and before such models can be seriously considered, concrete data far more robust than those summarized by Matson (1999) are required.

NORTHERN AND SOUTHERN SOUTHWEST DISTINCTIONS

Undoubtedly, regional environmental conditions were a factor in Late Archaic subsistence strategies, as they were in the subsequent Formative period. Some researchers have made much of this where significant cultural distinctions between the northern and southern Southwest continued into the Formative. Were these differences due to environmental constraints or cultural diffusion from Mesoamerica? I feel that compelling arguments exist for recognizing that distinctions in early agricultural communities between the northern and southern Southwest are largely due to distinct ecological characteristics in both regions (e.g., Huckell 1995; Wills 1995:234–242).

In this context, one really cannot consider early agriculturalists in the Southwest without addressing the use of irrigation canals, even if a true Wittfogelian hydraulic civilization model cannot be verified. Conventional wisdom was that the use of irrigation canals was largely confined to the southern region. This practice reached an apex with the elaborate Hohokam irrigation canal systems, and there are claims for earlier, if more primitive, such systems during the "Early Agriculture" period in the south (e.g., Ezzo and Deaver 1998; Mabry 2000b). Recently, however, a convincing argument also has been made for early agricultural canals on the Colorado Plateau, at the sites of K'yana Chabina and K'yawa:na'a Deyatchinanne near Zuni (Damp et al. 2002). From my perspective, there is no reason to be surprised by such technological enhancements, whether in the north or the south. But it is a mistake to assume that canal technology is necessarily part of an early agricultural "package," presumably associated with actual population movements, as Damp et al. (2002:675) seem to suggest. Many authors working with early agriculturalists in the Southwest apparently still view the issue as an all-or-nothing proposition and do not account for the huge range of variability that might have accompanied the spread of agriculture in the Southwest. I maintain that these concepts do not necessarily have to involve, in all cases, actual population movements.

The point here is that the available evidence indicates a remarkable degree of diversity in the Southwest at the time of (and before) the introduction of cultigens. This was not a universal adoption but rather apparently occurred in different parts of the Southwest in different manners and at different times. In the southern region, agriculture apparently became more important far more quickly than it did in the north. On first observation, this may seem an odd adaptation, given the natural abundance and richness of the Sonoran Desert (see Nabhan 1985). Many models of agricultural introduction are based on ecological stress and environmental marginality (e.g., Gebauer and Price 1992), although Smith (1995:213) believes that most early experiments with agriculture, on a global basis, occurred in relatively rich environments. In any case, marginality is a condition that might well fit the northern Southwest better than the south, despite initial impressions to the contrary. Consequently, models of cultigen introduction that posit either a substantial and quick impact (see Fish et al. 1992a, 1992b) or an initial low impact on Late Archaic society (see Minnis 1992:122; Simmons 1982, 1986; Smith 1995:203) may both be correct. One must simply accept the fact that the Southwest is a huge area and that unicausal explanations may be inappropriate (see Smith 1998:1651).

Despite recent research, the enigma remains that much of the earliest evidence for cultigens still occurs in the north, and apparently at sites that do not reflect great sedentism, if one accepts the Chaco, Chinle, and Black Mesa data. It may be that Haury's (1962) argument for an initial northern "highland corridor" diffusion for agriculture deserves more credibility than many recent researchers would have one believe. Assuming a diffusion of cultigens (although not necessarily of people) from Mesoamerica, it seems logical that the first occurrence in the Southwest should be in the southern regions. It is not. Thus researchers must not automatically assume that cultigens were first used in the southern Southwest and then diffused to the north. This remains an unresolved issue, and given the tantalizing developments in the southern Southwest, earlier dates may come to light there. But this has not yet occurred, and it would be a mistake to categorically state that "a sedentary way of life...seems to have been an important element in early experiments with domestication" (Smith 1995:213). Such may have been the case in the southern Southwest, but it does not appear to be so in the north, where the earliest evidence for the use of cultigens occurs.

CONCLUSIONS

This chapter is not the place to discuss these indisputably complex issues. The point is that a few years ago we would not even be having these kinds of discussions—the data and theoretical framework simply were not there. This represents an exciting trend in contemporary southwestern archaeology, where divergent opinions, supported with hard data, have become common. Such debate is healthy and is the way science progresses. Don Graybill would have approved.

We now realize that the Southwest is not simply one or two homogeneous environmental zones, as earlier studies assumed. Indeed, even the often-used northern and southern dichotomies are undoubtedly too simple. We understand that economic adaptations will differ across shifting environmental landscapes, and the introduction of cultigens will be affected by this pattern. Many pertinent issues still require resolution, and they can best be investigated by the sort of multidisciplinary regional perspectives that Don Graybill promoted. I would like to conclude with certain topics that might profit from more investigation.

First, we need a fresh look at the issue from a more comparative basis. One might argue that Formative southwestern cultures represent Neolithic societies and that the Late Archaic is functionally equivalent to the Late Natufian (as reflected by the San Pedro phase) and early Near Eastern Pre-Pottery Neolithic (Cienega phase) of the Near East. Although no reasonable researcher would make direct parallels, there are undeniable similarities, and these precise terms were freely bandied about in a symposium on the Late Archaic at the sixty-fourth annual meeting of the Society for American Archaeology in Chicago (Vierra 1999). Several broadly framed questions may be asked in making such comparisons. For example, what are the actual similarities in terms of settlement patterns and degree of sedentism? How did the lack of domesticated animals in the Late Archaic, as contrasted with the Pre-Pottery Neolithic, affect cultural trajectories? Are such comparisons even appropriate, considering that the Near East witnessed a true in situ domestication process, what Minnis (1992:121) calls "pristine domestication," whereas in the Southwest, cultigens had their origins elsewhere in Mesoamerica? More relevant comparisons might be between the Near East and Mesoamerica, and yet it might be useful, from a broad framework, to consider such comparisons as examples of how cultigens can affect subsistence adaptations in arid environments (see Fish and Fish 1991; Mabry 2000b).

Second, there appear to be clear distinctions between the northern and southern Southwest. This is an issue that needs to be more fully investigated and may partly be a perceptual problem. We should begin by asking, is this a reflection of cultural reality, or is it due to research bias? It is important to realize that in recent years, the southern Southwest has witnessed a flurry of research activity related to large-scale projects. It might well be that a disproportionate amount of attention has been paid to the south in comparison with the north. I suspect, however, that these pattern contrasts are indeed accurate and that they are primarily due to the distinct biogeographic differences between the north and the south. Keep in mind, however, that a greater degree of precision between simply north and south is going to be necessary for any true resolution.

Third, we need to focus on the variety of domesticates involved. Maize is always mentioned, but what about the other two major resources—squash and beans? A fruitful pursuit might be examining this constellation of resources. There is limited evidence for the early introduction of these cultigens, although squash is present as early as ca. 1000 B.C., again in the north (Simmons 1986:79). Several questions are pertinent: how important were the nutritional components of these resources, and did they arrive as a "package" or were individual species separately introduced? What was the mechanism for introduction? Perhaps, since ceramics were lacking, squash gourds were used to carry beans and maize. A related issue involves possible local domestication of some plants (e.g., amaranth) and their relationship to imported tropical plants, such as maize (Mabry 2000b; Minnis 1992).

Fourth, precise chronology remains an issue. This is the sort of problem with which Don Graybill was very concerned; he was well known for his precise dendrochronological research. Unfortunately, the types of data required for this sort of research—that is, trees—simply are not usually preserved at Archaic sites. This is one reason that high-frequency resolution of both chronological and environmental issues (see Dean 1988a) for the Archaic will remain elusive. Dating becomes an even more important issue with the realization that the earliest domesticates and sedentary villages in Mesoamerica may, in fact, not be as old as once thought (MacNeish and Eubanks 2000; Matson 1999:10; Piperno and Flannery 2001; Pope et al. 2001), thereby shortening the transmission time from there to the Southwest. More direct dates on cultigens are clearly needed in order to fine-tune chronological variation and to better understand their antiquity and regional distribution. The large suite of radiocarbon determinations on both cultigens and charcoal from the Santa Cruz Valley sites (Mabry 1997:5) goes a long way toward providing precise regional chronologies, and future research must ensure that adequate funding is available for such studies. Also needed are comprehensive and thoughtful analyses of the available radiocarbon determinations throughout the entire region, such as provided by Smiley (1994).

Fifth, climate remains a big question. Paleoenvironment was another of Don's great concerns. It was here that he was most critical of archaeologists for their overreliance on the "men in the white coats"—that is, our awe of so-called hard sciences. While recognizing the absolute necessity of natural sciences to archaeological interpretation, Don was always amused that archaeologists tend to jump on whatever bandwagon is current. He was the first to admit that the overprecision characterizing many natural sciences was a seductive attraction to archaeologists, who are used to working with marginal data. Don candidly acknowledged that it was easy to become lost in precisely dissecting the proverbial tree while not seeing the forest. He constantly pointed out that human beings were, after all, the focus of archaeology, something that some archaeologists seemed to have forgotten.

An acute problem in southwestern agricultural origin studies is determining the overall climatic regime—was it dry and marginal or wet and not so marginal? Of course, we first have to agree when cultigens were introduced. We also need better paleoenvironmental data. Or do we? Conflicting climatic interpretations already exist, depending on what data sources one uses (Dean 1988a:165). That is, pollen might yield one environmental scenario, and packrat middens another. Don would have argued that the most sensitive climatic indicators were dendroclimatological. Indeed, he would probably have been correct (see Dean 1988a, 1988b; Dean et al. 1985), although packrat middens also have yielded impressively precise data (see Betancourt et al. 1990; Spaulding 1991; Van Devender and Spaulding 1979). But if the necessary information is not preserved at individual sites, we must rely on proxy data, which can lead to problems. Nevertheless, the types of painstaking environmental reconstructions summarized by Fish and Fish (1994:88–90) are beginning to provide a better ecological mosaic from which to evaluate Late Archaic adaptations. Given the nature of the archaeological and environmental data for the Archaic, however, it is unlikely that studies such as the innovative streamflow reconstructions developed by Don (Graybill 1989) or the precise high-frequency-resolution dendroclimatological model of Dean and his colleagues (Dean 1988a, 1988b) will be forthcoming for the Late Archaic. On the other hand, Don would have encouraged squeezing the maximum amount of information out of marginal data sets.

There is a remarkable amount of environmental diversity within the Southwest, and this may have shifted significantly in the past. Several authors (e.g.,

Fish and Fish 1994:88–90; Wills 1988b:51–58) have commented that environmental variables critical for agricultural success include both regional and topographical diversity and that these have not remained constant through time. For example, Late Holocene climatic differences between north and south can be attributed to decreased winter precipitation and the irregular effects of summer monsoons in the southern desert regions. Additionally, winter-dominant precipitation occurs to the west and summer-dominant precipitation occurs to the east, spring and fall frosts occur at upper elevations, and summer heat is intense in low basins. Individual regions in the Southwest have significant seasonal and annual variability, and early agriculture was undoubtedly affected by these variables. Only the concept of marginality—a unifying theme in southwestern archaeology—seems consistent. Yet we seem unable to achieve a consistent definition of marginality (see, e.g., Moore and Winters 1980); this greatly amused Don.

Finally, there has been an explosion of information over the past several years, yet much of it is published in reports receiving limited circulation. Detailed synthetic reviews need to make their way into mainstream journals so the profession as a whole can better evaluate these exciting data.

Overall, though, I believe "the Don," as he was known to many of us, would be impressed by recent events in Archaic archaeology in the Southwest, particularly as they relate to a truly multidisciplinary approach to understanding the introduction and integration of cultigens. This is an exciting time for research on this complex issue. I can visualize that familiar gait of Don's, or hear that distinctive clinking of ice cubes, as he contemplated a new site, or a new series of radiocarbon dates, or a new technique for assessing domestication. That twinkle would be in his eyes, and he'd say, "How do you really know?"

3

Prehistoric Human Response to Landscape Change in the American Southwest

Michael R. Waters

To fully understand prehistoric human behavior and the changes that occurred in prehistoric human cultures, it is necessary to reconstruct the environment and landscapes within which people lived and interacted. Such a reconstruction should include the biological environment (floral and faunal communities), physical setting (landscape), and climate of an archaeological site or sites at the time of occupation. Because the biological environment, landscapes, and climate change through time, these reconstructions should be not only static, showing what the environment was like at any specific time, but also dynamic, showing how the environment changed through time. When coupled with archaeological data, environmental reconstructions provide a powerful explanatory tool to address such questions as why people chose particular areas for occupation and use, why certain localities were abandoned while others remained occupied, and why certain adaptive strategies were chosen and why they changed. In this chapter, I examine how landscapes affected people and their behavior.

LANDSCAPE RECONSTRUCTION

The landscape is the physical platform on which all people, animals, and plants live and interact. It includes the landforms created by erosion and deposition, and the geomorphic processes that created these landforms. It is important to recognize that the landscape is not static but dynamic and changing. The landscape one sees today may not resemble the prehistoric landscape. Thus it is essential to reconstruct the landscape at the time a site was occupied as well as before and after occupation. Placing prehistoric human occupation in the context of a dynamic and evolving landscape helps us elucidate human-land interactions and perhaps identify causal factors that may explain culture change. In short, by reconstructing the environment, we put human behavior into its environmental context (Butzer 1982).

In any landscape, a portion is relatively stable and remains unchanged while other areas are constantly changing over the same period of time. This fact allows us to divide the landscape into two major components: constant landscape elements and dynamic landscape elements. Whether components of the landscape are considered constant or dynamic depends on the temporal scale of investigation. Given that the time frame of interest to southwestern archaeologists is typically the late Quaternary, the constant portions of the landscape from an archaeological perspective would be the geographic features that require millions, hundreds of thousands, or tens of thousands of years to change, such as the location of valleys, mountain ranges, passes, and basins and the stabilized (inactive) depositional landforms that did not change significantly during the late Quaternary, such as Pleistocene river terraces above a floodplain and Pleistocene segments of alluvial fans. The dynamic components are those landscape elements that were active and changed during the late Quaternary, such as a floodplain where a river changed its position and actively flooded adjacent lowlands, sand dunes that migrated downwind within a dune field, or a colluvial slope that has been aggrading. Through time, the dynamic portion of the landscape is characterized by alternating conditions of stability, deposition, and erosion. For example, a stabilized area may become an active area of deposition or erosion, or an area undergoing deposition may become stabilized. Furthermore, different conditions may characterize different segments of the landscape at any given time. For example, a floodplain may be undergoing arroyo cutting while deposition is occurring on the bajada and on an adjacent dune field.

To decipher human-land interaction, we must reconstruct both the dynamic and constant elements of the landscape. The key to reconstructing the landscape is preserved in the sediments, soils, and erosional contacts making up the archaeological site matrix and the regional stratigraphy surrounding the site (see Waters 1992). Sediments provide information on depositional (geomorphic) processes and the specific environments that were present, soils preserve evidence of landscape stability, and erosional unconformities record episodes of landscape erosion (degradation). Stratigraphic and geochronological investigations interpret the superposition of sediments, soils, and erosional contacts to define the spatial and temporal record of landscape aggradation, stability, and degradation. This evidence is first used to differentiate between the constant and dynamic elements of the landscape and then to define the landscape history of the dynamic component in greater detail.

Reconstructions of the prehistoric landscape should be both synchronic and diachronic. Synchronic reconstructions define the geomorphic processes and configuration of the landscape that existed at and around a site at the time of occupation, placing the site in its prehistoric landscape context. Site-specific synchronic reconstructions of the landscape are vital to the interpretation of site function and subsistence activity, while regional synchronic reconstructions provide an understanding of how coexisting landscape components affected the spatial patterning of settlements and activity loci. For example, the position of a prehistoric site, as well as the regional pattern of late prehistoric sites, may reflect the distribution of arable land and the hydrologic conditions of the region.

Diachronic landscape reconstructions at the site or regional level place prehistoric people in the context of an evolving landscape. The landscape is reconstructed at different points in time, and then archaeological settlement data are compared with the corresponding diachronic landscape reconstructions. In this way changes in the spatial distribution of geomorphic processes or changes in the landscape are correlated with the changing loci of human activity through time.

Synchronic and diachronic reconstructions of the landscape are crucial to the archaeological interpretation of prehistory. In many cases the landscape has changed considerably since the site was last occupied. It must be recognized that prehistoric people occupied a landscape that may not correspond to the configuration of the existing one. For this reason, the proper interpretation of human behavior requires that prehistoric sites be placed in the context of both static and evolving prehistoric landscapes.

CAUSES OF LANDSCAPE CHANGE

The dynamic parts of the landscape are best seen as being in a state of dynamic equilibrium with periodic adjustments (Bull 1991; Ritter et al. 1995). For example, a river may occupy its floodplain for 500 years with no change in channel morphology. This channel is adjusted to the average stream discharge, and the channel morphology does not change even though there are periods of below-average and above-average streamflow. Thus this stream is in a state of dynamic equilibrium. Periodically, however, the stream will undergo a major adjustment; for example, the channel may downcut into its floodplain and eventually stabilize at a lower elevation and create a new floodplain. Once the stream has adjusted to the new set of hydrologic and geomorphic conditions, a new dynamic equilibrium is established.

These periodic adjustments to the landscape are the result of changes to an external or internal variable or a combination of variables (Bull 1991, 1997; Ritter et al. 1995). A delicate balance or equilibrium exists between landscapes and the processes that create and maintain them. A landscape in equilibrium has definite parameters that represent the limits of a balanced system. These limits of equilibrium are known as *thresholds* (Bull 1991; Ritter et al. 1995; Schumm 1977). Changes to external or internal geomorphic variables or both may stress the landscape system beyond its defined limits of stability and send the landscape into disequilibrium. External variables include climate, tectonic stability, vegetation, and human activity. If any of these variables change and cause the parameters of a landscape system to become exceeded and ultimately sent into disequilibrium, then an extrinsic threshold has been crossed. For example, if irrigation ditch construction or change from a dry to a wet climate results in arroyo cutting, then a change in an extrinsic variable caused the geomorphic system to be sent into disequilibrium. However, if external variables remain relatively constant and the geomorphic system becomes unstable and changes, then an internal or intrinsic threshold has been exceeded. Intrinsic thresholds are reached as a result of gradual change within the landscape. For example, as the valley floor slope increases by normal deposition year after year, a slope threshold may eventually be reached and exceeded, triggering channel entrenchment (gullying) with no change in climate or other external factors (Bull 1991, 1997; Ritter et al. 1995; Schumm 1977). Whenever thresholds of equilibrium are exceeded, the system is sent into disequilibrium and eventually a new equilibrium is established.

Major adjustments to a landscape (e.g., channel downcutting or widening), as a result of changing geomorphic conditions, occur infrequently and usually very rapidly. These are considered low-frequency environmental landscape changes, that is, changes that occur at intervals greater than a human generation (ca. 20 years). Conversely, when the landscape is in a state of dynamic equilibrium, geomorphic thresholds are not crossed even though flooding and droughts may occur. These periodic intervals of flooding or low streamflow do not trigger major changes to the landscape. Landscapes in a state of dynamic equilibrium are affected by high-frequency geomorphic processes.

HIGH- AND LOW-FREQUENCY LANDSCAPE PROCESSES AND PEOPLE

Human cultures are adapted to high-frequency environmental processes, those that operate over short periodicities (i.e., year-to-year or season-to-season variability), because people perceive this as stability (Dean 1988b). In order for a culture to maintain itself over time, its adaptation must be flexible enough to handle expected, but unpredictable, high-frequency variations in environmental variables. Low-frequency variations that occur infrequently, with periodicities greater than one human generation, may have more severe consequences for people (Dean 1988b).

The human response to low- and high-frequency environmental changes depends on a number of variables. Two of the most important are prehistoric subsistence patterns and how close the population is to the carrying capacity of the land (Dean 1988b). Both low-frequency changes, such as arroyo entrenchment that fundamentally alters the landscape, and high-frequency changes, such as a catastrophic flood within the thresholds of the landscape system, probably would have had minimal impact on mobile hunters and gatherers. Likewise, such changes to the landscape would probably have had minimal impact on small populations of agriculturalists if the population was below the carrying capacity of the land. These people could cope with environmental changes that caused crop failure by drawing on stored surpluses from previous years, increasing their use of wild resources, and relocating fields to unentrenched segments of the floodplain or onto the bajada.

In contrast, if population is at or near the carrying capacity of the land and people are dependent on an intensive agricultural subsistence, then low- and high-frequency changes to the landscape become critical threats to survival. Low-frequency landscape degradation (e.g., entrenchment of the floodplain, leading to abandonment of fields) or high-frequency drought or catastrophic flooding that destroys canal systems will lower the carrying capacity of the land and thus the number of people who can be sustained. If the carrying capacity is reduced to the point that people can no longer be supported by the extant subsistence system, a rapid human response is required to stave off disaster, especially if food losses cannot be replaced by wild foods, crop yields from other field areas, or storage. Thus, subsistence patterns and population are key factors in the response of people to environmental changes. If they strain the supportive capacity of the habitat, both low-frequency changes (especially landscape degradation) and high-frequency changes (depending on the timing and severity) can have serious consequences for large populations.

In the following sections, two case studies show how prehistoric people responded to low- and high-frequency landscape changes in the American Southwest during the late Holocene.

HUMAN RESPONSE TO LOW-FREQUENCY LANDSCAPE CHANGE ON THE SANTA CRUZ RIVER, SOUTHERN ARIZONA

The Hohokam occupied southern Arizona from around A.D. 300 to about 1450 and were skilled agriculturalists who constructed extensive canal networks along the perennial Salt and Gila rivers (Doyel 1991a; Fish and Fish 1994). To the south, in the Tucson Basin, these farmers used a combination of dry and floodwater farming and local ditch and canal irrigation. In the Tucson Basin, the landscape is dominated by the Santa Cruz River, now an ephemeral arroyo, which flows down the central axis of the basin. Flanking the Santa Cruz River floodplain are Holocene-age alluvial piedmonts (bajadas) and Pleistocene-age alluvial fans that abut the surrounding mountains.

The constant elements of the landscape included the mountains, Pleistocene alluvial fans, and Holocene bajadas extending from the adjacent mountains; the dynamic element of the landscape was the Santa Cruz River floodplain. Low-frequency changes to the landscape and hydrologic conditions on the floodplain influenced the position of Hohokam settlements and agricultural strategies. This is well illustrated along the 15-km segment of the Santa Cruz arroyo traversing the San Xavier Indian Reservation south of Tucson. Here the location of settlements is well documented for six separate Hohokam phases dating between A.D. 800 and 1500 (Doelle et al. 1985). The geomorphic history of the landscape is also well established through study of the alluvial stratigraphy (Freeman 2000; Haynes and Huckell 1986; Waters 1988a, 1988b;

Waters and Haynes 2001). When the settlement patterns are superimposed on the corresponding landscape reconstructions, it becomes clear that the changing configuration of the floodplain influenced Hohokam use of the riverine environment (Figure 3.1).

Before 50 B.C., the channel of the Santa Cruz River was entrenched. Over the next thousand years the channel filled with alluvium, and by A.D. 950 the floodplain had stabilized. The last 150 years of aggradation, from A.D. 800 to 950, coincided with the Rillito phase, when settlement was characterized by occupation at five primary villages and a number of hamlets on the western side of the floodplain. Floodplain stability continued into the early Rincon subphase (A.D. 950–1000), and the settlement pattern closely resembles that during the Rillito phase, but the number of hamlets and the intensity of occupation at primary villages increased. The floodplain during both the Rillito and early Rincon was characterized by a broad, sandy surface that was probably traversed by a shallow channel or draw. Discharge across most of the floodplain would have been ephemeral; however, small isolated seeps may have been present. This environment was well suited to floodwater farming. Crops could have been planted on the floodplain, where they would have been watered during overbank flows, and along the margin of the floodplain, where gullies draining the bajada emptied out onto the floodplain. Villages were located immediately adjacent to the floodplain next to major washes entering it but above areas of active flooding. Additional *ak chin* (floodwater) farming (Nabhan 1986) could have been pursued on the gully-mouth fans on the bajada to the west.

By the middle Rincon subphase (A.D. 1000–1100), arroyo cutting occurred in the central portion of the floodplain and extended southward. Channel entrenchment led concurrently to the formation of an arroyo-mouth fan and sand dunes to the north. The arroyo destroyed arable land in the south but in the process created arable land to the north. *Ak chin* farming was possible on the gully-mouth fan and dry farming on the sand dunes. These landscape changes appear to have triggered settlement reorganization. During the middle Rincon, the number of primary villages decreased from five to one, and settlements generally shifted northward. By the late Rincon subphase (A.D. 1100 to 1150), a cienega (spring-fed marsh) developed in the northern part of the floodplain and arroyo cutting in the southern portion of the floodplain continued to destroy arable land. In response, Hohokam settlements shifted to the northern and eastern edges of the floodplain.

In the subsequent Tanque Verde phase (A.D. 1150–1300), the landscape remained much like that of the previous 150 years. The cienega expanded and stabilized, and the arroyo channel in the southern portion of the floodplain began to backfill. During this time the number of settlements continued to decline on the western side and increase on the eastern side of the floodplain. New primary villages were established around Martinez Hill and the dune complex, and the number of hamlets also increased.

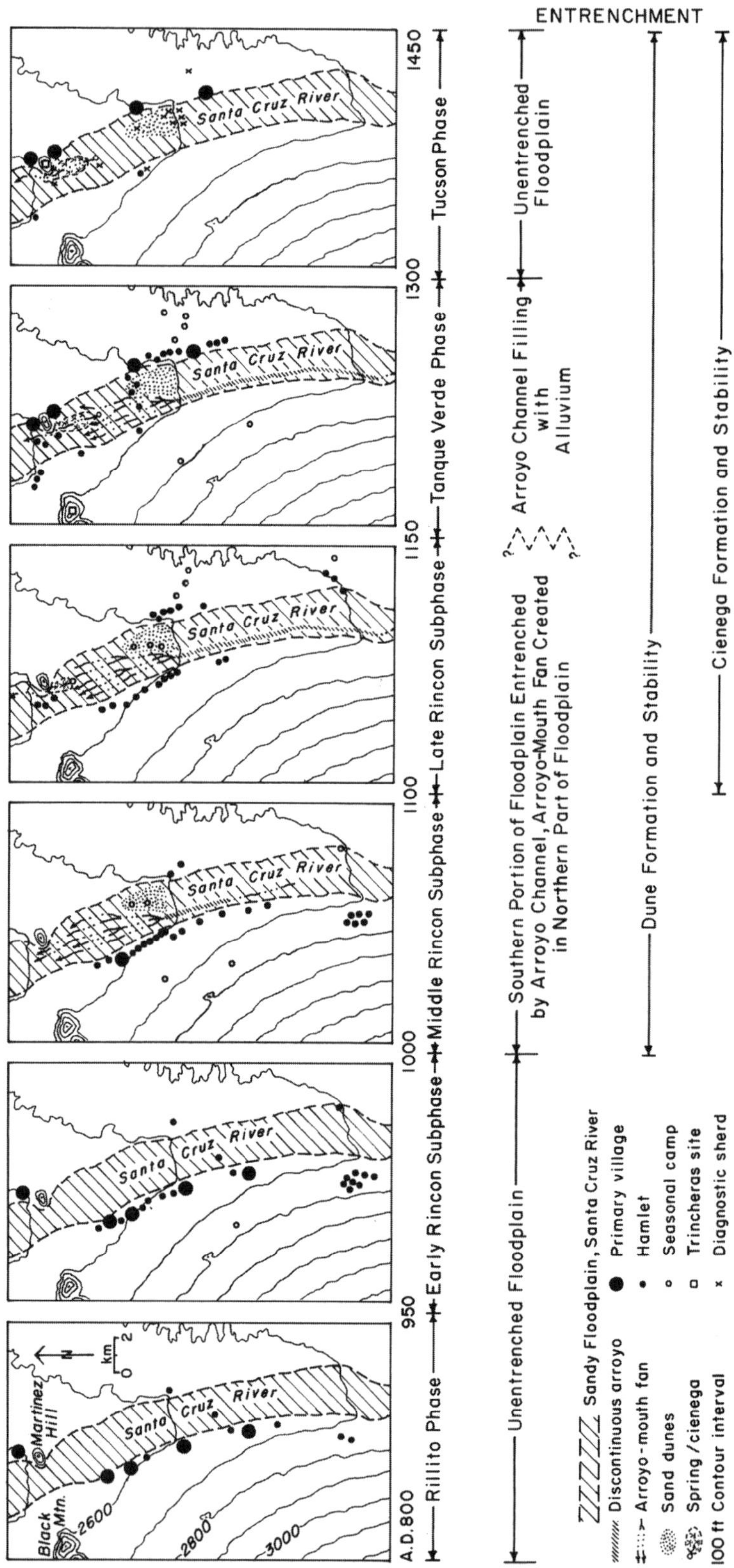

FIGURE 3.1.
Landscape changes along the San Xavier reach of the Santa Cruz River, Arizona, from A.D. 800 to 1450, and their effect on Hohokam settlement patterns.

The settlement shift to the northern and eastern sides of the floodplain, completed by the Tanque Verde phase, was likely a response to the destruction of arable land in the southern and western parts of the floodplain and the creation of environments suitable for farming to the north. Major villages were established next to the cienega and the sand dunes, and smaller sites occurred around and within these two environments, which provided optimal areas for farming in the newly created landscape. Water could have been drawn from the cienega to fields situated on its margin, and dry farming could have been pursued on the sand dunes. Floodwater farming could have been conducted on the gully-mouth fan and along the undissected portions of the floodplain-bajada interface on the eastern side of the floodplain. Clearly, the unentrenched northern and eastern portions of the floodplain offered favorable agricultural conditions.

By the beginning of the Tucson phase (A.D. 1300–1450), the arroyo had filled with alluvium, and the floodplain was no longer entrenched, while the cienega and sand dune environments remained. The population appears to have concentrated into the primary villages established during the Tanque Verde phase. A few small sites are present, and diagnostic sherds of the Tucson phase are found on activity loci on the floodplain.

Major entrenchment of the Santa Cruz floodplain followed the Tucson phase around A.D. 1450. This widespread environmental degradation would have again rendered the floodplain unfarmable, and following the entrenchment, the large primary villages were abandoned. Occupation is not documented along the San Xavier reach of the Santa Cruz River until sometime after A.D. 1650 (Ravesloot 1987), by which time the arroyo channel was largely filled.

The landscape and settlement history of the San Xavier reach of the Santa Cruz River shows dramatically how low-frequency environmental changes can affect prehistoric people (Figure 3.1). During periods of landscape stability, the Rillito and early Rincon phases, settlement patterns did not change. In this period, high-frequency geomorphic changes occurred on the floodplain (periods of high and low streamflow), but these variations were within the tolerances of the Hohokam adaptive system and consequently did not affect the people who used the river. The low-frequency landscape changes at the start of the middle Rincon and lasting into the Tanque Verde phase—floodplain entrenchment, development of an arroyo-mouth fan, and the emergence of a cienega and sand dune field—dramatically influenced the Hohokam agriculturalists. As a consequence of these landscape changes, the Hohokam relocated their settlements and fields. However, it is interesting to note that the Hohokam did not abandon this area but instead adapted to the new environmental surroundings. This indicates that their subsistence, social systems, and population were sufficiently flexible to adjust to the dramatic changes in the landscape, probably because they had enough food from upland floodwater fields, storage, and expanded use of wild plant resources.

Following a short period of stability in the Tucson phase, a major low-frequency environmental change occurred (arroyo cutting) from which the Hohokam did not recover. Floodplain entrenchment was so severe that it rendered the floodplain unfarmable. This time the Hohokam reacted by abandoning their large villages and hamlets, perhaps dispersing into small homesteads across the landscape. In this case, Hohokam subsistence, social systems, and population were not flexible enough to adapt to a landscape change of this magnitude.

Devastating, low-frequency changes to the floodplains of other streams used by the Hohokam farmers in southern Arizona appear to have been an important factor in the Preclassic to Classic period transition (Waters and Ravesloot 2000, 2001). At this cultural juncture there was channel downcutting and widening along every major stream used by the Hohokam: the Gila River (Huckleberry 1993; Waters and Ravesloot 2000, 2001), the Salt River and Tonto Creek in the Tonto Basin (Waters 1998), and the San Pedro River (Waters and Haynes 2001). It appears that this synchronous landscape catastrophe triggered a cultural response that accelerated social, political, demographic, and other changes already under way at the end of the Preclassic. Interestingly, there is no consistent low-frequency landscape event (such as channel downcutting) in all these drainages at the end of the Classic period that might help to explain the demise of the Hohokam. Thus, other factors or a combination of factors, such as internal or external human and cultural variables or even high-frequency geomorphic or climatic changes, were important elements associated with the cultural changes seen at the end of the Classic period. A human ecological approach to the interpretation of prehistory (Butzer 1982) allows us to recognize which variables (environmental and/or cultural) may be important in explaining changes seen in the archaeological record.

HUMAN RESPONSE TO HIGH-FREQUENCY LANDSCAPE CHANGE IN THE TONTO BASIN, CENTRAL ARIZONA

The Tonto Basin is about 90 km northeast of Phoenix in the rugged, semiarid Transition physiographic zone of central Arizona. This area was occupied in the Classic period (ca. A.D. 1150 to 1450) by late prehistoric agriculturalists known as the Salado (Fish and Fish 1994; Rice 1992). The basin is surrounded by mountains that are adjacent to an extensive set of Pleistocene alluvial terraces, pediments, and fan surfaces. Flowing through the middle of the Tonto Basin are Tonto Creek and the Salt River, both of which have narrow Holocene floodplains. Alluvial stratigraphic studies show that three Holocene terraces occur along these streams (Figure 3.2; Waters 1998). Terrace 3 lies about 10 m above the modern streambed and formed from around 2000 to 400 B.C. Terrace 3 is poorly preserved and is not a dominant landscape element of the basin. Terrace 2 lies about 4.5 m above the active beds of both streams and was formed between approximately 400 B.C. and A.D. 1000. Terrace 1, which lies

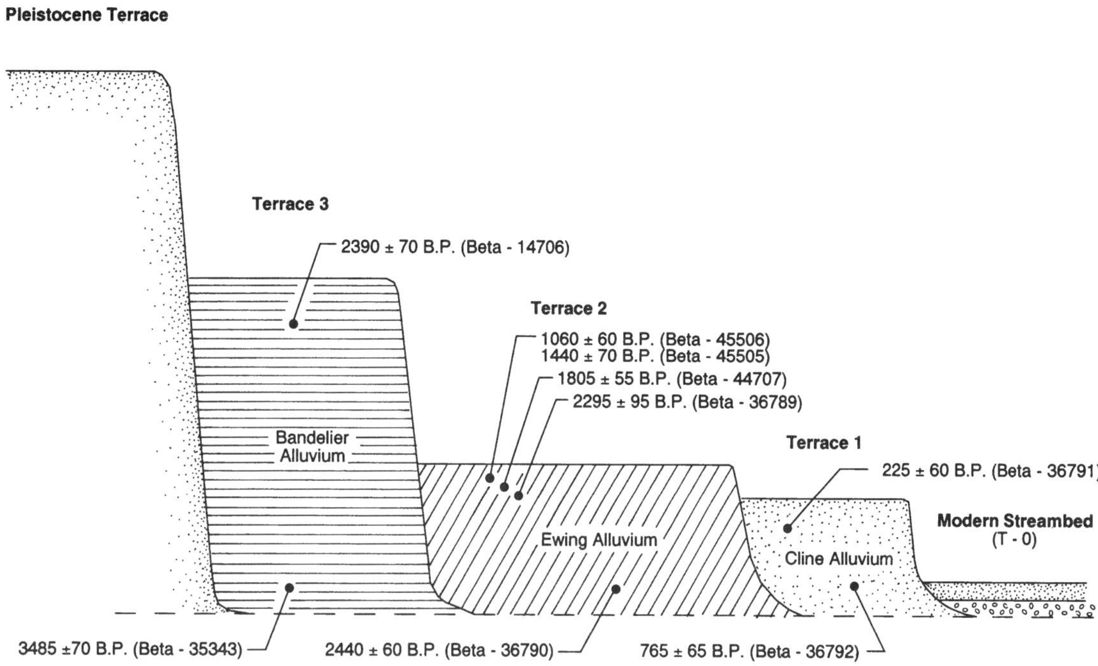

FIGURE 3.2. Generalized late Quaternary stratigraphy of the Tonto Basin, Arizona. Radiocarbon ages associated with the alluvial units are shown.

about 3.5 m above the modern streambeds, formed from around A.D. 1250 to 1800. Terraces 1 and 2 are extensive along both Tonto Creek and the Salt River.

When the Salado occupied the Tonto Basin, the constant elements of the landscape would have included the Pleistocene surfaces surrounding the basin and Holocene terrace 3 (Figure 3.3). These landscape elements would have been stable surfaces undergoing no deposition, except for minor accumulation of alluvial and colluvial sediments in a few localized areas, and minor erosion.

The dynamic portions of the Salado landscape included terrace 2 and the then-active floodplain (the alluvium that would eventually become terrace 1; Figure 3.3). Terrace 2 would have been present throughout the basin and for the most part was stable. However, this surface was sometimes flooded during catastrophic flood events. Terrace 1 did not exist during the Classic period. Instead, at that time the floodplains of both rivers were aggrading. The streambed would have been sandy and gravelly and only 2 m below the surface of terrace 2.

This reconstruction provides the landscape context to examine the location of sites and activity areas during the Classic period (Figure 3.3). For the most part, major settlements were located on the stable Pleistocene surfaces of Tonto Creek and the Salt River. Here settlements were protected from flooding and overlooked the floodplain. These locations also maximized the total irrigable acreage on the valuable and limited farmland of terrace 2. In addition to irrigation agriculture, the Salado practiced dry farming and floodwater farming on the Pleistocene surfaces (Van West and Altschul 1994). These strategies,

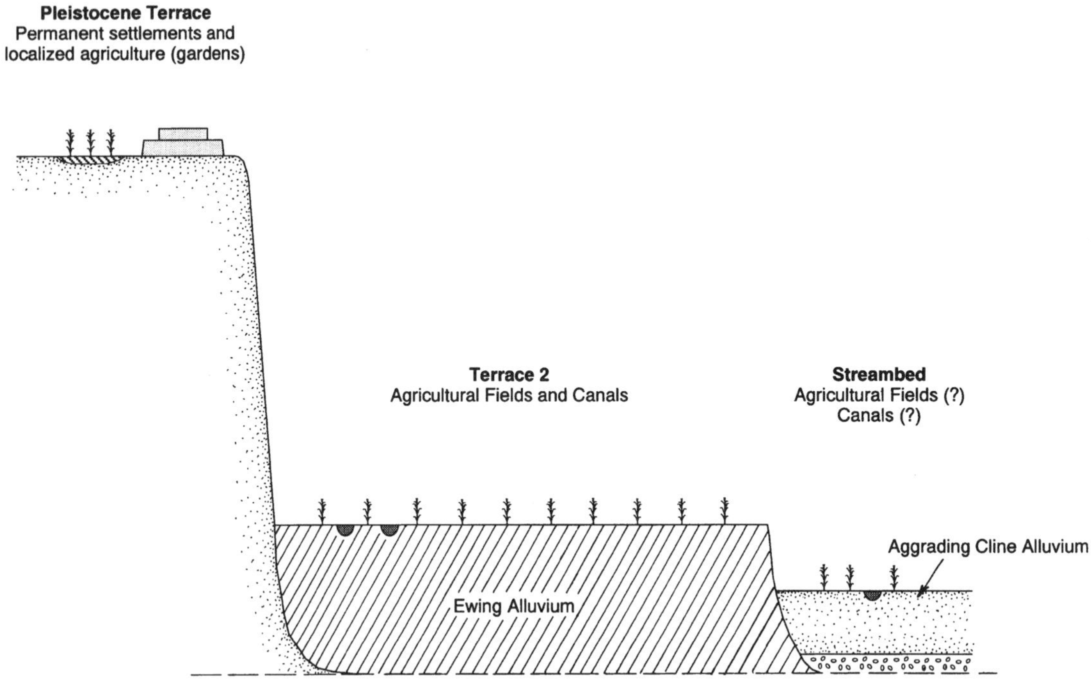

FIGURE 3.3. Reconstruction of the Salado landscape along a typical segment of either Tonto Creek or the Salt River. The Holocene Terrace 3 is not shown because in most places it was eroded away before the Salado period.

however, were secondary to irrigation agriculture along Tonto Creek and the Salt River.

Canal irrigation was practiced on both the Salt River and Tonto Creek during the historic and prehistoric periods (Ciolek-Torrello et al. 1994; Waters 1998; Welch 1992). The aggrading floodplains of Tonto Creek and the Salt River were potential agricultural fields, but the floodplain sediments were coarse-grained, low in organic matter and nutrients, and highly permeable. More important, crops planted in such a setting would have been highly vulnerable to the slightest high flow. Thus, the floodplains would have been poorly suited to farming, which would have been a high-risk venture. Terrace 3, though it had some suitable cropland, was too high above the prehistoric streambed (8 m) for the Salado to have brought water, and few remnants of the surface were available for use.

Terrace 2, only 2 m above the active streambed, was the ideal location for crops (Figure 3.3). It would have been easy for the Salado to bring water onto this surface with their irrigation technology. Further, at 2 m above the channel, crops would have been protected from the frequent high streamflows within the channels of Tonto Creek and the Salt River that occurred in the late summer. The terrace 2 soils were texturally and chemically suited for crops. One drawback to canal irrigation was the limited amount of available farmland. Holocene alluvium occupied narrow areas adjacent to the streams. It is estimated that only about 930 ha (2300 ac) of terrace 2 surface would have been available for farming in the entire Tonto Basin (Waters 1998). Thus, the limiting factor in canal irrigation farming was not water but land.

ANNUAL STREAMFLOW

An important factor affecting prehistoric canal irrigation in this area was annual streamflow. Along the Salt River in the Phoenix Basin, the Hohokam adjusted their agricultural practices to the seasonal variations in streamflow, especially the occurrence of large floods that caused varying magnitudes of damage to canal systems and ultimately dictated the timing of their operation (Graybill 1989; Nials et al. 1989). Streamflow along the Salt River and Tonto Creek in the winter and spring is highly variable and unpredictable. If large floods occur, they typically come in late winter or early spring. Conversely, the streamflow regime is most predictable and uniform in the summer months of May, June, and July. However, flooding may again occur near the end of summer. Graybill and others suggest that in most years, winter–spring streamflow damaged the canal systems to some extent, and repairs had to wait until the end of the spring flood season; in that case, the canals would not have been operational until the end of April or beginning of May. Further, the potential floods of August could again render the canal system inoperable. Thus, the maximum season in which crops could have been irrigated with little fear of canal damage and consequent interruption of the water supply during plant growth extended from the end of April to at least the end of July. In some years, the irrigation season could have been extended into August. This pattern of flooding and associated damage to canal systems likely precludes the possibility of double cropping (Nials et al. 1989). Since the Salt River streamflow in the Phoenix Basin is similar to the pattern of streamflow in the upper Salt River and Tonto Creek, these observations probably hold true for the Tonto Basin as well.

Next we must consider the pattern of annual discharge and large floods and the resultant damage to canal systems in relation to the irrigation season during the Classic period (i.e., could the available land be farmed every year?). This examination is possible because Don Graybill (1989), using dendrochronological data, reconstructed the streamflow along the segment of the Salt River that traverses the Phoenix Basin. For the Phoenix Basin, Graybill examined the annual streamflow data from the Salt River in terms of recurrence intervals above and below the mean annual streamflow. Next he noted that during the historic period, there was a correlation between periods of high annual discharge and years of flooding. Based on this correlation and Graybill's (1989) streamflow reconstructions, Nials and others (1989) established a hierarchy of the probable effects of recurrent flows and associated flooding on the prehistoric canal systems in the Phoenix Basin (i.e., the extent of damage to the canal systems as a result of flooding associated with flows of various recurrence interval ranges on an annual basis and thus the calculated loss of part or all of an irrigation season). Fortunately, Graybill provides reconstructions of streamflow for the Salt River and Tonto Creek in the Tonto Basin. However, because there is only vague information on the impact of damage to historic canal systems due to flooding in the Tonto Basin, the same hierarchy

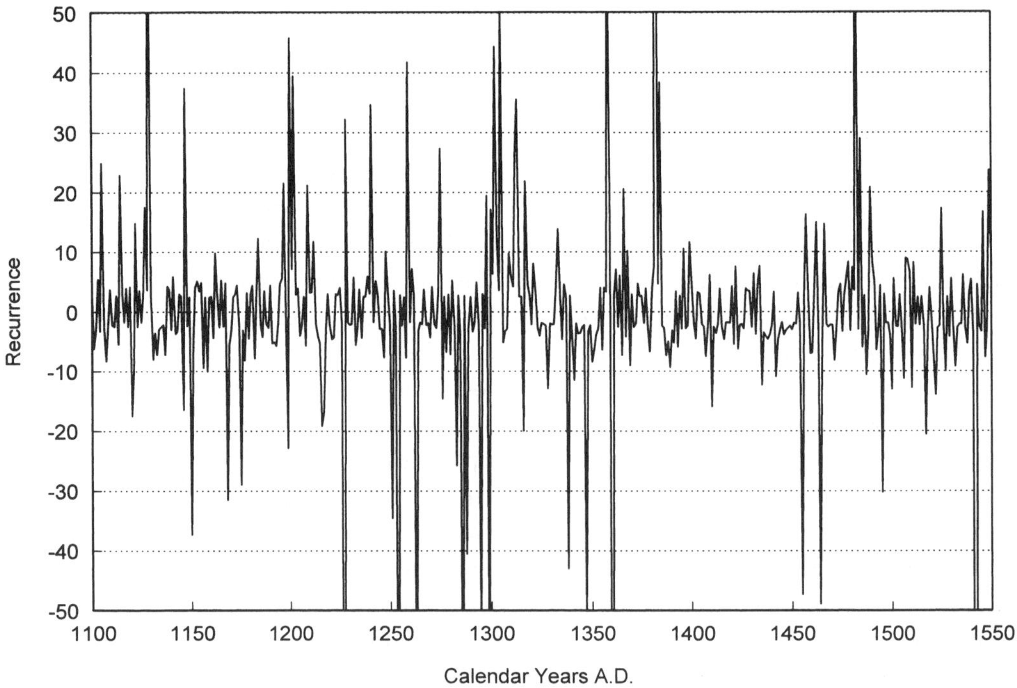

FIGURE 3.4. Recurrence of streamflow values above and below the mean for the Salt River from A.D. 1100 to 1550.

of recurrence intervals and potential damage proposed by Nials and others (1989) is projected onto the Salt River and Tonto Creek in the Tonto Basin.

The annual streamflow reconstructions for the Salt River during the Classic period (A.D. 1150 to 1450; Figure 3.4) show that annual streamflow was within an acceptable range of variance for 256 years, or 85.3 percent of the time (RI 5 or below). Likewise, along Tonto Creek, streamflow was within an acceptable range for canal irrigation for 259 years, or 86.3 percent of the time (Figure 3.5). Minimal damage would have occurred to canal systems during these years, and repairs could have easily been made, thus more than likely ensuring a full growing season. Flows with a recurrence interval of 6 to 10 years occurred 18 years out of 300 years, or 6.0 percent of the time, along the Salt River. Along Tonto Creek, flows of this magnitude occurred 21 years out of the 300 years, or 7 percent of the time. Canals during some of these years may have been inoperable for a portion of an irrigation season. Flows with a recurrence interval from 11 to 20 years would have caused enough damage to the canal systems to result in the loss of part of an irrigation season. This would have happened 9 years out of the 300-year record, or 3.0 percent of the time, along the Salt River. Flows of this magnitude would have occurred 14 years, or 4.7 percent of the time, along Tonto Creek. Flows with a recurrence interval between 21 and 50 years happened along the Salt River 14 years out of the record, or 4.7 percent of the time. During these 14 years, the canal systems would have been inoperable for a part of one irrigation season and possibly longer. However, along Tonto Creek, similar damage would have occurred only 4 years, or during 1.3 percent of the record. Three extreme flows, exceeding a recurrence interval of 50 years,

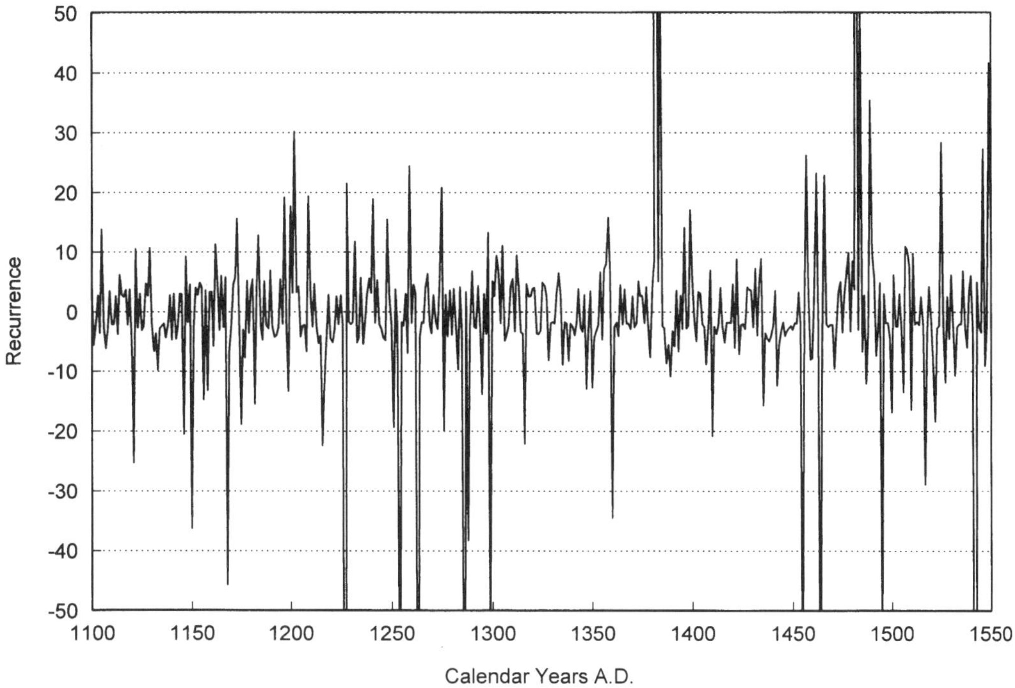

FIGURE 3.5. Recurrence of streamflow values above and below the mean for Tonto Creek from A.D. 1100 to 1550.

occurred during the 300-year record, or 1.0 percent of the time, along the Salt River. These three extreme streamflow events occurred on the Salt River in A.D. 1305, 1358, and 1382. Similar extreme flows occurred along Tonto Creek in A.D. 1382 and 1384 (0.7 percent of the time). These periods would have resulted in catastrophic destruction of the canal systems, rendering the systems inoperable for a year or more. In summary, the canal systems along the Salt River would have probably functioned normally 85.3 percent of the 300 years from A.D. 1150 to 1450, with partial or total loss of the growing season 14.7 percent of the time. Similarly, the canal systems along Tonto Creek would have functioned normally 86.3 percent of the time over the same period, with partial or total loss of the growing season 13.7 percent of the time.

DISCUSSION

The rise, growth, and collapse of the Salado in the Tonto Basin between A.D. 1150 and 1450 can be examined in this environmental context. As we will see, it was not only the magnitude of streamflow events that was important to prehistoric people but the timing and duration of these events.

Before the start of the Classic period (ca. A.D. 1150), the population of the Tonto Basin was dispersed into small hamlets situated on the Pleistocene alluvial piedmont and adjacent to the Salt and Tonto streams (Ciolek-Torrello et al. 1994; Elson and Gregory 1995). This trend lasted until the latter part of the thirteenth century. Starting around A.D. 1280, a change occurred; people constructed a regional network of ceremonial platform mounds around which habitations were loosely clustered. This pattern existed until A.D. 1320, when the platform mounds and adjacent communities were abandoned and people were consolidated into fewer but larger platform mound and pueblo settlements (Ciolek-Torrello et al. 1994; Elson and Gregory 1995). This first period of mound building and abandonment, from A.D. 1280 to 1320, is called the Roosevelt phase, which may have begun as early as A.D. 1250 and ended as late as 1350. This settlement pattern shift is most dramatically documented in the eastern portion of the Tonto Basin (Rice 1992), which included the Pyramid Point mound, Meddler mound, Pillar mound, Pinto mound, Schoolhouse mound, and the Griffin Wash site complex and their associated satellite communities. With the exception of Schoolhouse mound, these sites were abandoned around A.D. 1320 (Elson and Gregory 1995; Rice 1992). Other mounds and associated communities that flourished during the Roosevelt phase were abandoned around the same time.

The abandonment at A.D. 1320 was followed by the aggregation of people into fewer but larger settlements (Ciolek-Torrello et al. 1994; Elson and Gregory 1995). Some earlier Roosevelt phase sites, such as Schoolhouse mound, were expanded, while others, such as the Cline Terrace mound, were established at new locations. This new period of aggregation is known as the Gila phase, which lasted until A.D. 1450 or perhaps a little later. It is unclear whether the population of the basin remained stable, increased, or declined. Examination of the archaeological record for the eastern portion of the basin shows that although there is evidence for expansion, all the people who abandoned the Roosevelt phase communities could not have congregated into the Schoolhouse site (Lindauer 1996). The question remains whether some people left the basin or whether they migrated to the central and western portions of the basin and helped establish new Gila phase communities. The Salado flourished for roughly 130 years, until around A.D. 1450, when there appears to have been a sudden abandonment of the area.

Two events stand out in the Tonto Basin. The first is the creation and collapse of the Roosevelt phase platform mound communities between A.D. 1280 and 1320. The second is the creation and collapse of the Gila phase platform mound communities between A.D. 1320 and 1450. The growth and collapse of these communities may be tied to a variety of causes.

The climate and hydrology of the Tonto Basin appear to have been important factors in the growth and expansion of the Roosevelt mound communities. The last quarter of the thirteenth century in the Tonto Basin was a period of repeated drought. Calculation of Palmer Drought Severity Indices (PDSI) for the basin showed that the period from A.D. 1275 to 1299 had regular episodes of alternating wet and dry years with six major droughts, occurring in A.D. 1276, 1283, 1286, 1288, 1295, and 1299 (Altschul and Van West 1992; Van West and Altschul 1994). Van West and Altschul (1994) suggest that because of the unpredictable nature of crop yields during this time, dry farming and floodwater farming on the alluvial piedmont, which previously had been a viable subsistence strategy, was no longer economically tenable. They suggest that independent farmers on the piedmont abandoned all but the most productive areas and focused on the floodplains of the Salt River and Tonto Creek, where irrigation agriculture could be pursued. Thus, the start of the Roosevelt phase was a period of intensification of irrigation agriculture on the floodplains of both Tonto Creek and the Salt River.

The hydrologic record shows that the period from A.D. 1250 to 1298 was characterized by very predictable streamflow and minimal flooding—ideal conditions for canal irrigation agriculture. In addition to the local movement of people, Van West and Altschul (1994) suggest that immigration occurred into the Tonto Basin from the Colorado Plateau, where the Great Drought (A.D. 1275–1300) was severely affecting farmers. Immigration into the Tonto Basin has been confirmed by the construction of new rooms and compounds and the presence of nonlocal artifacts within them at Saguaro Muerto, Griffin Wash, Meddler Point mound, and Schoolhouse mound (Elson and Gregory 1995). Although the number of people who migrated to the Tonto Basin is unknown, the irrigation systems within the basin were surely enlarged to increase the carrying capacity of the fields to feed these displaced people. Thus, climate change appears to have triggered displacement of people, and local hydrological factors favored the intensification of irrigation agriculture. This may have led to greater social organization, cooperation, and ritual behavior, as suggested by the platform mounds.

However, 20 years after the creation of a new settlement pattern and social order dependent on irrigation agriculture, the hydrologic conditions changed dramatically on the Salt River. The period from A.D. 1298 to 1317 was a 32-year interval of repeated flooding, which would have severely impacted the people irrigating the floodplain of the Salt River. During this period (1.5 generations), there were 4 years (A.D. 1298, 1300, 1303, 1309) of moderate flooding (recurrence interval 10–20 years), when enough damage was done to canal systems to cause the loss of part of the irrigation season; 4 years (A.D. 1302, 1312, 1313, 1317) of intense flooding (recurrence interval 21–50 years), when the canal systems were damaged so severely that they would have been inoperable for a part of one irrigation season and possibly longer; and one year (A.D. 1305) of catastrophic flooding on the Salt River (recurrence interval of over 50 years), during which significant damage would have occurred to the canal systems, resulting in the loss of the growing season. Interestingly, along Tonto Creek high streamflows occurred less frequently and were of lower magnitude. During the same 32-year period, high streamflow occurred only five times, in A.D. 1298, 1302, 1303, 1305, and 1312. All these episodes of streamflow were within a recurrence interval of 10 to 20 years and thus caused only moderate damage to canal systems. The collapse of the Roosevelt phase mound community system and the timing of the hydrologic changes are probably more than coincidental and may have been a critical factor in the collapse of the first mound system around A.D. 1320. The damage to the canal systems, the potential loss of part or all of the growing season, years of low or no harvests, and the future uncertainty of crop yields during these years would have created stress on people dependent on irrigation farming. This would have been compounded by the lack of attention paid to storage of surplus crops during the Roosevelt phase. Lindauer (1992, 1995) notes that there was only a small amount of storage capacity within the Bass Point mound and little or none in Roosevelt phase compounds. These stresses must surely have contributed to the collapse of the Roosevelt phase mound system.

Adding to the climatic and hydrologic factors of the Roosevelt phase collapse were undoubtedly social stresses. Conflict may have erupted between the immigrants and the indigenous peoples and between other groups in the Tonto Basin. There is evidence of conflict within the basin at the end of the Roosevelt phase. At site AZ U:8:221, a small three-room compound, three human bodies lying face down were found on the floor of an intentionally burned room (Whittlesey 1994). They appear to have died as a result of blows to the head (Turner et al. 1994). Further, Roosevelt phase structures are commonly burned throughout the Tonto Basin.

The abandonment at the end of the Roosevelt phase is most dramatic along the Salt River in the very easternmost end of the basin (Rice 1992). Again, this timing is probably no coincidence. Flooding here likely would have been more severe than at any stretch of the Salt River because at this point the river leaves its confined bedrock gorge to flow out into its broad alluvial basin. Floods immediately downstream of the gorge would have been very intense and may explain the abandonment of this area and the near-absence of occupation here since A.D. 1320. Interestingly, although there were some periods of flooding on Tonto Creek during the first quarter of the thirteenth century, they were substantially less intense than those along the Salt River. Thus, Tonto Creek may have been viewed as more stable than the Salt River for irrigation agriculture and may account for the decline in population of the Salt arm of the Tonto Basin and the increase in occupation along Tonto Creek at the start of the Gila phase, when people remained in the basin only to use the best and most protected parcels of land along the lower and middle portions of the Salt River (i.e., downstream of Schoolhouse mound).

After A.D. 1320, the people in the Tonto Basin amassed into fewer but larger nucleated settlements. These larger settlements had platform mounds and adjacent pueblos. The reason for this nucleation is unclear, but it may have stemmed from the need to deal with the unpredictable nature of the streamflow along both the Salt River and Tonto Creek, as well as defense. Many Gila phase sites are now submerged under the waters of Lake Roosevelt, and it is unknown whether population increased, decreased, or remained stable during this time. This renewed period of aggregation of people corresponds with a period of quiescence on the Salt and Tonto streams.

From A.D. 1320 to 1380, streamflow in the Tonto Basin became less variable and more predictable—once again, an ideal time for irrigation agriculture.

Streamflow was typically near normal values, and moderately high streamflows with a recurrence interval of 11 or more, during which part of the irrigation season would have been lost, occurred in A.D. 1333, 1358, 1359, and 1366 along the Salt River and in 1358 along Tonto Creek. The 18-year period from A.D. 1381 to 1399 was characterized by several years of high flow on both the Tonto and Salt arms. Large floods that could have substantially damaged the canal systems occurred four times along the Salt River and Tonto Creek (A.D. 1382, 1384, 1396, 1399). However, there was plenty of labor that could be mobilized to repair any damage to the irrigation systems. None of the floods appear to have significantly impacted the development of the Salado. The 70-year period from A.D. 1384 to 1457 would have been the best time for canal irrigation in the Tonto Basin during the Classic period. This was a time of very predictable discharge with only four minor episodes of flooding, in A.D. 1409, 1422, 1431, and 1434, and only three periods of minor drought. During this period, lasting 3.5 generations, the Salado flourished. Sites increased in size; the population probably grew internally, and additional immigrants may have entered the basin. For example, Grasshopper Pueblo and adjacent pueblos in the headwaters of the Salt River were abandoned by A.D. 1400 (Fish et al. 1994). Fish and others (1994) suggest that some of these people may have migrated into the Tonto Basin. As the population increased through the late Classic period, more and more demand was placed on the small and finite amount of irrigated land in the Tonto Basin. Undoubtedly, during these good years population increased to the maximum carrying capacity of the land.

The collapse of the Gila phase mound communities and the abandonment of the Tonto Basin is poorly documented. According to the existing archaeological data, the Gila phase communities were abandoned around A.D. 1450 or shortly thereafter. The large platform mound communities were abandoned, and it appears that the people left the basin. This abandonment appears to have occurred suddenly. Analysis of 480 burials revealed that the Gila phase occupants were in good health (Ravesloot and Regan 1995; Regan et al. 1994). Also, the macrobotanical remains show no evidence of a shift in the late Classic from heavy dependence on agricultural foods to wild plant resources. There was a heavy dependence on maize until the abandonment (Dering 1996).

The reason for the abandonment is unclear. Late Holocene geology indicates no low-frequency landscape catastrophe at the end of the Classic period in the Tonto Basin that may have rendered the agricultural land unfarmable. There is no evidence of major channel entrenchment that would have eroded the agricultural fields or left them too high above the level of the river to get water to them. On the contrary, there was channel aggradation; deposition occurred on the floodplain, and terrace 2 was stable. Further, terrace 2 appears to have remained a suitable place for agricultural fields (i.e., the soils were acceptable for plant growth, good drainage prevented salinization of fields, and crops would have been protected from most high flows). Also, there was plenty of water in the Salt River and in Tonto Creek to irrigate these fields, even during the driest years. Therefore, we must look elsewhere for an explanation of the Salado collapse.

The two most important variables in the disappearance of the Salado may have been population growth and the limited carrying capacity of the land. The population undoubtedly expanded during the 3.5 generations from A.D. 1384 to 1454, when annual crop yields were at their maximum. During this time, the Salado population probably expanded to the maximum carrying capacity of the land and perhaps beyond. This growth placed the Salado in a precarious situation because they had become dependent on good and reliable annual harvests (as had occurred for 70 years).

Disaster struck in A.D. 1455, when a severe drought occurred and lowered streamflow on both the Salt River and Tonto Creek, followed by flooding in A.D. 1457 and 1462, drought and lower streamflow again in 1464, and more flooding in 1466 along both streams. Thus, over this 11-year period, the agricultural stability that had lasted for 3.5 generations was severely disrupted. If harvests declined (which they probably did) and population was at its maximum, this was undoubtedly a period of stress. If that was not enough to send the Salado into disequilibrium, additional high streamflow and potential flooding occurred on both Tonto Creek and the Salt River in A.D. 1477, 1478, and 1480, followed by very severe flooding and damage to the irrigation systems in A.D. 1482, 1484, and 1489 and minor flooding in 1490 followed by drought in 1495. Extreme flooding during these years would have rendered the canal systems potentially inoperable for part if not all of a growing season. This period of streamflow instability and drought may have caught the Salado off guard even though they appear to have amassed enough storage at the larger sites, such as Schoolhouse, to provide food to withstand one to two years of poor harvests (Lindauer 1996). They may have been made complacent by the good years and allowed their population to expand to the productive limits of the land. The reduction in crop yields and increased uncertainty of yields from one year to the next may have led to social stresses, political strife, and disputes over land tenure. The changing environmental variables that affected the food supply may have intensified existing and ancient animosities between people in the basin. Internal and external migration of people into the Tonto Basin had brought a diverse group of people together in a small area that had only limited agricultural potential. Disputes between various groups may have begun long before A.D. 1455, and the environmental instability that disrupted food production may have been the factor that brought underlying hostilities to the surface. Thus, a combination of environmental and social factors appears to have been responsible for the final collapse of the Salado.

The geomorphic and settlement history of the Tonto Basin provides a good example of how changes in high-frequency geomorphic processes affected prehistoric people. For the most part, the social and subsistence systems of the occupants of the basin were flexible enough to deal with the highly variable discharge of the Salt River and Tonto Creek. However, with time, as the population grew and became more diverse and more dependent on agricultural products, the people of the basin inched their way into a precarious situation. It appears that the frequency of high-magnitude flooding coincided with the time during the middle of the fifteenth century when the population reached the maximum carrying capacity of the land. Repeated crop failures over a short period of time created a shortfall of food that could not be made up from storage, upland crop yields, and wild foodstuffs. These resources were just not enough to sustain the entire population of the Tonto Basin for long. In the end, there was just too much strain on the cultural system because of crop failures and cultural diversity, and the cultural system collapsed. In sum, the collapse of the Salado occurred when the population had grown to the limits of the environmental system and cultural diversity was at its maximum, and was triggered when high-frequency discharge events and droughts occurred over a very short period after a long period of stability.

CONCLUSIONS

Understanding high- and low-frequency landscape changes is crucial for understanding prehistoric human behavior. The record of high- and low-frequency landscape changes is preserved in the stratigraphy surrounding most archaeological sites. How prehistoric people responded to landscape changes is visible in the archaeological record. To elucidate relationships between the landscape and people, researchers should conduct geoarchaeological studies along with archaeological investigations, allowing prehistoric sites to be placed into geomorphic context in a dynamic prehistoric landscape.

ACKNOWLEDGMENTS

Funding for the research along the Santa Cruz River was provided by the Arizona Department of Transportation (ADOT) and the Cultural Resource Management Division of the Arizona State Museum. Funding for the research in the Tonto Basin was provided by the U.S. Bureau of Reclamation and the Office of Cultural Resource Management, Arizona Sate University. I thank David Doyel and Jeff Dean for offering comments on this chapter.

4

Cross-cultural Perspectives on Prehispanic Hohokam Agricultural Potential

Suzanne K. Fish and Paul R. Fish

In the southwestern United States, archaeologists often link the trajectories of indigenous societies to environmental change, especially as driven by climate. Perhaps more than in other, similarly arid world areas, this linkage reflects the dominance of a regional paradigm of agricultural marginality, coupled with the availability of highly detailed paleoclimatic records. Our notions of the relationship between prehispanic cultivators and their environment are influenced, however, by expectations rooted in the agricultural traditions of temperate climates and industrialized technologies. Early southwestern farmers are thus primarily seen as facing daunting challenges to an agricultural lifestyle (Fish and Fish 1994). Tim Kohler (1993:269) underscores the pervasiveness of these perceptions of marginality in his observation that "the effectiveness of various ecological approaches for understanding culture change in these environments (which on a large scale of comparison, are marginal for agriculture and sensitive to climatically induced change and human impact) has allowed these positions to remain more dominant than they are, for example, in Europe."

Our studies with traditional farmers represent an attempt to approach the interplay between agriculture and environment through eyes unconstrained by comparative concepts of marginality. Instead, we sought viewpoints that, like those of farmers in the past, were shaped by rich, multigenerational experience in the opportunities and risks of southwestern agricultural practice. By "traditional farmers," we mean those who incorporate persisting indigenous agricultural traditions and who cultivate using technologies that are relatively nonmechanized and small scale compared with industrialized operations. The social institutions of farmers are integral to these agricultural traditions, which typically include risk-sharing and cooperative techniques.

Archaeological science allows us to track well-defined environmental variables with increasing precision at multiple temporal and spatial scales and to examine how trends in these measures articulated with the rate and direction of cultural change. Such current concern with long-term monitoring of environmental conditions and their consequences is not without parallels among members of the region's traditional farming societies. Oral traditions of many tribal groups in the U.S. Southwest transmit knowledge of environmental events and their impacts, frequently encoding proper and improper human response. Likewise, traditional farmers in northern Sonora are able to recite a calendar of destructive flood events over the last century (Sheridan 1988:55). Record keeping of environmental episodes and periodicities beyond individual memory is particularly well exemplified by Pima calendar sticks. Using symbols inscribed for successive years, the keepers of these calendric devices were able to recount recurrent flood events on the Gila River over many decades (Russell 1975:36–66).

Longitudinal records of critical environmental variables are invaluable for understanding and interpreting the histories of southwestern farmers. However, we lack similarly focused and systematic insight at the human scale of assessment and decision making by individuals and their immediate social units in the cultural context of farming societies. These processes are more complex than the articulation of key variables, and it is the cumulative outcomes of such decisions in the past that register archaeologically detectable change. Moreover, Robert Netting (1993) has convincingly argued that strategies at these organizational levels are the determinants of the success and long-term viability of traditional smallholder agriculture.

Consultation with traditional farmers provides insights into the practitioner's dynamic role in mediating the relationship between agriculture and environment. We do not propose that even those consultants who have the most direct continuities with ancestral populations replicate the knowledge, outlook, and decisions of their predecessors. Rather, they illustrate which variables are deemed critical by farmers using similar techniques in similar environments and provide a glimpse into conceptual frameworks for monitoring and applying these variables in making agricultural decisions. Consultant deliberations also reveal the classificatory systems of traditional agriculturalists and exemplify the role of both economic and non-economic motivations.

Our immediate archaeological objectives in consulting with traditional farmers were (1) to enhance our ability to reconstruct and interpret farming practices and outcomes associated with particular sets of agricultural remains and (2) to improve our assessments of the prehistoric productive potential of particular locales. We also took seriously a responsibility to share archaeological knowledge with Native American communities. All the tribal consultants had long-standing personal and professional interests in traditional and prehistoric agriculture and, as cultural preservation advocates, were well suited to communicate their observations to others. It was not until the final study of the three discussed in this chapter, however, that the focus turned from evaluating prehistoric agriculture per se to also understanding the manner in which migrating farmers assessed and responded to new environments at

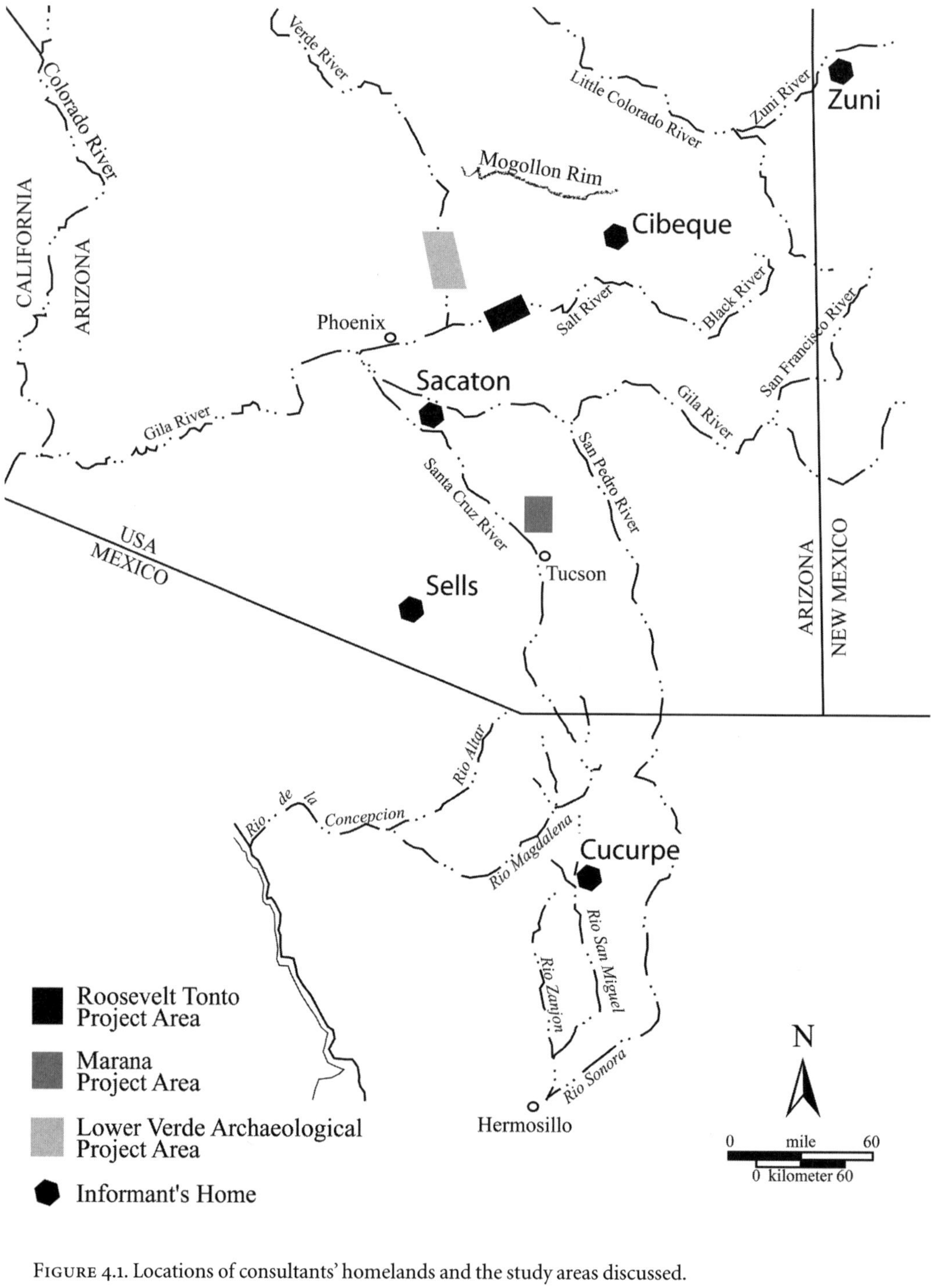

FIGURE 4.1. Locations of consultants' homelands and the study areas discussed.

their destinations. The processes by which migrating farmers established productive strategies under such circumstances have parallels in the reactions of consultants to unfamiliar locations.

Farmers who maintained traditions of indigenous cultivation participated as consultants for three archaeological study areas in southern and central Arizona (Figure 4.1). Each study shared overlapping research objectives and methodologies, and three of the six consultants participated in more than one project. The initial study by the authors in 1986 focused on agriculture in the Hohokam Marana Community in the northern Tucson Basin (Fish et al., eds. 1992:41–52). The second, by John Welch in 1992, was part of a broad-based examination of changing land use in the Tonto Basin (Welch and Ciolek-Torrello 1994:51). The final and most extensive consultation, by the authors in 1994, involved agricultural settings in the Horseshoe Basin of the Lower Verde Valley (Fish and Fish 1997:239–248). In each case, consultants included traditional farmers living relatively near the study areas and cultivators from more distant localities with environmental and technological similarities.

CONSULTATION METHODOLOGY

The "ideal" consultants for analyzing factors governing the prehistoric agricultural potential of each study area would (1) embody a maximum degree of cultural continuity with the former occupants, (2) be conversant with techniques suggested by the archaeological remains, and (3) have personal farming experience in the local environment. Consultants meeting these criteria were not available; farmers maintaining indigenous practices had long been absent from the three locations. Consequently, an alternative methodology of comparing and contrasting responses by members of agricultural traditions in surrounding regional sectors was followed. Consultants included Sonoran mestizo, Zuni, Apache, Pima, and Tohono O'odham farmers.

Consultant observations were not intended to replace insights from the extensive literature on southwestern traditional farming (e.g., Doolittle 1984, 2000; Muenchrath et al. 2002; Nabhan 1986) or experimental studies with traditional techniques and crops (e.g., Muenchrath 1995; Nabhan 1984) but represented an additional and geographically specific avenue for exploring alternatives and decision making in areas lacking clear ethnographic models.

In view of differences in consultants' homeland environments, techniques, and experiences, their judgments and strategies can be construed as only generalized analogues for those of prehistoric agriculturists in the study area. Eliciting exhaustive or overly detailed evaluations would not have increased the accuracy or applicability of responses and, indeed, might have fostered misleading conclusions. Each consultant was encouraged to address topics on which he felt able to comment meaningfully. Many of the same issues were raised in each case, but the studies did not pose standardized queries. To obtain a level of comparability in responses, consultants visited a designated set of locales representing a range of environmental situations. Some locales contained prehistoric agricultural features, and some were considered probable field settings for nearby settlements.

BACKGROUND OF CONSULTANTS

Delores Lewis and Joseph Joaquin, Tohono O'odham from the vicinity of Sells in southwestern Arizona, were familiar with floodwater (*ak-chin*) farming from ephemeral streams in low-elevation desert basins. Archie Russell, a Pima, spent his youth on a farm that had been irrigated from the Gila River near Sacaton in central Arizona. Glenn Cromwell, an Apache, cultivated an irrigated field along Cibecue Creek and a runoff field near Oak Creek in the mountainous zone just below the Mogollon Rim of east-central Arizona. Jerome Zuni farmed irrigated and runoff plots near Zuni, New Mexico, on the Colorado Plateau. Adalberto Cruz, a farmer from Cucurpe along the Río San Miguel in northwest Sonora, was familiar with irrigation from a perennial stream and with floodwater techniques on alluvial fans and in ephemeral tributaries. All consultants had active interests in the agricultural practices of previous generations and in archaeological indications of farming in their homelands. Each gave his permission to be identified and quoted.

The success of farmers employing nonmechanized technologies is dependent on an intimate understanding of local conditions (e.g., Wilken 1987). A critical common denominator in the Southwest is the need to command the knowledge necessary to ensure delivery of supplemental water at the proper stage in the growing season. Key environmental variables in the areas studied differ significantly from those in the consultants' homelands, including the seasonal distribution of rainfall and temperature, soils, topography, and hydrological regimes. However, the largest intervening distances do not exceed the scale of well-documented population movements in the past, as from the Four Corners area to the Rio Grande Valley (e.g., Kohler 1993:293–297) or from the vicinity of Kayenta into southeastern Arizona (e.g., Woodson 1999). Moreover, migrations into the Verde Valley have been posited from both highlands to the north and broad, flat basins to the south.

Elevation above sea level is an easily compared and relevant measure for farming. In descending order, the elevation is 1,885 m at Zuni, 1,640 m at Cibecue, 930 m at Cucurpe, 750 m at Sells, 670 m at the central mound site in the Marana Community, approximately 650 m at Horseshoe Dam on the Lower Verde and at Roosevelt Dam in the Tonto Basin, and 460 m at Sacaton. Obviously, frost as a limiting factor to agriculture would differ widely over such a range in elevations and latitudinal position from north to south (Figure 4.2). Elevation trends are further magnified or ameliorated locally in consultants' homelands by factors such as slope aspect and cold air drainage.

Hydrology likewise differs in the consultants' homelands and the study

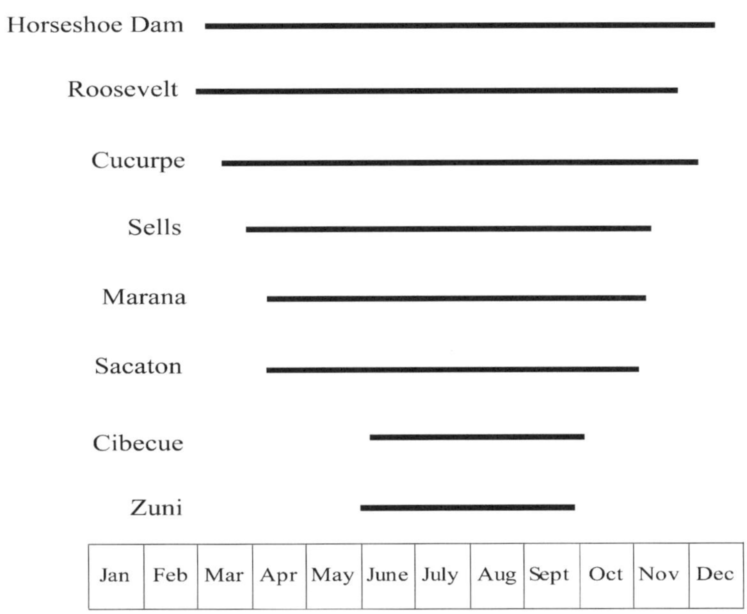

Sites	90% Probability After	90% Probability Before
Horseshoe Dam	March 10	December 16
Roosevelt	March 3	November 20
Cucurpe	March 15	December 1
Sells	March 29	November 6
Marana	April 10	November 3
Sacaton	April 10	November 1
Cibecue	June 7	October 1
Zuni	May 30	September 24

FIGURE 4.2. Probabilities for early and late frost in consultants' homelands and the study areas. Arizona and New Mexico data are from the National Weather Service's Western Regional Historical Climate Summaries, www.wrcc.dri.edu. Cucurpe, Sonora, data are from Durrenberger and Murrieta 1978:69 and Hastings and Humphrey 1969.

areas. Peak summer flows in major watercourses near Marana, Sells, and Cucurpe reflect relatively local rainfall. In each of the other areas, major watercourses draw on more distant upland watersheds and sustain peak flows in the spring. Before damming, the Gila River near Sacaton carried substantially more water than the Santa Cruz River in the Tucson Basin, upper Salt River in the Tonto Basin, and lower reaches of the Verde River, but at notably lower gradient. The largest prehistoric Salt and Verde irrigation networks would have been matched only along the Gila, although by the smaller systems there. The drainage systems of the Río San Miguel at Cucurpe, the Zuni River, and Cibecue Creek are smaller networks than those of the Gila, Verde, and Santa Cruz. On the other hand, tributaries and mesa edge situations near Cibecue and Zuni offer opportunities for farming reminiscent of those in the Tonto and Verde study areas.

The amount and timing of rainfall is a pivotal variable. Annual rainfall in

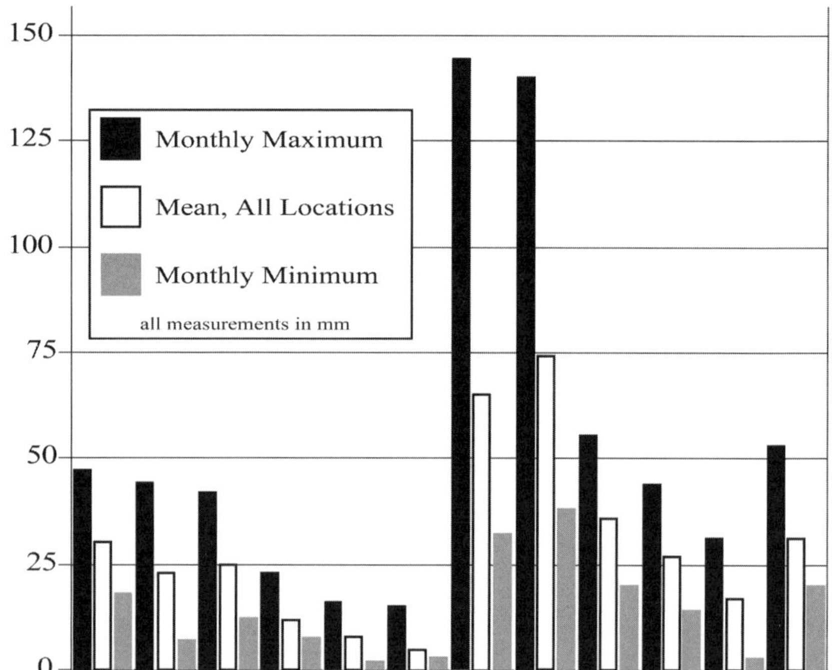

	Jan	Feb	Mar	Apr	May	June	July	Aug	Sept	Oct	Nov	Dec	Annual
Horseshoe Dam	42	37	39	14	6	7	40	54	31	36	30	41	377
Roosevelt	47	44	43	18	8	8	37	50	33	28	31	52	399
Cucurpe	33	7	12	9	10	10	144	140	55	29	3	24	476
Sells	20	15	19	7	2	4	72	66	33	26	16	36	316
Marana	29	29	19	10	4	3	49	56	22	17	11	28	277
Sacaton	21	19	21	8	4	3	32	38	20	14	16	27	223
Cibecue	47	40	42	23	12	15	62	78	47	44	31	41	482
Zuni	18	17	21	13	12	13	54	55	34	22	14	20	293
Mean, all locations	32	26	27	13	7	8	61	67	34	27	19	34	355

FIGURE 4.3. Monthly variability in precipitation in consultants' homelands and the study areas. Arizona and New Mexico data are from the National Weather Service's Western Regional Historical Climate Summaries, www.wrcc.dri.edu/. Cucurpe, Sonora, data are from Durrenberger and Murrieta 1978:69 and Hastings and Humphrey 1969.

the consultants' home areas varies by a factor of more than two, and seasonal distribution also differs (Figure 4.3). In yearly totals and seasonal proportions, for example, Lower Verde patterns resemble those of Cibecue, at a higher elevation. In July, the critical month for summer planting, Lower Verde precipitation is substantially less than at Cucurpe, Cibecue, Zuni, Marana, Tonto, and Sells; only the driest, hottest Sacaton locale receives lower rainfall in this month. At two-thirds the annual total, summer rainfall at Cucurpe is high; the average during the July–September growing season is almost 300 mm, while the next highest, at Sells, is only 171 mm.

CONSULTANT EVALUATIONS IN THE TUCSON BASIN

Understanding the diversified agricultural strategies within the 146-km², Early Classic period (A.D. 1150–1300) Marana Community has been a research goal for more than two decades (e.g., Fish 1995, 2000; Fish et al., eds. 1992). Environmental variation translated into mixed farming practices from valley floor to mountain flanks: moderate-scale Santa Cruz irrigation, floodwater fields on alluvial fans and the floodplains of larger drainages emerging from the Tortolita Mountains, and a host of rockpiles and linear rock alignments trapping overland runoff. Some prehistoric techniques such as upland floodwater cultivation and midbasin rockpile fields were not duplicated in postcontact times. The first of the three consultant studies was undertaken here with the aid of Gary Nabhan and Thomas Sheridan, who had previously benefited from the knowledge and advice of the expert farmers Delores Lewis and Adalberto Cruz. The Marana study initiated collaborative approaches to evaluate agricultural remains and the potential of archaeological landscapes.

Floodplains of large washes were judged to be agriculturally desirable by both Lewis and Cruz on the basis of vegetation indicators. Cruz noted the size of mesquite trees in identifying good agricultural land. William Doolittle (personal communication 2002) has noted that contemporary Sonoran Desert farmers almost universally recognize improved crop conditions in the vicinity of mesquites and leave them in fields, perhaps because these trees facilitate nitrogen fixation in the soil. Lewis observed the size and intense green of palo verde trees and the presence of catclaw acacia on wash floodplains to judge the potential for hand-dug wells (the authors subsequently encountered water at a depth of less than 2 m during the annual June drought). Both informants commented on ironwood concentrations on slopes as indicating the localized absence of frost and therefore the possibility of early crops grown with winter rains.

Although Cruz and Lewis agreed on the desirability of cultivating floodplains along major washes, Lewis was much more willing to consider smaller drainages with comparatively limited arable land. These situations were compatible with his experience in Papaguería, where, in the absence of rivers, all farming employed floodwater techniques and was primarily oriented toward subsistence. Cruz, on the other hand, was an irrigator from the perennial Río San Miguel and a rancher attuned to growing cattle feed in a market economy and only limited food crops. His primary labor was directed toward highly productive fields along the river and only secondarily toward floodwater cultivation on tributaries.

Neither consultant favored midbasin slopes for cultivation, deeming them "too dry" for farming and without opportunities to divert floodwaters to improve soil and moisture conditions. The extensive prehistoric fields with rockpiles and alignments were outside their farming experience, and they could not suggest possible crops. While recognizing the mulching effects of the rock features and their role in intercepting overland runoff, they believed moisture to be insufficient for annual crops. When told that archaeological evidence suggested agave cultivation, Cruz favorably compared the features with similar rock plantings in his son's ornamental landscape business in Tucson.

Both consultants considered conditions for floodwater farming on alluvial fans to be favorable. The relative coarseness of soils along the channels helped them judge the rate and force of flow; Lewis carefully noted the extent of plant debris left behind by flooding and runoff events. Their notions of optimal field placement at points coinciding with natural water spreading along drainages, in conjunction with diversion structures in upslope watershed areas, conform with descriptions of Tohono O'odham fields of the late nineteenth century (e.g., Bryan 1929; Nabhan 1986). Large fans would require greater effort, involving the construction of upstream devices to concentrate runoff and longer ditches to deliver water to the fields.

In some locations, both consultants considered the soil too sandy to achieve high yields. They suggested that finer-grained and more water-retentive mediums could be created by diverting water to deposit fine silts for one to several years prior to cultivation. These improvements might have encouraged further mediations such as ditches extended to sequential fields, as in the incremental expansion of floodwater field systems described by Doolittle (1984) along the Río Sonora.

Cruz and Lewis designated portions of the Santa Cruz floodplain adjacent to the Tucson Mountains as certainly the most productive land in the Marana Community. They also identified there some stretches of the river, where underlying bedrock kept flows near the surface, as the location where canals should head. However, recent river conditions are so dramatically altered that neither consultant was willing to suggest more specific farming strategies. Living fencerows of trees like those planted along the river at Cucurpe, Sonora, for soil conservation (Sheridan and Nabhan 1978) were considered, but Cruz believed more information about river conditions in the past was necessary to evaluate this possibility.

CONSULTANT EVALUATIONS IN THE TONTO BASIN

A second collaboration with traditional farmers was occasioned by a study of prehistoric rural settlement and changing prehistoric land use around Lake Roosevelt in the Tonto Basin of east-central Arizona (Welch and Ciolek-Torrello 1994:51). In this Statistical Research project, John Welch (1994:229–251) visited the study area with Glenn Cromwell, an Apache farmer, and Sonoran Adalberto Cruz. Again, Thomas Sheridan facilitated Cruz's participation. The consultants visited locations associated with various types of farming, including riverine irrigation, upland spring-fed irrigation, *ak chin* or floodwater farming, and slope systems of small-scale terraces and check dams.

Cruz and Cromwell both identified a number of soil, water, slope, and vegetative indicators for promising situations, frequently agreeing on means of assessment. For instance, a lack of depth, apparent low fertility, and the presence of salts suggested to them that successful crops such as corn were improbable in locations with check dams and terraces. Although it was clear that these devices were constructed to control runoff, they both concluded that the purpose must have been to control erosion or to increase the production of natural vegetation rather than to grow typical annual crops.

The consultants expressed variable perceptions of agricultural potential and appropriate technologies based on their differing homeland experiences. For example, Cromwell, the Apache farmer, identified an upland area of Civilian Conservation Corps erosion-control alignments surrounding springs as having the highest potential of visited locales (Welch 1994:233). The Sonoran farmer preferred locations for riverine irrigation, where both water and soil fertility could be better managed, despite the greater need for infrastructure and heightened flood risks to fields and canals. He believed that living fencerows would be worthwhile for stabilizing field edges along the long, narrow floodplain of Tonto Creek but not along the Salt, and that the larger river could not be similarly "tamed" by planting trees.

Cruz, a small-scale rancher as well as a farmer, believed a mixed cattle and cropping enterprise would be necessary to successfully deploy his technological repertoire in the Tonto Basin, although Cromwell did not share this conviction. To Cruz, cattle were necessary to loosen the soil, adding to the erosional sediment load, thereby increasing fertile deposits on irrigated fields. However, neither consultant systematically discounted possibilities of areas not suitable for canal or floodwater irrigation. Philosophically, Cruz observed that we can reconstruct only a fraction of the knowledge of prehispanic peoples and may never be certain what they knew about what to grow where and when (Welch 1994:234).

CONSULTANT EVALUATIONS IN THE VERDE VALLEY

The final collaboration with traditional farmers was conducted as part of a larger interdisciplinary study of prehistoric agriculture by Statistical Research (Homburg and Ciolek-Torrello 1997) on the Lower Verde River north of Phoenix in central Arizona. Jeffrey Homburg accompanied the authors and consultants, and Karen Adams also participated in some of the visits. This more extended consultation followed the same general methodology as in the previous instances. Traditional farmers included Apache Glenn Cromwell, Tohono O'odham Delores Lewis, Tohono O'odham Joseph Joaquin, Pima Archie Russell, and Puebloan Jerome Zuni.

Archaeologists designate most Lower Verde archaeological remains Hohokam or Sinagua on stylistic grounds. There is little agreement on the identity of postcontact Sinagua descendants (for discussion, see Whittlesey 1997:672–688). Piman-speaking groups, usually regarded as having the closest cultural ties to the Hohokam, did not live in the Verde Valley during the postcontact era. Subsequent records of cultivators in upstream sections of the valley do not describe occupational or agricultural patterns that would account for the full range of archaeological remains. Spanish explorers encountered *ranchería* peoples (e.g., Bartlett 1942), but their cultural affiliation with later groups is unclear. Both Yavapai and Apache farmed on a small scale in the valley and uplands, in some cases until early in the twentieth century. These groups were not strongly dependent on agricultural production, however, and had not practiced traditional farming for decades by the time of the archaeological study (Morris 1971:42). There is no historical record of indigenous peoples farming in the Horseshoe Basin study area at any time.

In the absence of groups closely allied culturally or geographically with the Horseshoe Basin study area, Native American consultants were sought from surrounding homelands, in keeping with the posited origins of prehistoric migrants to the Verde Valley. Indeed, the archaeology of the Verde Valley strongly suggests this area's role as a cultural crossroads and a destination for external populations and influences (e.g., Whittlesey 1997). Appropriately, the consultants' backgrounds encompassed upland zones as well as low, hot basins. Their reactions on entering the Verde Valley are highly relevant to migration scenarios in which newcomers had to rapidly assess local conditions and respond as subsistence farmers.

The consultants examined a series of situations representative of the study area (Figure 4.4). Johnson Ranch offers both relatively substantial riverine irrigation and river terrace/mesa edge floodwater farming. The McCoy Mesa site contains a series of stone check dams. The Cow Wallow site is on an alluvial fan that appears suitable for floodwater farming. Locations at the margins of the circumscribed floodplain of Lime Creek are potentially amenable to small-scale irrigation and floodwater diversion. All consultants visited these locations, with the exception of Russell at Lime Creek. Only Cromwell observed the extensive rockpiles and cobble terraces at the Mullen Mesa site, and only Zuni viewed the rockpile and terrace field at the Crash Landing site.

Assessing Johnson Ranch

At Johnson Ranch, approximately 150 acres of sandy loams were planted at the time of the study under gravity irrigation on the first terrace of the Verde River, and about 40 additional acres were cultivated with runoff. Mindeleff (1896:238–239) described and Midvale (1946) mapped a prehistoric canal two miles long that headed in approximately the same location as a modern one. These canals and other historic remnants demonstrate that the path of canals shifted in the area over time. A non-irrigated field is on a colluvial slope at the base of McCoy Mesa. It appears to have been outside the bounds of prehistoric irrigation, as is currently the case. Within and uphill from this field, Midvale (1946) recorded stone terrace and grid alignments and diversions in washes draining the mesa.

Lewis and Joaquin commented that the soil in the irrigated field was

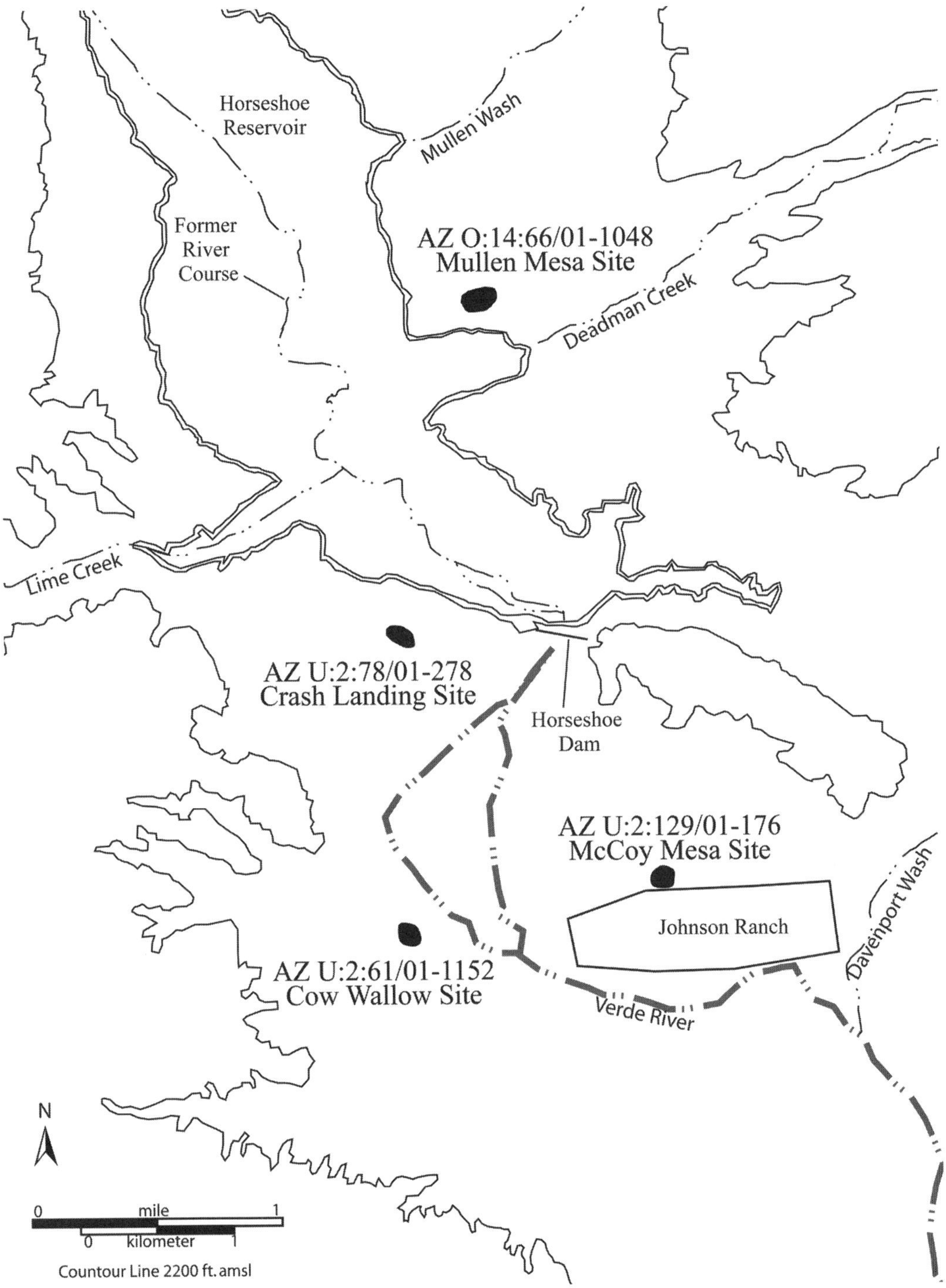

FIGURE 4.4. Locations visited by consultants in the Horseshoe Basin study area.

of good quality for farming, although a little sandy. The presence of desert broom and the size and vigor of mesquites are indicators of farming potential. The tendency of a crust to form on the soil surface, inhibiting the emergence of seedlings, is probably a consequence of plowing. An alternative would be to flood the field first and then poke holes for seed. Low earthen berms could have been constructed to help spread the water in an irrigated field like this one. Corn should be watered at germination, when the plants are between 1.5 and 2 feet tall, and again when the ears are forming. Additional watering would only encourage vegetative growth. Corn and beans could be grown here. Rotation of the crops every two years would be a good practice.

According to Russell, the soil in the irrigated field was good and showed no sign of alkali accumulations. The dirt clods are relatively big and would have to be broken up for planting. The amount of sand is no problem; its presence would help to limit the size of clods. However, when soil is overly sandy, seedlings may sprout too quickly or be damaged by alkali content. Mesquites at the field edges are a sign of good agricultural soil.

Cromwell found the soil to be slightly sandier than optimal: this type of soil requires more water and heats up so much it can kill young plants. On the other hand, if planting is done early in the year, high soil temperatures are an advantage (an early planting would be from February to March in Cibecue, but he estimated February at Johnson Ranch by comparing it with Oak Creek). Sunflowers growing at field edges indicate fertile soils. Corn, beans, and squash could be grown. Because of the warm, sandy soil, Cromwell would plant corn early, spacing plants 10 to 12 inches apart. He would intersperse more corn during the summer to have longer, staggered harvests. He might alternate corn and squash in a given row in sequential years. Beans would be planted in the sandier parts of the field because they mature in less time and could take advantage of the higher water infiltration rates. He noted that if the soil is very sandy, irrigation may be required every three or four days. Cromwell thought this field also could be cultivated without irrigation, using surface runoff. Cottonwoods and willows would be left along the edge of the river or could be planted to serve as a buffer against flooding.

The soil in the field had about the same texture as the soil on Zuni's farm. Since it is deep, it should drain well and not become salty. He tasted the soil to confirm the lack of salt. He mentioned corn, beans, and melons as crops that would do especially well. Zuni would irrigate every 10 days and thought that the water would not settle at this frequency of flooding (settling was considered disadvantageous, apparently because of waterlogging). An optimal field size here would be about five acres, using traditional irrigation methods. He suggested that long, continuous brush structures following the natural contours should be constructed to guard against flooding.

Zuni commented that he never needed to fertilize at Zuni even though the river was currently dammed (damming causes localized deposition of suspended nutrients behind the dam and should make downstream flows less effective for fertilization than prehistoric water supplies). However, he noted that weeds also might be burned off and the ashes mixed in and that portions of fields sometimes could be left fallow to increase fertility.

All consultants noted the high sand content in the Johnson Ranch irrigated field. Although they drew somewhat differing conclusions about the effect on crops, none found this amount of sand an important obstacle to farming. Sandy flood deposits could become a hindrance in any irrigable area along the Verde. The consultants' consensus as to the desirability of this land demonstrated their tacit agreement that such difficulties would not offset the advantages and sustainability of irrigation. Techniques for combating flooding included the construction of brush structures to baffle flows and preserving or planting trees along the river edge.

The relatively large watershed, substantial gradient, and narrow floodplain of the Verde presents a threat to irrigated production as great or greater than in any of the consultants' homelands. Steeper gradients than in the Phoenix Basin heighten risks of washouts where canals must cross tributaries. An elaborate stone aqueduct across a side drainage recorded by Midvale (1946) reflects the investment needed to extend canals laterally in the Verde Valley. The continuous repair of delivery systems must have been a routinely accepted part of irrigated farming in prehistoric times. The consultants' comments identify the trade-off as high productivity without the need to fertilize or practice extensive fallowing. At the same time, the Verde's hydrological volatility likely is responsible for a low archaeological visibility of previously irrigated land.

Zuni made the significant observation that the position of the irrigated field at Johnson Ranch would have permitted prehistoric watering by directed slope runoff and by canal irrigation. Dual sources of supplemental water would have enhanced productivity and predictability. When flooding disrupted canal segments near the active floodplain, reduced cropping still might have been possible with the use of runoff. Such runoff farming also might have fared better than irrigation during episodes of river downcutting.

Lewis and Joaquin judged the soil in the upper, non-irrigated field at Johnson Ranch to be even better than in the adjacent irrigated land. Lewis thought it had a better balance of clay and sand. Both were impressed by the large mesquites within and around this field. In clearing a mesquite bosque, they also would leave a few trees to provide shade and to help spread out the moisture and make it available for crops. The Tohono O'odham farmers noted that the adjacent washes had large watersheds (a fact confirmed by topographic maps) and that it would be necessary to use small, side-ditch diversions to direct flows onto field areas. Water draining off the adjacent uplands also would have to be managed with brush structures to protect these side diversions, simultaneously trapping water and sediment.

In Russell's opinion, soil in this field was about the same quality as that in the irrigated field. He noted that sunflowers in the vicinity were healthy and green, a sign of good soil. Although this Pima farmer never practiced floodwater farming himself, he thought that relatively high rainfall would be necessary for success here.

Like the Tohono O'odham farmers, Cromwell found the soil in the upper field even better than in the irrigated field because there was less sand. The field setting is very similar to many Apache plots at mesa edges. If new to the area, Cromwell would cultivate this field first and later develop irrigation downslope. He also would emphasize upper field cultivation during seasons when floods threatened the irrigation system. In comparison with fields at Cibecue, this field seemed at less risk from washouts. Side diversions would spread out the water draining the uplands. Cromwell would use rocks to protect parts of the field threatened by gully erosion. Mesquite brush could be used to slow the water; sediment deposited as a consequence could be used to fill in low areas. In addition, he would dam the drainage at the eastern field edge with rocks so it would fill with sediment until it became sufficiently shallow for easier diversion.

Zuni's first impression was that a field like this would be good for growing medicinal plants. It would also be possible to grow the same crops as in the irrigated field. He recalled a productive Zuni field like this, irrigated from a tributary. As at Zuni, brush structures could slow down the water and spread it across the plot. The soil seemed to have as high an organic content as that in the irrigated field. The large tree left standing in this field prompted him to observe that occasional trees might be left to hold soil, provide shade, and act as windbreaks.

The promise of the upper field was a matter of consensus for each of the consultants familiar with floodwater farming. Compared with the irrigated field below, an advantage was its lower flood risk. Russell, who knew only irrigation from the Gila, nevertheless recognized soil and plant indicators of good agricultural potential. His speculation that high rainfall would be needed to cultivate this non-irrigated field highlights a fundamental contrast between his home area and the Lower Verde: annual rainfall in the study area is more than half again as great as at Sacaton.

Opportunities to extend canals well off the active floodplain, and thus irrigate appreciable acreage as at Johnson Ranch, are rare along the relatively narrow and steep valley of the Lower Verde. In contrast, situations like that of the upper field are widespread along the outer margins of the valley floor and could have been watered by directing local runoff. Uplands flanking the valley and even the higher river terraces offer catchments for concentrating rainfall. Smaller labor forces than those required for irrigation could manage such water. Because major infrastructure would not have been needed to begin cultivation, even fairly numerous newcomers might have followed Cromwell's strategy of farming these situations before developing irrigation systems. On the other hand, if prime non-irrigated situations were already in wide use, newcomers might have been gainfully employed to expand existing canal networks.

Cobble and grid-like alignments for slowing and directing runoff are still evident at the upper edge of the non-irrigated Johnson Ranch field. In view of the fact that the consultants mentioned brush as a construction material more often than rocks, archaeologically visible remains probably represent only a fraction of such prehistoric devices. It is likely that the cumulative productivity of land in settings like the upper field approached the scale of production on the relatively localized irrigation systems of the Lower Verde.

Assessing McCoy Mesa

At McCoy Mesa on a high Verde terrace, 12 check dams were constructed across shallow swales that drain into a deeply incised wash. The cobble dams are generally 4 to 5 m wide, spaced at approximately 3-m intervals, and form distinct terraces that alter surface elevations by about 20 cm. Soils in the immediate vicinity are a somewhat deflated, cobbly mix of colluvium and alluvium. The larger wash empties directly onto the currently non-irrigated upper field at Johnson Ranch. Three cobble structures appear to be field houses associated with the check dams.

Even in such rocky soil, Lewis stated that farming is usually possible if small weedy plants are present, such as those in the vicinity of the check dams. Corn would require supplemental water to do well in this location. Lewis and Joaquin noticed that one unbreached check dam still collected organic detritus, indicating the passage of appreciable water and a continuing baffling effect by the feature. Such constructions would be useful in slowing the water that runs onto the Johnson Ranch fields. Rocks thus arranged might serve as direct controls on soil and water or as anchors for an additional brush structure.

Russell recognized the check dams as "dams of some sort." He believed they could have been used for small plantings or to slow water to prevent erosion. However, he doubted their effectiveness in protecting the Johnson Ranch fields because the water could still overflow and continue rapidly downstream toward them.

Cromwell also thought the check dams might be used for either erosion control or garden plots. It might have been necessary to add to their height over time as silt accumulated behind them. He suggested that another way to stabilize the stone dam structures would be to plant walnut trees. Cromwell found the soil sandier and less fertile than at Johnson Ranch.

Zuni noted that the check dam was constructed to slow the water and still serves that purpose. Plantings might or might not be made on the upstream terrace surface. At Zuni, terraces are used for planting early-season corn, but they are not usually in the form of check dams placed within drainage channels. He thought that the soil had more sand and gravel than at Johnson Ranch and was not as good.

Although none of the consultants was able to conclusively choose between two plausible functions of the check dam solely on the basis of its morphology, none rejected the possibility of farming in this location, more marginal than the previous fields. The Tohono O'odham consultants were particularly adept at noting physical indications of cultivation potential. The mention of trees was an unanticipated response; archaeological interpretations focus heavily on annual crops.

Experienced Native American farmers acknowledged more than one possible function for prehistoric stone features but could not make a definitive choice. Clearly, resolving the function of specific check dams would not necessarily resolve it for all of them. Such ephemeral remains as single-course check dams may not be sufficiently well preserved or archaeologically recognized to accurately reflect the extent of land modified in this manner. The McCoy Mesa check dams demonstrate that prehistoric farmers sometimes deemed it profitable to control the flow of water or to accommodate limited plantings in small, ephemeral drainages.

Assessing Mullen Mesa

The Mullen Mesa site consists of an extensive area of cobble alignments on a high river terrace. It is at the distal end of a broad alluvial fan, a landform known as Mullen Mesa. Soils are shallow, somewhat deflated, cobbly colluvium. Heaped cobble mulches or rockpiles are present, but well-defined cobble contour terraces are more numerous in an array that resembles Hohokam fields used to cultivate agave (e.g., Fish et al. 1992). No roasting pits are present, but the artifact assemblage includes tools usually associated with agave harvesting. A colony of the probable Hohokam cultivar, *Agave murpheyi*, was identified about 1 km to the northwest.

Cromwell doubted that corn could be grown in the terraces. The soil did not appear promising, and he believed that heated terrace rocks might injure immature corn plants. Nevertheless, farmers in hot, dry climates worldwide plant in terraces and other stone features. The moisture benefits of such features (Fish et al., eds. 1992) usually outweigh damage from heat. The Mullen Mesa terraces are among the most substantial and formally designed stone agricultural features in the study area. In light of the consultants' frequent suggestions to build brush features, the effort invested in these more durable constructions appears significant.

Assessing the Crash Landing Site

This rockpile field is located on a broad ridge of a Verde River terrace with cobbly to very cobbly clay loam soils. Extensive rockpiles are accompanied by linear terraces and rock rings (Figure 4.5). No roasting pits or tools usually associated with agave cultivation were found.

Zuni immediately recognized the linear terrace alignments as similar to ones near Zuni, where such features slow water, catch topsoil, and provide a mulching effect. The clayey subsoils should remain moist longer under the rocks than in adjacent areas. He thought that June (the time of the visit) was too dry for growing but that it might be possible to plant beans, squash, and

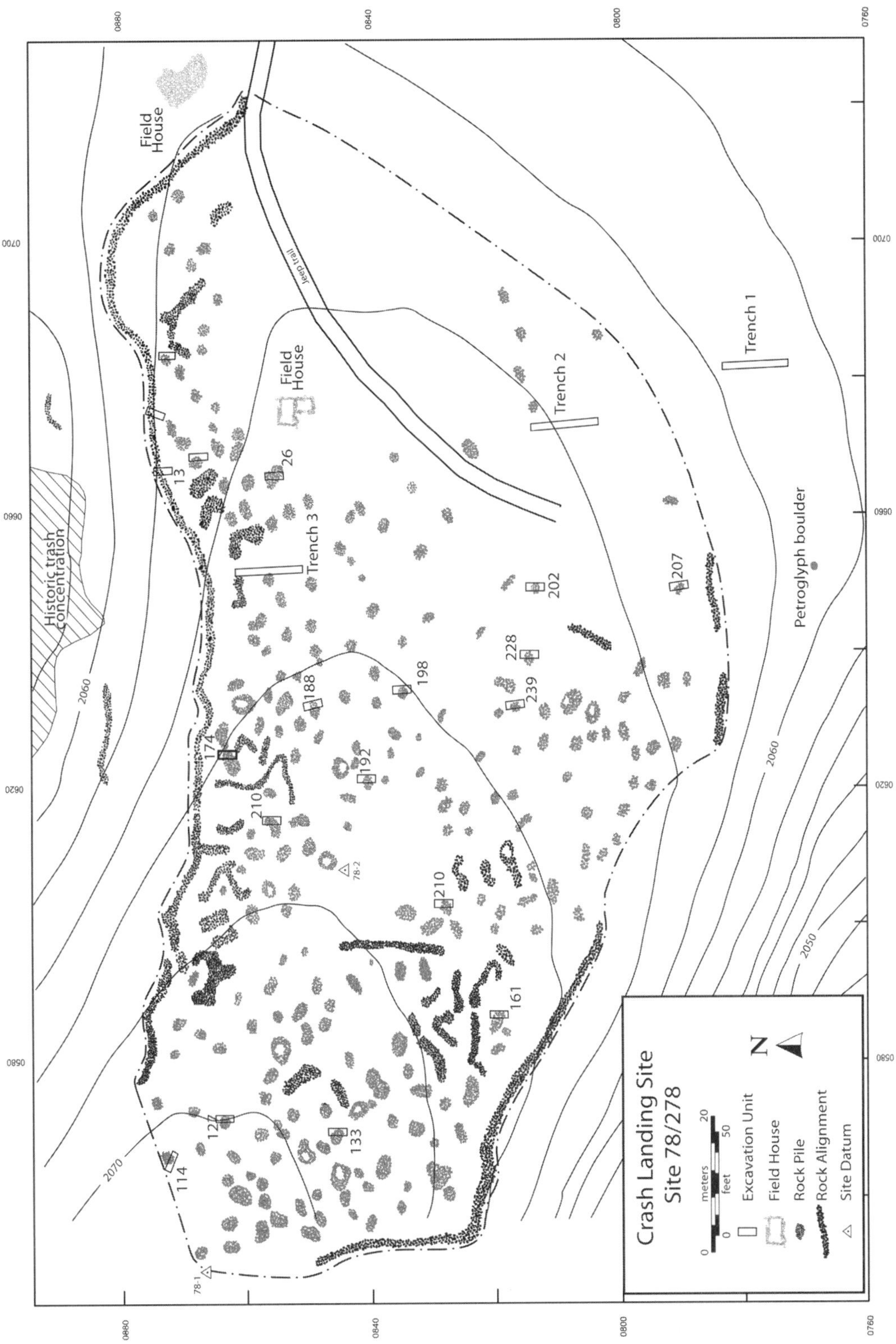

FIGURE 4.5. Rockpiles and linear alignments at the Crash Landing site.

possibly some corn during the summer rainy season. The Zuni consultant was not at a loss for agricultural possibilities, in spite of the fact that the predominant rockpiles were outside his experience. However, this repetitive prehistoric constellation of rockpiles and linear features throughout central and southern Arizona is not replicated in postcontact times.

Assessing Lime Creek

Lime Creek is a spring-fed tributary of the Verde River. With an extensive upland watershed, its flow is almost perennial along some segments, and ponds persist elsewhere as late as June. Irrigable pockets of alluvium are present, even after much bottomland was recently removed by destructive floods. Some of these alluvial pockets support dense mesquite bosques, and the consultants visited one well above the active reach of the current floodplain. Midvale (1946) recorded extensive terrace gardens with ditch irrigation near Lime Creek's confluence with the Verde.

The Tohono O'odham consultants viewed the Lime Creek locality only from an overlook above the floodplain. The water in this unincised creek made land just above the active floodplain worthy of farming. Diversions could be constructed so that floodwaters would deposit sediment for planting. Joaquin noted that this sort of cultivation might seldom be evident archaeologically because once farmers stopped tending the diversions, the plots soon would disappear without a trace. Both he and Lewis indicated that temperature would be an important consideration in planting decisions for such a setting. Joaquin stated that a farmer would have to "get used to the climate" before farming along the creek.

Cromwell associated the green leaves of mesquites with desirable conditions for farming. Although pea gravel covered most of the surface, he noted that the soil was less pebbly at depth and very moist. Before farming this area, he would place diversions on the higher terrace above in order to keep rocks from washing down onto the field area.

Near the mesquite bosque, Zuni believed that water could be obtained both by irrigation from the creek and from uphill runoff. The soil seemed relatively coarse, although he observed that it became finer just beneath the surface. At Zuni, squash is commonly planted in similar situations where runoff is a major water source. Corn and melons are also sometimes planted.

The consultants' reaction to the narrow floodplain of Lime Creek underscores the willingness of indigenous farmers to consider any source of sustained water for planting. Land beyond the reach of seasonal floods is limited, and even the protected pockets appear subject to high floods from the main channel and tributaries and from heavy runoff along the narrow, steep course. The mesquite bosque visited by the consultants is located sufficiently high above the active floodplain to merit a short canal. Zuni and Cromwell both recognized that runoff also could be concentrated in this area. This secondary water source could supplement irrigation but might require additional control structures.

Assessing Cow Wallow

The Cow Wallow site is at the toe of an alluvial fan near the mouth of a minor tributary to the Verde River. Fan sediments consist of a gravelly sandy loam. The locale supports an extensive mesquite bosque of very large trees.

Lewis and Joaquin saw the large size of mesquites as a sign of productive farmland. To prepare this area for farming, they would chop down the mesquites and remove the stumps. Planting locations would be selected by watching the water flow downslope over the fan. Where channels are present, brush diversions should be constructed to spread out the water. Initial fields would be cleared at the optimal *ak chin* location where fan drainages become shallow and water spreads over the surface naturally. For newcomers to the area, a crop would be possible in this situation in the first year following arrival. As time passed, more fields could be developed upstream. The first fields would be planted in the best natural situations. Level land would then be created and expanded by the use of diversion deposits to correct uneven topography and improve the soil.

The mesquites impressed Russell, even though he had not farmed in this type of non-irrigated setting. He thought that riverine irrigation should be possible at the toe of the alluvial fan near the river. The large size of the mesquites suggested good farmland to Cromwell. However, he was concerned that the soil became sandier downslope where the mesquite trees were larger. Water could easily be diverted from the many small tributaries of the main wash on the slope. Zuni likewise noted the floodwater potential of the drainages forming the alluvial fan. He thought it should be possible to grow corn, beans, and squash. At the same time, the slope seemed fairly steep and the soil fairly gravelly in some places. The best farming situation would be farther downhill where fan drainages neared the river floodplain. Brush structures there could spread out the water.

The vigor of modern vegetation impressed the consultants and indicated abundant water for floodwater fields. Those with floodwater farming experience could immediately discriminate the more favorable locations on the fan surface. Both the ease of diverting water and desirable characteristics for fields figured in these decisions. The improvement program suggested by the Tohono O'odham farmers illustrates that the consultants did not consider existing conditions of soil and topography to be immutable. A cautionary lesson for archaeologists is that observable states of environmental variables such as soil types may not accurately register the ultimate, modified potential of localities for farming.

Alluvial fans are acknowledged as desirable locations in the archaeological and ethnographic literature of the Southwest. Surfaces of fans along the Lower

Verde are fairly steep compared with those in broader and flatter basins, but evidently not to a degree to exclude floodwater cultivation. Like Cromwell at Johnson Ranch, Lewis and Joaquin could outline a strategy of incremental improvements to alluvial fans for newly arrived migrants.

CONCLUSIONS AND IMPLICATIONS

In spite of the widely differing home environments of the Sonoran, Tohono O'odham, Pima, Apache, and Zuni farmers, none hesitated to consider the possibility of cultivating any locale where water was available. Their evaluations of the Verde, Tonto, and Tucson study areas converged in many instances. They generally agreed on the qualities of soils; there was overlap again in the recognition of indicator plants as signs of fertility. Consultants often concurred on the risks of flooding and washouts, offering similar remedies of vegetation buffers and brush, earth, and stone constructions to counter these destructive forces. They similarly shared many methods for intercepting, concentrating, and delivering water.

Differences in assessments of agricultural potential arose from farming experiences as well as from variability in homeland environments. The Sonoran, Tohono O'odham, Apache, and Zuni farmers could visualize the outcomes of floodwater methods and suggest strategies for progressively expanding and improving fields by modifying drainages, soils, and topography. When they assessed prehistoric stone features that affected runoff on more marginal land, their interpretations of functions and agricultural outcomes were more varied. Manipulation of both channelized floodwaters and surface runoff was beyond the experience of the Pima farmer. He lived in the broad, flat valley of the Gila River and knew only irrigation from large canals. On the other hand, only this consultant was familiar with indigenous diversion structures of the scale necessitated by major rivers like the Verde and with the construction of canals traversing several miles. Only his tradition included social institutions for coordinating the use and maintenance of extensive canal networks.

All the consultants lacked local knowledge about environmental variables that they deemed critical for agricultural success. They could only speculate on seasonal proportions and timing of rainfall and its variability from year to year. They did not know the usual timing of peak flows and floods in watercourses or the seasonality and duration of surface water in tributaries. The interval of frost danger and the topographic vagaries of cold air drainage were also unknowns. More important, they could not predict the combinations of these factors that would promote efficient investments of seed and labor and help them to balance yields against risks.

The consultants' reactions bring into focus the kinds of challenges that would have faced individuals and groups entering the Lower Verde in the past. The first years of farming would have been risky indeed unless previous occupants shared their intimate and hard-won knowledge of local conditions.

Immigrants from distances beyond the reach of homeland stores would have been particularly vulnerable. They could not have afforded to observe Verde precipitation and frost for several years before planting. Likewise, they would have lacked detailed information on the distribution and timing of wild resources with which to supplement crops.

The logistics of relocation would have had further implications. Seed for initial planting must have been either carried along with all other possessions or acquired from new neighbors. Crop varieties best suited to homelands might be less suited to the temperature and rainfall regimes at destinations. Southwestern environments pose a risk of periodic failures and low yields. Indigenous agriculturalists of postcontact times uniformly articulated the goal of storing large fractions of harvests. Migrants who could not transport cumbersome subsistence supplies would have been dependent on the generosity of previous occupants during times of greatest common need. Because of their unfamiliarity with the local climate and hydrology, these incoming farmers would have been more liable than previous residents to make miscalculations. Furthermore, they could have less readily countered agricultural disasters by replanting if they were restricted to finite quantities of transported seed. Reliance on neighbors again was a likely outcome.

Under what conditions might these sorts of dependencies have been acceptable to both newcomers and resident farming societies? How might interactions with incoming households be envisioned if previous residents were called on not only to share productive land but also to provide a safety net for the first years? These questions should be addressed in investigations of population movements in the Verde Valley and as part of a renewed interest in abandonment and migrations in southwestern archaeology generally.

The Verde study suggests scenarios in which interactions could have been peaceful and beneficial to all concerned. Newcomers might have joined the existing labor pool in programs of agricultural intensification that involved structural investments. Seasonally concentrated labor would have been essential in diverting the Verde with its steep gradient and high volume of flow. Remains of aqueducts show that elaborate constructions also were necessary where canals crossed side drainages in this rugged valley profile. Stone features and modifications on slopes above the floodplain are alternative forms of labor investment for bringing additional increments of land into cultivation. Up to some threshold, increasing population and labor may have added to the overall productivity and diversity of the subsistence systems.

The agricultural logistics of relocation are ultimately embedded in the social institutions of farmers. Coexistence or cooperation between previous and newly arrived populations would have required major adjustments in territoriality, tenure of arable land, subsistence exchange, and related interactions. The responses of the Native American consultants in the three study areas underscore the fact that the conceptual and technical skills of entering farmers would not have been sufficient to ensure immediate agricultural success at destinations, the less so to the extent that environments diverged from those of homelands and entailed novel kinds and scales of technology. The agricultural decisions and solutions of farmers are fundamental issues for archaeologists seeking to understand migration processes in the ancient Southwest.

As exemplified by the consultants in this study, traditional southwestern farmers carry on agricultural traditions that enable them to define and monitor aspects of environment critical to productive and sustained harvests. They command classificatory systems for detecting the states of key variables and conveying the operational principles of cultivation. These conceptual frameworks provide the basis for immediate and long-term strategies that simultaneously reflect the cultural modes and motivations of farmers. Traditional farmers perceive their environments not as agriculturally immutable but as subject to purposeful modifications that, under differing conditions, may or may not be judged worthwhile. Such perspectives based in agricultural practice offer archaeologists an alternative to paradigms of regional marginality. These viewpoints of farmers challenge us to understand the nonstatic nature of prehistoric agriculture and its dynamic role in the changing configurations of prehistoric southwestern societies.

ACKNOWLEDGMENTS

Gary Nabhan and Thomas Sheridan helped plan the initial Tucson Basin study and thereby influenced subsequent field collaborations. We gratefully acknowledge logistical assistance and advice from Karen Adams, Jeffrey Altschul, William Doolittle, T. J. Ferguson, Jeffrey Homburg, and John Welch. At the heart of the study is the collegial spirit, professional interest, and farming expertise of the consultants, Glenn Cromwell, Adalberto Cruz, Joseph Joaquin, Delores Lewis, Archie Russell, and Jerome Zuni. The Tucson Basin study was funded by the National Science Foundation (BNS-8498141). The Bureau of Reclamation sponsored the Statistical Research projects in the Tonto and Horseshoe basins. The authors gratefully acknowledge Matthew E. Hill for assistance with drafting illustrations. We thank the volume editors for their extreme patience as they waited for us to finish this chapter.

5

Long-Term Streamflow Reconstructions, River Channel Morphology, and Aboriginal Irrigation Systems along the Salt and Gila Rivers

Donald A. Graybill, David A. Gregory, Gary S. Funkhouser, and Fred L. Nials

CANAL IRRIGATION HAS BEEN AN IMPORTANT FEATURE OF HUMAN adaptation to the desert environments of southern Arizona for at least three thousand years. Canals dating between 1200 and 800 B.C. have been discovered in the Tucson Basin, making them the earliest such features yet documented in North America (Mabry 1999; Muro 1998b). The impressive and well-known irrigation systems of the later Hohokam culture brought water to fields along the Salt and Gila rivers for over a millennium, and when Europeans first arrived in the seventeenth century, they found the Pima irrigating fields along the middle reaches of the Gila River. Today cotton and other economically important crops are entirely dependent on irrigation.

Variation in streamflow and resulting effects on the morphology of river channels are the most significant environmental variables faced by irrigation agriculturalists in southern Arizona (Graybill 1989; Graybill and Gregory 1989; Graybill and Nials 1989). In this chapter, we present long-term reconstructions of Salt River and Gila River streamflow, review the effects of variation in discharge on river channel morphology, and explore the implications of both high- and low-frequency variation (Dean 1988a) for aboriginal irrigation agriculture along the two rivers.

DENDROCHRONOLOGY AND PALEOCLIMATE

The long-term streamflow reconstructions presented here were made possible by increases in the quantity and quality of long, climatically sensitive tree-ring chronologies over the last 35 years. Between 1962 and 1975, the Laboratory of Tree-Ring Research, funded by the National Science Foundation, conducted a systematic reanalysis of thousands of tree-ring samples from archaeological contexts (Robinson et al. 1975:1–2). One result was identification of archaeological tree-ring samples suitable for dendroclimatic analysis. These and other data from ongoing research were used to develop a 25-station network of individual chronologies built from composites of proximate archaeological and living-tree samples (Dean and Robinson 1978). Subsequent research has strengthened and expanded the network and lengthened individual chronologies (Graybill 1987; Graybill and Rose 1992), and the tree-ring series used for the current study derive from these efforts.

As part of the Southwest Paleoclimate Project, this network of chronologies has been used to develop both relative and quantitative reconstructions of climatic variability in the Southwest (Dean and Robinson 1977; Dean and Funkhouser 1995). Since the late 1980s, the Southwest Paleoclimate Project has produced quantitative reconstructions of seasonal and annual precipitation and temperature (Dean and Funkhouser 1995), the Palmer Drought Severity Index (PDSI; Palmer 1965), and streamflow (Graybill 1989; Graybill and Meko 1995; Graybill and Nials 1989; Meko and Graybill 1993).

These successful applications of dendrochronological research resulted in part from parallel advancements in understanding of the mathematical and biological principles that govern tree growth–climate relationships and the procedures and methods used to study those relationships (Cook and Kairiukstis 1990; Fritts 1976, 1991; Graybill 1982; Hughes et al. 1982). Annual growth in arid-site conifers is primarily a process of new food balance over time, and the width of the growth increment produced in any one year is determined by the condition of the tree at the beginning of the growing season. This in turn is governed by the relative amounts of food production and food consumption that have occurred over the year or so prior to growth. Much of the initial annual growth period relies on food reserves, whereas current climatic conditions that affect growth processes become most important later in the growth period.

Prior and current climatic conditions can affect the width of annual growth increments in a variety of ways. Ring widths of southwestern arid-site conifers in stressful lower-forest-border settings primarily reflect climatic conditions affecting food balance during the winter preceding growth and, to a lesser extent, during the preceding or current summer. In essence, the trees integrate the variables of temperature and precipitation that in turn limit or enhance growth. These two variables can act singly or in concert at various times of the year to affect growth processes. Because the foremost factor limiting growth in arid, lower-forest-border sites is effective moisture, tree-growth records are dominated by climatic conditions that produce moisture-related stress.

DENDROCHRONOLOGY AND DENDROHYDROLOGY

Several pioneering studies (Smith 1981; Stockton 1975; Stockton and Jacoby 1976) employed chronologies from living trees in the Colorado River Basin to retrodict stream discharge. These studies provided empirical and statistical support for the premise that annual tree-ring series can be used to extend both annual and seasonal streamflow records because the two respond in like fashion to climatic stimuli. Tree-ring data are from the upper forested reaches of the river basins where most hydrologic input occurs. The effective moisture recorded therein is moderated by evaporation, infiltration, and the biological processes of the trees themselves, and the reduction of precipitation to streamflow is moderated by the same processes in these areas. Relationships between tree growth, precipitation, temperature, and runoff have been established for watersheds of both the Salt and Gila rivers (Graybill and Meko 1995; Graybill and Nials 1989; Meko and Graybill 1993; Smith 1981:23–45), with the effects of temperature found to be negligible in all cases. Thus a predictable relationship between tree-ring series and gaged streamflow records can be developed.

The nature and strength of the predictability of this relationship are established by various statistical regression models. When this procedure provides adequate results, the tree-ring series can be used in the manner described below to extend streamflow records back into times before gaged records were kept (Stockton and Boggess 1980). The studies cited above were of considerable importance in creating the conceptual and methodological basis for a previous reconstruction of prehistoric Salt River streamflow (Graybill 1989; Graybill and Nials 1989; Nials et al. 1989) and for the long-term reconstructions of Salt and Gila river streamflow reported here. The terms *streamflow* and *discharge* are used synonymously throughout this chapter.

TREE-RING DATA AND CHRONOLOGY DEVELOPMENT

Conifers in the upper forested reaches of the Salt and Gila river basins seldom exceed 300–400 years in age, but concerted searches have been made for sites with older trees in settings likely to produce drought-sensitive records. In several instances, sites selected for sampling were either on or near archaeological sites that also produced datable tree-ring series. This activity was primarily accomplished by the Archaeology Research Program of the Laboratory of Tree-Ring Research, as part of the previously cited effort to develop long, drought-sensitive series (Dean and Robinson 1978). Living tree-ring chronologies used here are based on sites collected or re-collected between 1985 and 1989 by the senior author. Tree-ring samples were dated and measured to the nearest 0.01 mm following standard procedures (Cook and Kairiukstis 1990; Stokes and Smiley 1968). Measurement accuracy was spot-checked, and each series was subjected to a battery of computerized cross-checks (Holmes 1983). Each series was plotted and examined for outliers and aberrations.

Accurately dated and measured ring-width series are time series that may have at least five different kinds of signals in the frequency domain, as follows.

For any individual specimen, let

$$R(t) = C + B + D_1 + D_2 + E,$$

where $R(t)$ is the measured ring width in year t; C is the macroclimatic signal common to trees at a site; B is the biological growth curve as a function of increasing tree age; D_1 is the tree-disturbance signal unique to a single specimen or tree due to random events that affected its growth; D_2 is the tree-disturbance signal common to most or all specimens in a stand due to fire, insect damage, or other disturbance; and E is the random growth signal unique to each specimen (Graybill 1982:29).

In chronology development, every effort was made to maximize the macroclimatic signal and remove or minimize the other variables in the equation. Disturbance factors were minimized by site and tree selection or by rejection of samples with anomalous growth characteristics. Nonstationary growth characteristics were removed by fitting either a straight line or a negative exponential function to each ring-width series (R_t) and calculating tree-ring indices (I_t), as follows:

$$I_t = R_t / Y_t,$$

where Y_t is the expected annual growth determined by curve-fitting (Fritts 1976). This removes most nonstationarity in ring width and variance resulting from tree age (Fritts 1976). Tree-ring index chronologies were then computed as averages of the indexed series by stand (Graybill 1979, 1982). This averaging process reduces localized noise in a tree-ring series by minimizing error variance due to individual tree or sample variation, thereby emphasizing the common macroclimatic signal.

Because there are usually fewer older than younger trees at a site, sample depth normally decreases toward the beginning of a chronology. As a result, the strength of the common signal may not be well approximated in the early part of a chronology. In order to avoid instability in the modeling process resulting from low sample depth, where there may be nonstationarity, we fit models for each chronology beginning in the year where subsample signal strength (SSN) was at least 90 percent (Wigley et al. 1984). In most cases this represented a starting sample depth of four or five individual tree-ring series. For sites with living trees, it was possible to use about twenty series per year for each chronology. Sample depths for archaeological chronologies are sometimes problematic, especially near their beginning and ending decades, where samples are scarce. The stringent criteria applied in specimen selection for those chronologies, however, ensures that reliable reconstructions can be based on them despite low sample depths during some short intervals.

Using these procedures, and guided by Jeffrey Dean and William Robinson of the Laboratory of Tree-Ring Research, we constructed archaeological tree-ring chronologies by site or region. Selection of individual series was directed at maintaining as much long-term (low-frequency) and short-term (high-frequency) climatic variation as possible. Therefore, an attempt was made to include only those series which were at least 100 years long and for which descriptive statistics indicated high growth variation and climatic sensitivity from year to year (Rose et al. 1981). Fourteen final chronologies were created by averaging the tree-ring indices of individual series meeting these criteria. Locations of the chronologies in relation to the respective watersheds are shown in Figure 5.1, as are gaging stations that provided streamflow data

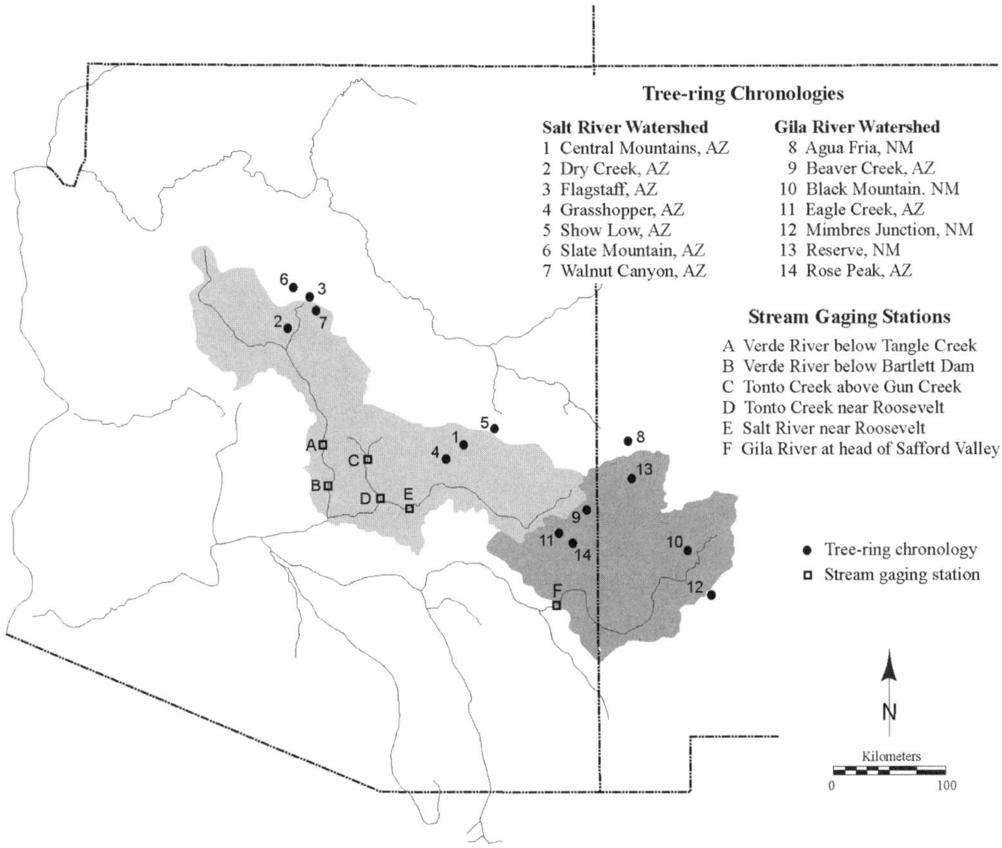

FIGURE 5.1. Locations of tree-ring stands and streamflow gaging stations.

(see below). Data for the fourteen chronologies are provided in Table 5.1, along with results of additional standardization measures discussed below.

Further transformation of tree-ring time series derived from the standardization process described above is necessary if the series possesses a significant persistence (autocorrelation) structure. Temporal persistence in a tree-ring chronology is present when successive values (observations) in the series are not independent, that is, when any value in the series can be partially predicted from one or more previous values (past observations). Autocorrelation in the annual values of a single or averaged set of series violates the assumption of independence in many statistical procedures, adds unknown bias to hypothesis tests, and results in unreliable variance estimates (Graybill and Rose 1992; Monserud 1986; Wonnacott and Wonnacott 1981). Since regression-based statistics are extremely sensitive to these factors, it is necessary to remove the autocorrelation. This is usually accomplished with procedures known as autoregressive moving average (ARMA) modeling (Box and Jenkins 1976). Autocorrelation is thought to be of nonclimatic or biological origin so its removal should not affect the actual frequency domain characteristics of past climate in subsequent reconstructions.

Statistically significant autocorrelation was found in all chronologies.

TABLE 5.1. Tree-ring stand information and time-series model data.

SITE NAME	SITE CODE	NORTH LAT.[a]	WEST LONG.[a]	ELEV. (M)[a]	SPECIES[b]	DATES	TIME-SERIES ANALYSIS PERIOD	BEST-FIT TIME-SERIES MODEL	% VARIANCE
Salt River Watershed									
Central Mountains, AZ[c]	CMT	34°00'	110°25'	1875	pp, pe, df, j	1040–1370	1040–1360	AR (1)	12.1
Dry Creek, AZ	DCP	34°54'	111°49'	1870	pn	1626–1988	1650–1986	ARMA (1,1)	18.7
Flagstaff, AZ[d]	FLA	35°15'	111°30'	2130	pp, pe, df	570–1370	735–1365	AR (1)	8.7
Grasshopper, AZ	GRS	34°04'	110°37'	1800	pp	1642–1985	1696–1985	ARMA (1,1)	9.9
Show Low, AZ	SGP	34°15'	109°56'	2070	pp	1596–1986	1622–1986	ARMA (1,1)	30.5
Slate Mountain, AZ	SMP	35°30'	111°50'	2195	pp	1590–1986	1658–1986	AR (1, 2)	42.9
Walnut Canyon, AZ	WCP	35°10'	111°31'	2060	pp	1309–1987	1532–1987	ARMA (1,1)	25.5
Gila River Watershed									
Agua Fria, NM	AFN	34°15'	108°37'	2200	pe	1403–1987	1510–1987	ARMA (1,1)	24.3
Beaver Creek, AZ	BCR	33°42'	109°14'	2390	pp	1559–1986	1587–1986	ARMA (1,1)	21.7
Black Mountain, NM[e]	BKM	33°23'	108°14'	2710	pp, df, wp, wf	1196–1988	1276–1987	ARMA (1,1)	18.0
Eagle Creek, AZ	EPN	33°28'	109°29'	1690	pe	1580–1987	1639–1987	ARMA (1,1)	12.7
Mimbres Junction, NM	MJN	32°57'	108°01'	1950	pe	1661–1986	1661–1986	ARMA (1,1)	8.2
Reserve, NM[f]	RES	33°27'	108°37'	3017	pp, pe, df, j	532–1286	646–1286	AR (1)	8.4
Rose Peak, AZ	RPP	33°25'	109°22'	2320	pp	1650–1987	1699–1987	ARMA (1,1)	29.5

[a] Location and elevation of tree-ring sites are approximate.

[b] pe Pinyon pine (*Pinus edulis*) — pp Ponderosa pine (*Pinus ponderosa*)
df Douglas fir (*Psuedotsuga menziesii*) — wf White fir (*Abies concolor*)
wp White pine (*Pinus strobiformis*) — j Juniper (*Juniperus* sp.)

[c] CMT represents a combination of data previously published as Central Mountain North Chronology and Central Mountain South Chronology (Dean and Robinson 1978).

[d] LA represents an average of data from the 1-degree-by-1-degree geographic quadrangles referred to as N and O (Bannister et al. 1966).

[e] Four living-tree chronologies from Black Mountain, NM, were combined to form a single chronology, BKM (Meko and Graybill 1993:3).

[f] Dean and Robinson 1978.

Standard Box-Jenkins (Box and Jenkins 1976) protocol was used to examine the structure of these processes. Procedures for finding the best-fit model initially involved fitting several low-order AR, MA, and mixed ARMA models to each chronology. Several diagnostic evaluations of each fit were used as guidelines for final model selection. These included (1) the significance or nonsignificance of model parameters, (2) minimization of the residual sums of squares, (3) penalties for overfitting, and (4) determination of the presence of remaining significant autocorrelation in the residual (prewhitened) series computed for each model fit (i.e., the difference of subtracting model fit values from index series values). As in previous studies (Biondi and Swetnam 1987; Graybill 1985, 1989; Graybill and Rose 1992; Monserud 1986), modeling procedures usually showed that a first-order autoregressive moving average fit (ARMA [1,1]) was best for the living-tree chronologies and that a simple first-order autoregressive fit (AR [1]) was best for the archaeological chronologies (see Table 5.1).

The residual series produced from this filtering process are referred to as *residual tree-ring chronologies.* Seven residual chronologies were created and then averaged to form a single chronology for the Salt drainage (SALT_RES); similarly, seven residual chronologies were created and averaged to create a single chronology for the Gila drainage (GILA_RES) (Table 5.1).

STREAMFLOW DATA

Twentieth-century records of monthly streamflow that permit characterization of Salt River discharge in the Phoenix Basin include gaged records from immediately above Lake Roosevelt on the Salt River; from two stations on Tonto Creek above its confluence with the Salt; and from two stations on the Verde River, one below Bartlett Dam and the other below Tangle Creek, approximately 14.5 km upstream from Horsehoe Dam (Figure 5.1). Use of the Tangle Creek gage for post-1945 Verde discharge values avoids some errors that develop when the Bartlett flow record is corrected for net changes in reservoir content. Tonto Creek data were not used in an earlier reconstruction of Salt River streamflow because of potential inaccuracies in the gaged record (Graybill 1989:10-11); these data have since been reevaluated and judged to be adequate for inclusion here. For the Gila River, modern discharge data are from a station at the head of the Safford Basin (Figure 5.1). Although there has been modern diversion of water for irrigation above this station, no adjustments have been made for changes in diversions over time (see Meko and Graybill 1993:2–3). In all cases, monthly flow data were obtained from EarthInfo's (1990) compact disc version of the U.S. Geological Survey database. As with the tree-ring series, discharge records were examined for evidence of autocorrelation, and none was evident in any of the records. Data for each gaging station are presented in Table 5.2.

TABLE 5.2. Streamflow gage data.

GAGE NAME	USGS NUMBER	PERIOD OF RECORD	WATERSHED AREA (KM^2)
Salt River Watershed			
Verde River below Bartlett Dam, AZ	09510000	1895–1990	15,177
Verde River below Tangle Creek, AZ	09508500	1945–1990	15,594
Salt River near Roosevelt, AZ	09498500	1914–1990	11,163
Tonto Creek near Roosevelt, AZ	09499500	1913–1940	2,176
Tonto Creek above Gun Creek, near Roosevelt, AZ	09499000	1940–1995	1,748
Gila River Watershed			
Gila River at head of Safford Valley, AZ	09448500	1911–1990[a]	20,450

[a] Some values from another gage were used to fill in values for 1930s ; see Meko and Graybill 1993.

Transfer Function Development and Verification

The transfer functions (calibration equations) developed here are equations that predict streamflow from a tree-ring series. Preliminary data analysis confirmed previous findings suggesting that many monthly and, consequently, seasonal and annual discharge series are not normally distributed and are curvilinearly related to the residual tree-ring chronologies (Graybill 1982; Meko and Graybill 1993). Log-10 transforms of these data resulted in series with better distributional properties and a more linear relationship to the tree-ring data. Correlation and response function analysis (Fritts 1976) of numerous seasonal and annual log-10 streamflow series with the appropriate residual tree-ring chronologies indicated that October through April seasonal and October through September annual (water year) streamflow transfer functions could be developed for both drainages. Additionally, reconstructions derived from these equations are comparable to those in previous studies and consistent with the hydrologic literature.

Multiple linear regression procedures (Draper and Smith 1981) were used to develop (calibrate) statistical models that describe streamflow–tree growth relationships. In each case, the dependent variable (predictand) was one of the log-10 seasonal or annual streamflow series described above. As noted above, the autocorrelation structure of each streamflow series was evaluated and found not to be significant. Therefore, the log-10 discharge series were not prewhitened. Independent variables (predictors) in each case included the relevant residual chronology values for years t (the current growth year) and $t-1$ (the prior growth year). Inclusion of the lagged predictors was allowed because the ARMA modeling process is not a perfect filter and may leave a

TABLE 5.3. Verification statistics and final equations.

VALUE	RECONSTRUCTION PERIOD	R^2	$R^2_{PREDICTION}$	FINAL EQUATION
Salt River log-10	October–April	.74	.72	5.853193 + (.874794*SALT_RES$_t$) + (.234794*SALT_RES$_{t-1}$)
Salt River log-10	October–September	.72	.70	5.999380 + (.704959*SALT_RES$_t$) + (.213923*SALT_RES$_{t-1}$)
Gila River log-10	October–April	.64	.61	5.203613 + (.940282*GILA_RES$_t$) + (.384189*GILA_RES$_{t-1}$)
Gila River log-10	October–September	.60	.57	5.427689 + (.735406*GILA_RES$_t$) + (.312354*GILA_RES$_{t-1}$)

small degree of autocorrelation in the prewhitened series. Since the timing of hydrologic seasons or years and tree growth is not identical, there can be possible carryover effects from one year to the next. In the calibrations performed for this study, each of the two regionally averaged tree-ring series were used for the independent variables (X_t) and (X_{t-1}) to predict annual or seasonal flow values (Y) in the regression equation

$$Y = a + bX_t + cX_{t-1} + e,$$

where a is a constant (y intercept), b is the regression coefficient for the current year, c is the regression coefficient for the prior year, and e is an error term.

Calibration and verification procedures were used to ascertain the predictability of Salt and Gila river discharge from tree-ring data. Calibrations were performed for seasonal (October–April) and annual (October–September) periods for each river, and the full calibration interval was A.D. 1914 to 1986 for each. Figure 5.2 illustrates the full-period actual and reconstructed values for both drainages.

Verification of the full-period calibration equations was accomplished by calculating their prediction error sum of squares (PRESS) and evaluating each ($R^2_{\text{prediction}}$) relative to the variance explained (R^2) by the regression equation (Montgomery and Peck 1992). This measure is derived by removing one observation from the calibration period, calculating a new regression equation with the remaining $n - 1$ observations (where n equals the total number of observations in the full-period calibration equation), and then using the new equation to predict the removed observation. This is repeated for each of the n observations and provides a useful statistic for determining how well the full-period regression equation predicts new values.

Table 5.3 illustrates results of the verification procedures described above and presents the final equations used with the residual tree-ring series for streamflow reconstructions. The verification procedures suggest that our models are satisfactory and indicate a somewhat higher degree of predictive capability for the October–April transfer functions than the October–September for both drainages. The final equations were developed with regressions that used all available data, ensuring that the fullest range of sample

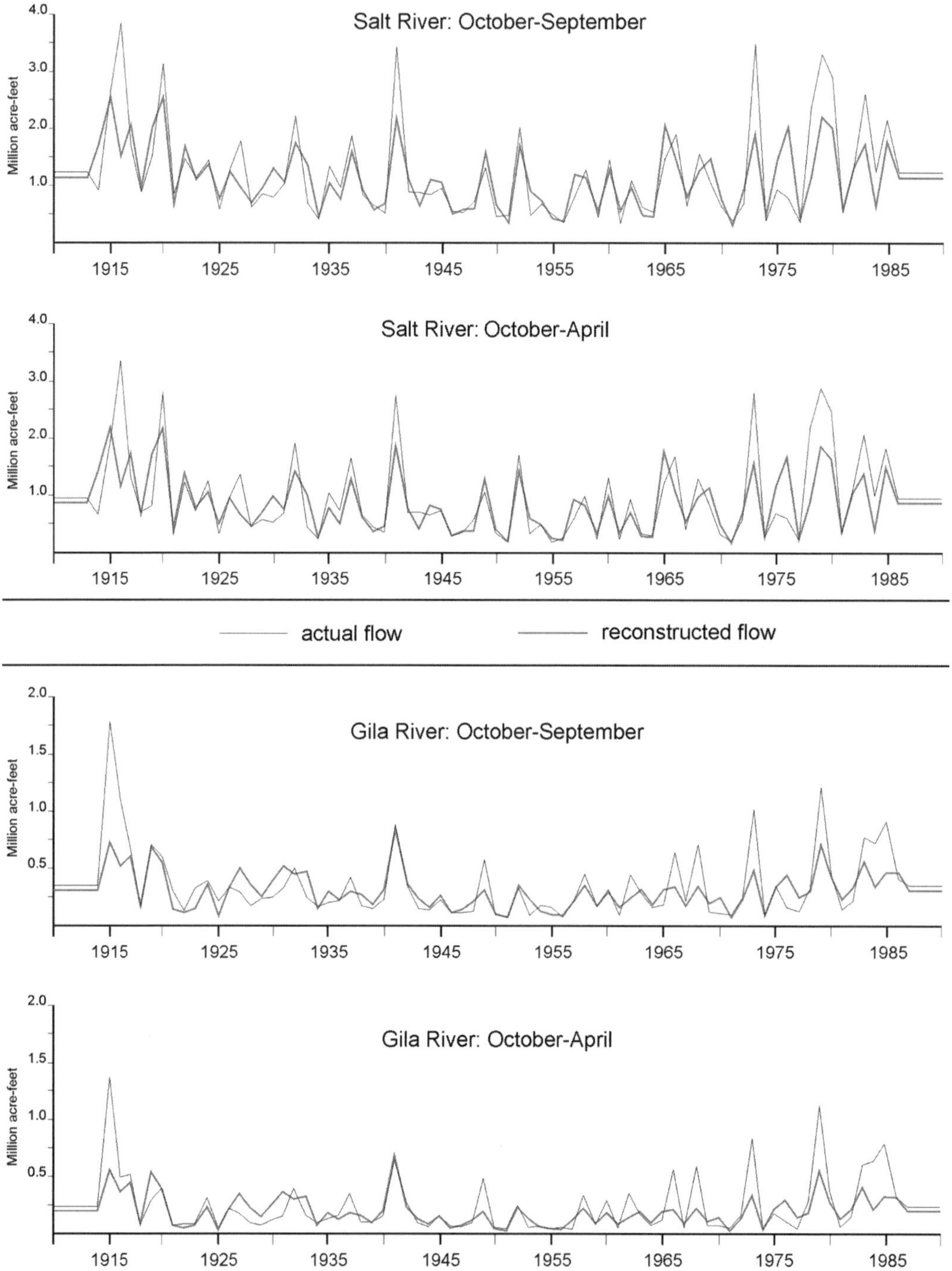

FIGURE 5.2. Actual and reconstructed Salt and Gila river streamflow, A.D. 1914–1986.

covariance might be included in model formulation. These equations were applied to the two residual tree-ring series (SALT_RES, GILA_RES) to reconstruct log-10 streamflow values for each river for the seasonal and annual periods. The reconstructed log-10 streamflow values were then converted back to thousands of acre-feet to create reconstructions covering the period A.D. 572 to 1988 for the Salt and A.D. 534 to 1988 for the Gila (see below).

Use of log-transformations in discharge reconstructions satisfies model assumptions but also introduces additional error. When log-acre-feet are back-transformed, the size of errors on high flows is magnified and a bias is built into the mean of the reconstructed values: "reconstructions from log-transform models must be considered to have a sliding scale of accuracy: low-flow years [are] more accurately reconstructed than high-flow years" (Meko and Graybill 1993:7; see also Graybill 1989:35–36). Thus the reconstructions presented below are highly accurate with reference to years having lower-magnitude flows but tend to underestimate the magnitude of higher annual flows.

Additional Considerations

The reconstructions presented here were created in large part to provide a basis for inferences concerning the geomorphic effects of discharge and the implications of changing patterns of discharge and channel morphology for the operation of aboriginal canal systems along the two drainages. Given these objectives, we must note some additional considerations, including the length and nature of the modern record and the geographic applicability of the reconstructions.

Both the reconstructions and inferences concerning geomorphic responses to specific discharge events and patterns are based on a relatively short modern record. Climate and precipitation patterns during the modern instrumented era are not necessarily representative of the entire period of reconstruction. In addition, these data encompass a period when human intervention may have significantly altered natural flow characteristics. Historic grazing, farming, logging, and urbanization, coupled with the almost ubiquitous presence of arroyos and gullies in valley bottoms, combine to increase both runoff and sediment yield in the Salt and Gila drainages. As a consequence, even without significant changes in climate, the peak discharge produced by a given amount and rate of precipitation under historic landscape conditions would be significantly larger than that produced under similar conditions in an ungullied prehistoric landscape. These factors must be considered in any direct comparison of discharge characteristics and geomorphic consequences of discharge patterns.

For the Salt River, the reconstructions provide valid estimates of annual flow between the Salt-Verde confluence and the Salt-Gila confluence. The combined area above the Salt River gages is approximately 86 percent of the 33,670 km^2 of the drainage basin above the confluence of the Salt and Verde near Phoenix. That portion of total flow that reaches the Phoenix area and is detectable by a sum of data from these gages is probably greater than 90 percent. The majority of flow is from snowmelt that occurs primarily above the gages. Also, because the area between the gaging stations and the confluence of the two rivers is relatively low and arid, inflow from it is not directly proportional to its size. Thus the Salt River reconstruction is appropriately used for inferences concerning the effects of variation in discharge and channel morphology on the well-known prehistoric Hohokam canal systems of the Salt River Valley (Graybill and Nials 1989; Nials et al. 1989).

For the Gila River, the current reconstruction applies directly to the Safford Basin, from immediately above Safford to the mouth of the San Carlos River. The Gila reconstruction does not include data from the San Carlos and San Pedro rivers, which have watersheds of 2,657 and 11,533 km^2, respectively. This unfortunately precludes quantitative assessments of the absolute *amount* of water present in the reach between Buttes Dam and the Salt-Gila confluence, where many of the principal prehistoric Hohokam settlements and irrigation systems along the Gila were located. However, given the effects of infiltration, evaporation, and possibly diversion of water for irrigation between San Carlos and the Buttes, it is likely that the amount of water reaching the middle Gila was actually *less* than the reconstructed values for the Safford Basin. In addition, annual *variation* in flow in the middle Gila likely mirrored that shown in the current reconstructions. Thus, with appropriate qualifications, it is possible to use the current reconstruction as a basis for inferences concerning the effects of variation in discharge and channel morphology on canal systems located along this archaeologically important reach of the river (see below).

STREAMFLOW PATTERNS AND CHANNEL MORPHOLOGY

As a prelude to presentation and interpretation of the two reconstructions, it is necessary to review the basis for inferences concerning discharge-related geomorphic processes that determine the hydraulic geometry of the river channel, and differences in characteristics of the two drainages that influence the particular operation of these processes. Streamflow retrodictions described below clearly show that both short- and long-term changes have occurred in cumulative annual discharge variation. It has long been recognized that streams adjust their hydraulic geometry to changing conditions of discharge and sediment load (Burkham 1972; Chorley et al. 1984; Graf 1988a, 1988b; Leopold and Wolman 1957; Mackin 1948).

Discharge and Channel Morphology

Relationships between discharge and hydraulic geometry are often described in terms of a qualitative equilibrium model, wherein the river adjusts its channel through time to prevailing discharge and load conditions. Schumm's (1969, 1977) characterization of *river metamorphosis* is perhaps the best-known and most widely applied of these models. In Schumm's equations, the relationships between channel morphology, slope, discharge, and sediment load are incorporated into a set of predictions regarding the direction of channel changes that occur when discharge and sediment load are modified by natural and/or human processes. Independent variables in Schumm's equations include some index of discharge and bedload, where discharge may be represented by monthly, seasonal, or annual mean discharge, total annual discharge, annual peak flood, or flood frequency (Chorley et al. 1984; Ritter 1978). Dependent variables are channel width and depth, meander wavelength, channel gradient, and sinuosity. Responses to changes in the discharge/sediment load ratio may vary not only between rivers but also between different reaches of the same river (Schumm and Brackenridge 1987; Huckleberry 1993:18–20). Schumm's model accurately predicts channel change for rivers in temperate semiarid and humid environments and was employed as an explanatory device in an earlier study relating retrodicted annual discharge and geomorphic changes in the prehistoric Salt River channel (Graybill and Nials 1989; Nials et al. 1989).

In recent decades, there has been an increasing realization that rivers in arid lands may not respond in precisely the same way as humid-region rivers and that long-term equilibrium concepts do not necessarily apply (Burkham 1972; Graf 1981, 1983; Stevens et al. 1975). For rivers in arid and semiarid environments, average discharges are considerably smaller than flood discharges, and these rivers often lack the stream power and bank-stabilizing vegetation to quickly adjust their channels to previous equilibrium forms (Graf 1983; Richards 1982, 252; Schumm and Lichty 1963; Wolman and Gerson 1978). Thus channel and floodplain morphology tend to be controlled by infrequent flood events, and the time necessary for a return to the quasi-equilibrium condition is markedly prolonged; in fact, this condition is often not regained prior to the next channel-changing event.

Additional factors not directly related to individual discharge events may influence channel position and/or morphology more significantly in arid-lands rivers. For example, drought in headwater areas causes increased erosion, leading to an increase in sediment load, which may in turn exacerbate the effects of floods and result in development of new channel patterns or persistence of existing ones. Drought–sediment yield relationships in arid lands may be more pronounced than those in humid regions, generally increasing the potential effects of drought in upstream areas on channel morphology in downstream reaches.

As a result of these factors, the process of stream metamorphosis in arid-lands rivers is more accurately represented as one of punctuated trends toward equilibrium, in which infrequent high-magnitude discharge events are the principal instigators of change. The degree and direction of changes wrought by such events may be influenced to varying degrees by the conditions immediately prior to their occurrence, and a return to near-equilibrium conditions following such events will be prolonged and usually incompletely realized before the next flood of sufficient magnitude to induce change.

Burkham's (1972) study of the Gila River in the Safford Basin showed conclusively that nearly 50 years passed before a return to near-equilibrium conditions following a cluster of large floods in the early 1900s. His findings regarding behavior of the stream channel were reinforced by Graf's (1981, 1983, 1988a, 1988b) important studies on both the Salt and lower Gila rivers. These studies show that channel morphology in both rivers may be best viewed in terms of two contrastive sets of events and conditions. Unusually large floods cause erosion of channel banks and development of broad, braided channels. This is followed by periods of relatively lower discharge, a gradual return to a narrower active floodplain, and development of a single meandering channel within the boundaries of the recently active braided channel, Graf's "compound channels" (Graf 1988a, 1988b). During the latter phase, channel avulsion within a relatively narrow active floodplain may be the dominant response to smaller, higher-frequency flood events. Secondary channels flanked by terraces may also occur during the latter phase (Graf 1988b:237). Figure 5.3 illustrates the dynamic relationships between discharge and the variations in channel morphology described above.

Several types of annual discharge events or patterns represented in the reconstructions are perceived as potentially significant for inducing major changes in channel position and/or morphology. These are (1) extremely high annual flows that may have included "catastrophic" flood events, (2) closely spaced clusters of high-magnitude annual flows, (3) larger annual flows or clusters of them preceded by prolonged periods of drought in headwater areas, and (4) prolonged periods of pronounced variability.

Comparison of Salt and Gila River Characteristics

The Salt and Gila rivers differ from each other in a number of characteristics that may influence the timing, rate, magnitude, and nature of channel adjustments. These factors are sediment load, channel substrates, and discharge, the last including annual flow values as well as the distribution of flow over the annual period.

Sediment load in the Gila is proportionally higher than in the Salt and includes more suspended load. Channel substrate sediments in the Salt are significantly coarser than the sandier, more easily eroded floodplain and channel bottom deposits in the Gila. Thus the Gila has a lower potential threshold for discharge events likely to produce changes in channel position and morphology.

Owing to differences in the topography and elevation of the two drainage basins, and thus to the manner in which climate affects precipitation and discharge, the Salt and Gila differ markedly in discharge volume. For the period A.D. 1914–1988, the portion of the Salt watershed included in this study produced an average of 42.7 acre-feet of flow per square kilometer annually, while the Gila watershed yielded only 17.3 acre-feet per square kilometer per year. Thus although the area of the Gila watershed studied here (20,450 km^2) repre-

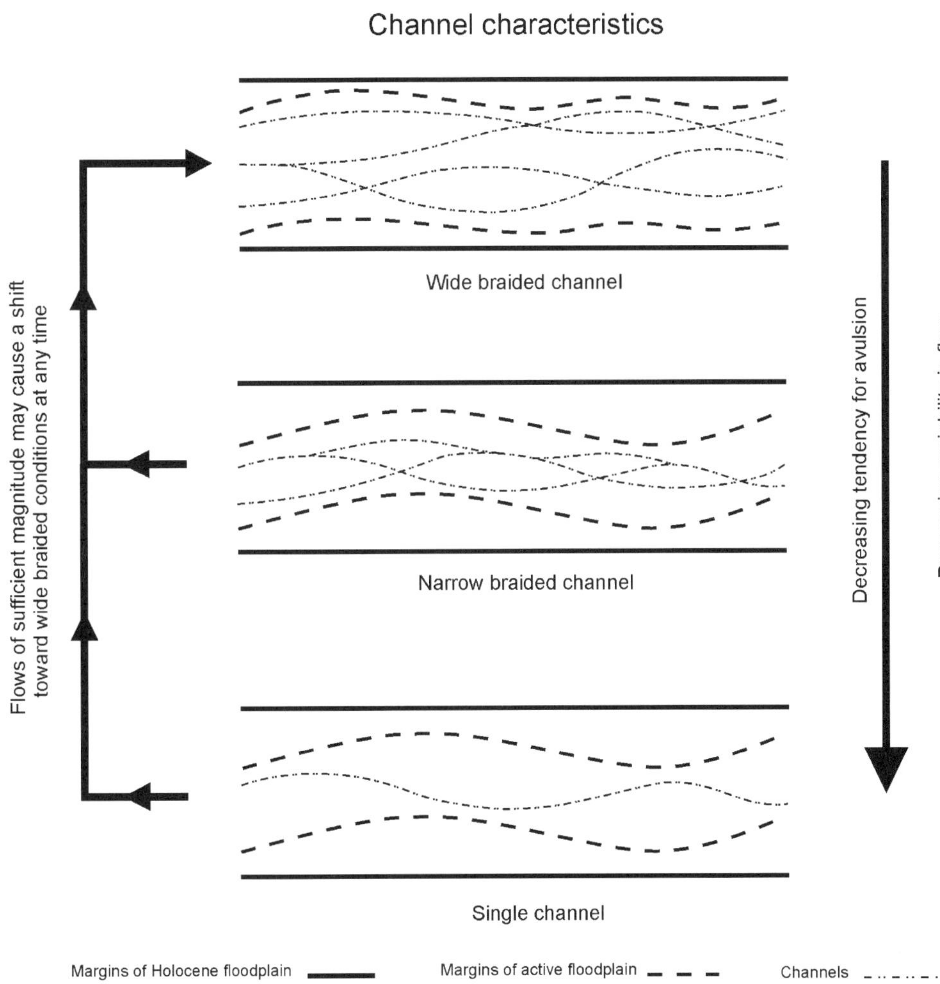

FIGURE 5.3. Dynamic relationships between discharge characteristics and channel morphology.

sents 71 percent of the area of the Salt watershed (28,933 km^2), the mean annual discharge of the Gila (353,929 acre-feet) is less than one-third that of the Salt (1,234,994 acre-feet). Hence base-flow conditions in the Salt may include flows as high as those occurring in average years in the Gila, and flood events in the Salt are of significantly greater volume than in the Gila.

For the same reasons, the distribution of discharge across the annual period is also markedly different between the two drainages. These differences are well illustrated with reference to modern data from two of the gaging stations used in the reconstructions: the Salt River near Roosevelt (1914–1997) and the Gila River at the head of the Safford Valley (1920–1997).

Annual discharge of the Salt River is determined primarily by winter precipitation in the upper reaches of the watershed (Graybill and Nials 1989; Hirschboeck 1985), and the patterned distribution of flow over the annual period consistently reflects this dominance (Figure 5.4). Flow increases steadily beginning in November, with maximum discharge usually occurring in

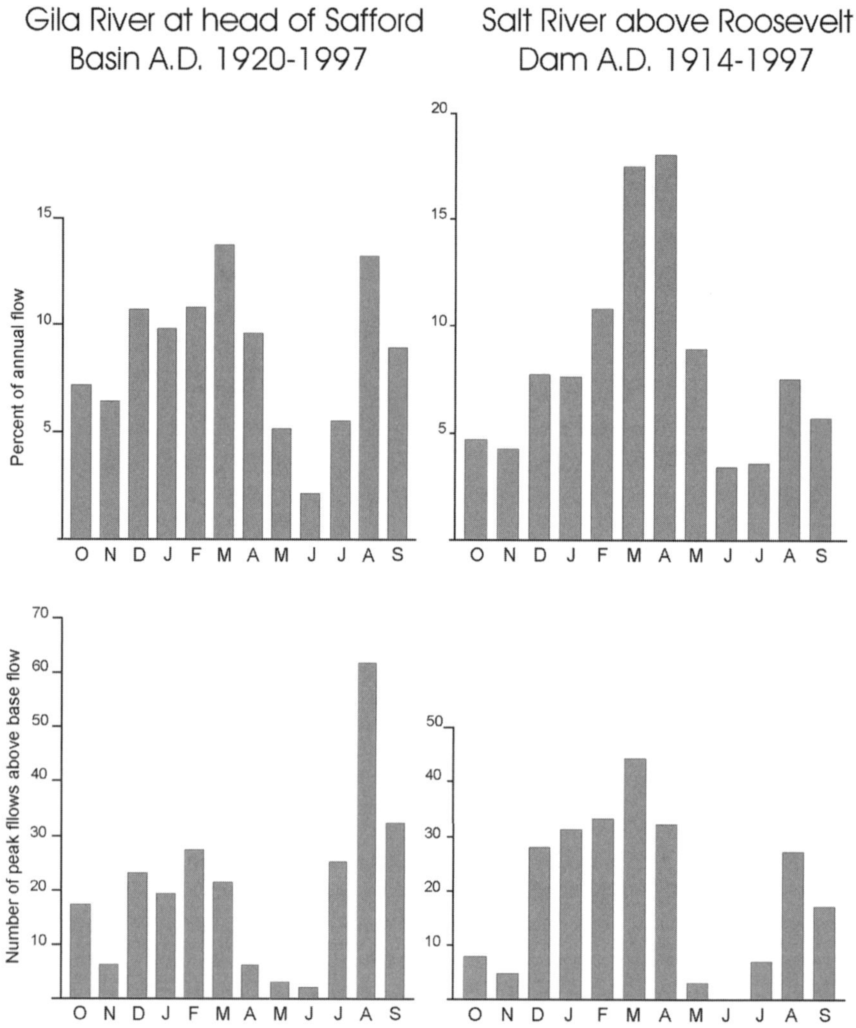

FIGURE 5.4. Modern Salt and Gila river flow characteristics.

March or April. It then declines rapidly to its lowest levels during June and July, followed by a slight peak in August and a return to lower flows from September through November. Most of the total annual flow occurs between November and April, and variation in annual flow is accounted for primarily by discharges occurring during these months. It is also during these months that peak flows above base flow as well as flood events most commonly occur, and there is a strong correlation between years having high annual discharge and years in which floods occur (Graybill and Nials 1989). May through October discharge contributes a much smaller percentage of the annual flow and is less variable, although there is a slight increase in peak flows above base flow in August and September. Thus flows of a magnitude sufficient to produce changes in channel position and morphology are most likely to occur between November and April (Graf 1983).

By contrast, Gila River discharge reflects a much more substantial contribution from a summer convective rainfall (monsoonal) component than Salt River discharge (Meko and Graybill 1993:605) and thus exhibits a markedly bimodal annual distribution. As with the Salt, the influence of winter precipitation is apparent: a substantial portion of annual runoff occurs from December to May, mainly because of rainfall at lower elevations during this period and snowmelt at higher elevations in April and May (Meko and Graybill 1993:606; Stockton 1975). Interestingly, flow in the Gila is more evenly distributed over the December–April period than in the Salt, and the highest mean discharge in the Gila is during March. Whereas flows in the Salt increase steadily from December to a marked peak in March and April, Gila flow increases in December, stays at a relatively even level until the peak in March, and has already begun to decline in April. These differences may be due to the fact that the portion of the Gila drainage represented is lower in average elevation and experiences somewhat higher temperatures. Thus throughout the December–April period, more precipitation reaches the river directly as a result of rainfall and less is retained as snowpack. As with the Salt, Gila flow declines to its lowest levels during June. However, July flow shows a marked increase over June, and the mean flow in August is nearly equal to that of March. Peak discharges above base flow in the Gila show the same bimodal distribution, occurring most frequently from December through March and from July through October. As with the Salt, there is a strong correlation between years having high annual discharge and years in which floods occur. Flows of a magnitude sufficient to produce changes in channel position and morphology are most likely to occur between December and April, but there is also the possibility of such events in July, August, or September, particularly with respect to localized floods.

Salt and Gila River Streamflow and Inferred Geomorphic Effects

Reconstructed annual discharges of the Salt and Gila rivers are presented and discussed below. Each reconstruction has been divided into multiple analytic intervals that contrast with one another in terms of several characteristics likely to have influenced channel morphology and change: mean flow; overall variability; the number, relative magnitude, and temporal spacing of high-magnitude annual discharges; and the number, severity, and duration of prolonged periods of low mean discharge characterized by low variability (droughts). Prolonged periods of low flow are defined here as intervals of eight years or longer during which the mean flow was approximately 0.5 standard deviations or more below the overall mean; in most cases, the maximum flows during these periods are less than the mean for the entire reconstruction. Reconstructed discharges for the two rivers are sufficiently different that the two sets of analytic intervals are not coeval, with five intervals identified for the Salt and six for the Gila.

Along with graphs of flow for each of the analytic intervals, high-magnitude annual discharges with a rank of 25 or greater are listed, as well as the reconstructed discharge for those years, their specific rank, the number of standard deviations above the mean represented, and their recurrence intervals (see below). Also provided for each analytic period is the number of years above and below the overall mean for the entire reconstruction, as well as the number of years in the interval having recurrence intervals of greater than 10 years and greater than 20 years. Prolonged periods of low flow within each analytic period are similarly listed, and the mean and standard deviation, the coefficient of variation, and the maximum and minimum discharges for each low flow period are provided. In the graphs for each interval, a line representing a 25-year centered moving average is also shown; this line may be taken as an approximation of average variation on the scale of a human generation (see Dean 1988a).

Inferred geomorphic responses of the two drainages are based primarily on studies of the effects of floods per se. Although more conservative, annual discharge data represented in the reconstructions can serve as a basis for inferring the relative potential for floods. As noted above, there is a high correlation between years having high-magnitude annual flows and years in which floods occurred in both drainages (Graybill and Meko 1995; Graybill and Nials 1989; Meko and Graybill 1993). Thus while the *magnitude* of floods in any given year is not linearly related to the magnitude of annual discharge and cannot be consistently predicted from annual flow data, high-magnitude annual flows serve as a proxy for the potential *occurrence* of flood events. The timing of large floods inferred on the basis of an earlier Salt River streamflow reconstruction (Graybill 1989; Nials et al. 1989) has been confirmed by recent studies of paleoflood events represented in slackwater deposits and dated by radiocarbon assays (Baker 1987; Ely and Baker 1985; Ely et al. 1993).

Recurrence intervals (*RI*) for reconstructed annual flows have been calculated using the formula $RI = n + 1/r$, where n is the number of years in the reconstruction and r is the magnitude rank of the annual flow in question. This formula is used commonly in hydrologic studies to calculate the recurrence intervals of floods and characterize flood probability magnitudes (10-year floods, 100-year floods, and so on; Bloom 1978). Use of this probability expression in relation to annual flow data provides a convenient way to isolate and characterize patterns of annual discharge (see Nials et al. 1989) but should not be confused with or substituted for flood recurrence intervals. Given this formula, annual flows with a rank of 25 have a recurrence interval of just over 50 years in both reconstructions.

Salt River Streamflow, A.D. 572–1988

Figure 5.5 illustrates reconstructed annual Salt River discharge from A.D. 572 to 1988, indicates the five defined analytic intervals, and presents box plots of discharge for each of the intervals. Tabular data for the entire period of reconstruction and for each interval are also provided.

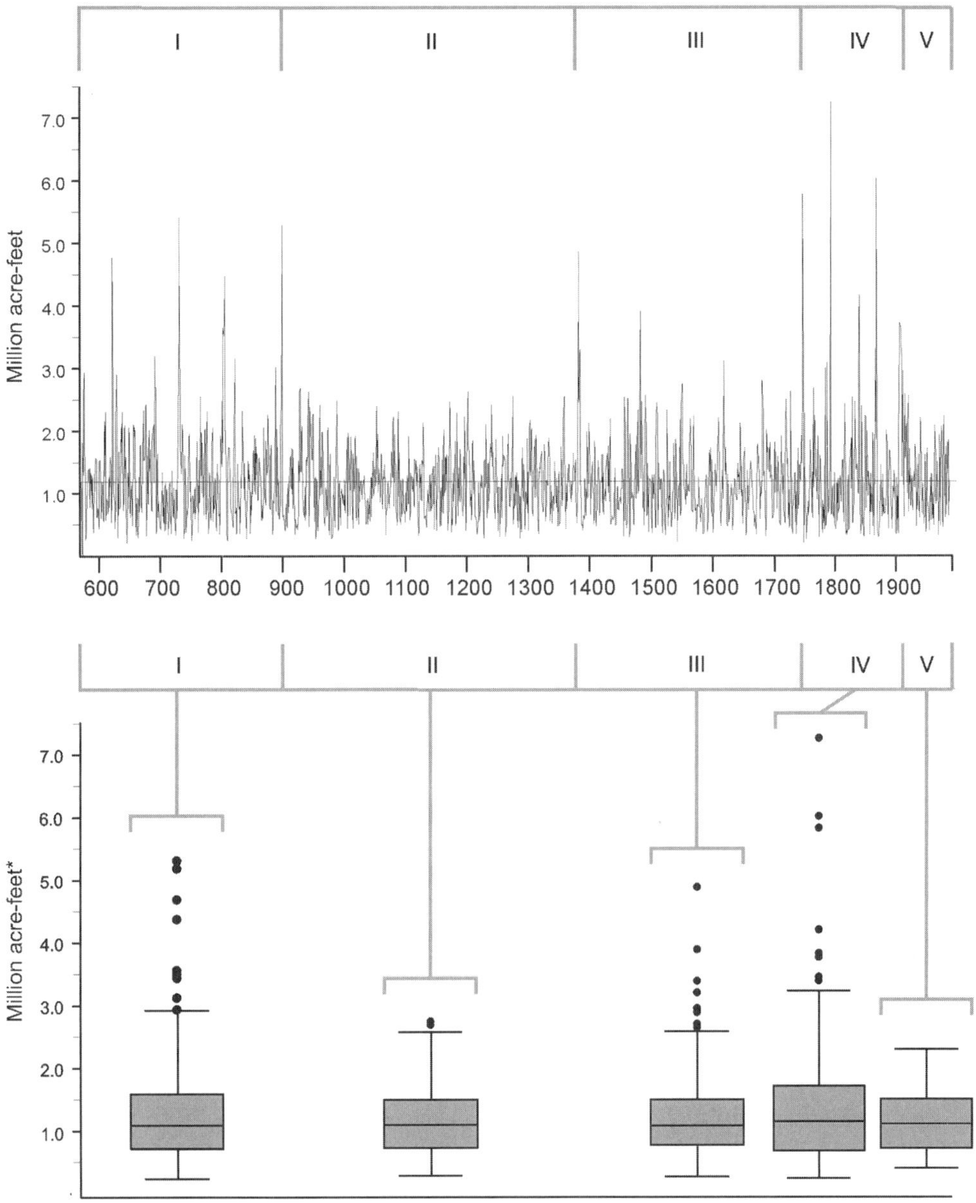

Interval	Dates	N Years	Mean	SD	CV	Minimum	Maximum	> Mean	< Mean	RI > 10	RI > 20
All	572-1988	1417	1.164030	.686980	.59	.185627	7.213954	603	814	n/a	n/a
I	572-899	328	1.220996	.767882	.63	.192305	5.377931	151	177	42	20
II	900-1380	481	1.108238	.512864	.46	.250801	2.655416	202	279	30	12
III	1381-1745	365	1.128684	.609600	.54	.201142	4.819558	143	222	34	16
IV	1746-1920	175	1.317849	1.040980	.79	.185627	7.213954	81	94	30	22
V	1921-1988	68	1.074232	.510182	.47	.314034	2.213421	26	42	5	0

*Box plots represent medians, quartiles, outliers, and extremes for flow values in each interval.

FIGURE 5.5. Reconstructed Salt River streamflow, A.D. 572–1988, showing identified intervals.

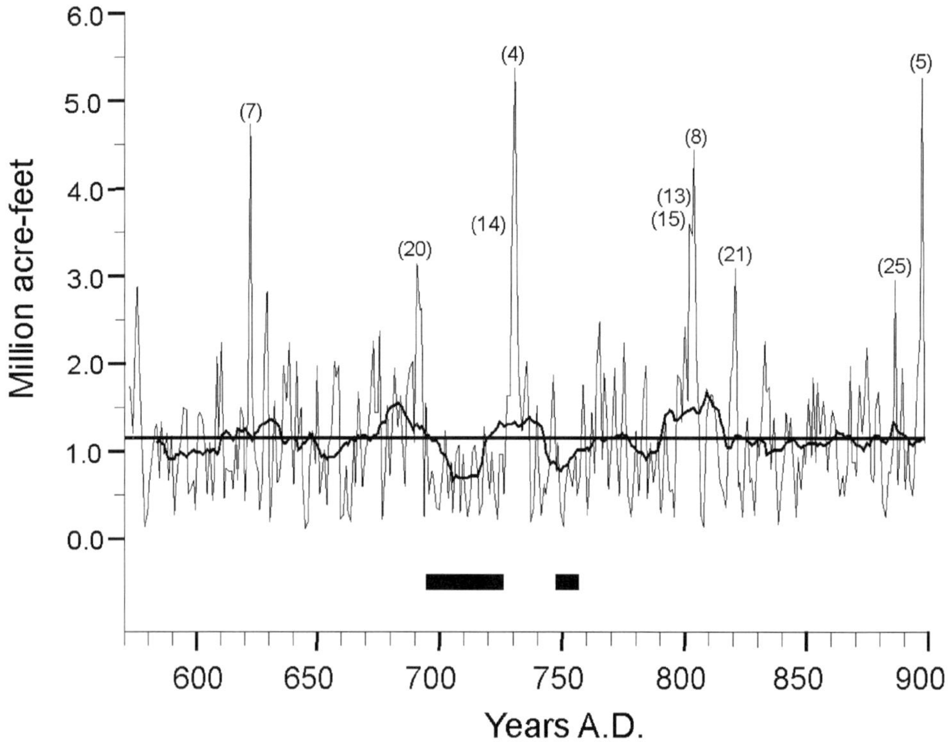

High magnitude annual flows (rank in parentheses)

Rank	Year	Flow	SD above mean	RI
4	731	5.377931	6.13	354
5	899	5.254196	5.95	284
7	622	4.743495	5.21	203
8	805	4.439728	4.77	177
13	803	3.626986	3.59	110
14	730	3.523021	3.43	101
15	804	3.477003	3.39	95
20	691	3.161202	2.91	71
21	822	3.122209	2.85	68
25	888	2.983407	2.65	57

Periods of low flow (bars)

Interval	Length	Mean	SD	CV	Minimum	Maximum
696 - 727	32 years	.762059	.303434	.40	.309711	1.344923
748 - 758	11 years	.739444	.287220	.39	.218767	1.171513

FIGURE 5.6. Salt River Interval I: A.D. 572–899.

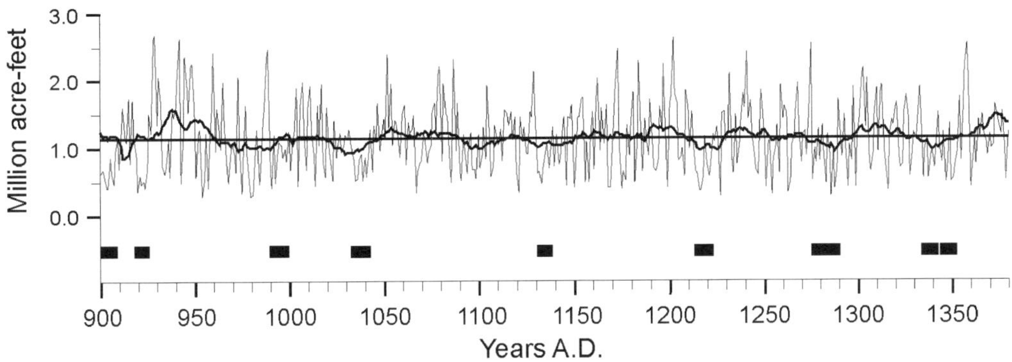

High magnitude annual flows
None

Periods of low flow (bars)

Interval	Length	Mean	SD	CV	Minimum	Maximum
900 - 908	9 years	.588960	.170989	.29	.373420	.901419
918 - 925	9 years	.603649	.246363	.42	.332132	1.103416
990 - 999	10 years	.820553	.246756	.30	.375491	1.177294
1033 - 1042	10 years	.720243	.188966	.26	.482811	.988115
1131 - 1138	8 years	.730496	.188316	.26	.427579	1.029170
1214 - 1222	9 years	.612658	.232751	.38	.321053	1.151386
1276 - 1289	14 years	.791870	.315998	.40	.257768	1.296755
1335 - 1342	8 years	.808710	.202665	.25	.502216	1.094052
1344 - 1353	10 years	.791342	.303987	.38	.448757	1.313529

FIGURE 5.7. Salt River Interval II: A.D. 900–1380.

Salt River Interval I: A.D. 572–899. The initial 328 years of the reconstruction exhibit relatively high mean discharge, numerous high annual discharges, and high variability (Figure 5.6). Ten of the largest 25 reconstructed annual discharges occur during this interval, including the fourth, fifth, seventh, and eighth largest. One of these, in A.D. 899, marks the end of the interval (5.3 million acre-feet; rank = 5). Clusters of high annual discharges are represented by the flows of A.D 730–731, A.D. 803–822, and A.D. 888–899. Two periods of sustained low flow are identified, A.D. 696–727 and A.D. 748–758, accounting for 13.1 percent of the years in this interval. The first of these is the longest period of low discharge identified in the entire reconstruction and came shortly after the high-magnitude flow of A.D. 691. Given the relatively short intervals between high-magnitude annual flows, this was a period of stream-channel disequilibrium. As a result, a single-channel condition and position would have obtained only for brief periods, and a pronounced tendency toward wide, braided channels is probable. Channel avulsion would have been a common phenomenon, and the potential for radical shifts in the channel was high.

Salt River Interval II: A.D. 900–1380. The lengthy Interval II is characterized by low mean flow and low variability (Figure 5.7). None of the largest 25 reconstructed annual discharges occurred, and this interval displays the lowest variability of the five identified intervals. Nine sustained episodes of low discharge are identified but account for only 18 percent of the years in this interval. The first two periods of low flow immediately follow the high-magnitude discharge of A.D. 899. These two nine-year intervals (A.D. 900–908 and 918–925) have the lowest mean discharges of any identified for the entire reconstruction. The remaining six periods of low discharge are more or less evenly spaced across the interval, occurring between 50 and 100 years apart. Discharge stability would have been accompanied by a more pronounced tendency toward channel-within-channel conditions and a gradual trend toward development of equilibrium channels. Although flood events undoubtedly occurred, the majority were probably of relatively low magnitude. As a consequence, radical avulsion would have been minimized and less frequent than in Interval I.

Salt River Interval III: A.D. 1381–1745. The beginning of Interval III (Figure 5.8) is marked by the largest annual discharge in 480 years (A.D. 1382, rank = 6) and is characterized by moderate variability overall. With the exception of A.D. 1382 and three other high-magnitude annual discharges, this interval is quite similar to the preceding one. Six periods of prolonged low flow are identified, accounting for 22.7 percent of the years in the interval. Broad, braided channels accompanied by bank erosion and potentially radical avulsion probably would have developed with the high discharges at the beginning of the interval. A single-channel pattern may have developed in the low-flow periods between subsequent high-magnitude annual flows.

Salt River Interval IV: A.D. 1746–1920. Interval IV begins with the third largest annual discharge for the period of reconstruction and is characterized by frequent high-magnitude discharges and high variability (Figure 5.9). Eleven of the 25 largest reconstructed annual flows occurred during this interval, including the first-, second-, and third-ranked annual discharges. The rank 3 flow marks the beginning of the interval. Three clusters of high annual flows occurred, A.D. 1787–1793, A.D. 1866–1868, and A.D. 1906–1909. Four periods of sustained low discharge are identified, and three of these follow immediately after high annual flow events or clusters of them (A.D. 1750–1757, 1794–1811, and 1870–1883). These periods of low flow account for 30.3 percent of the years in this interval, the highest proportion of any of the five analytic intervals. Discharge characteristics of Interval IV would have promoted channel widening through bank erosion, wide braided channels would have developed, and avulsion would have been common and potentially radical.

Salt River Interval V: A.D. 1921–1988. The remaining 68 years of the reconstruction form Interval V (Figure 5.10). None of the largest 25 reconstructed annual flows occurred, and no prolonged periods of low flow are identified. This is the interval during which all modern studies that allow reconstruction of

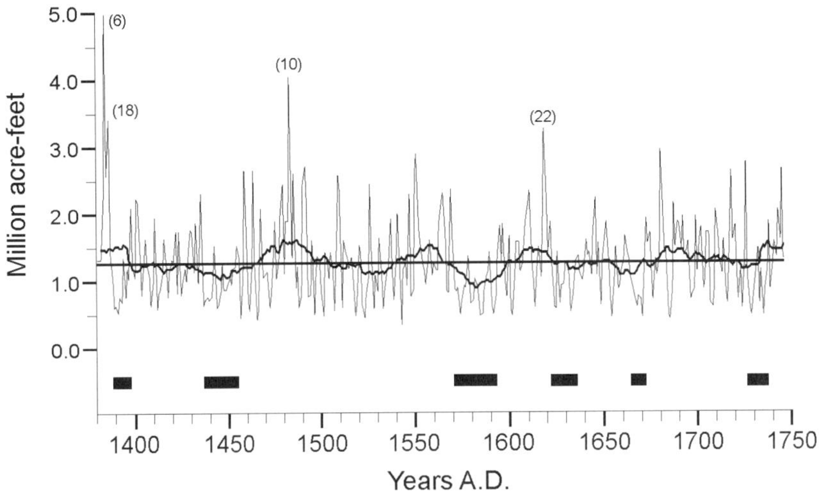

High magnitude annual flows (rank in parentheses)

Rank	Year	Flow	SD above mean	RI
6	1382	4.819558	5.32	236
10	1482	3.881364	3.96	142
18	1384	3.259460	3.05	79
22	1618	3.087523	2.80	64

Periods of low flow (bars)

Interval	Length	Mean	SD	CV	Minimum	Maximum
1386 - 1395	10 years	.709654	.256931	.36	.385733	1.197809
1435 - 1452	18 years	.768489	.242927	.32	.466255	1.379572
1570 - 1592	23 years	.696179	.251863	.36	.330081	1.274237
1622 - 1634	13 years	.778071	.280884	.36	.371590	1.270856
1664 - 1671	8 years	.625210	.179723	.29	.315719	.918253
1727 - 1737	11 years	.747066	.305604	.41	.325749	1.313016

FIGURE 5.8.
Salt River Interval III: A.D. 1381–1745.

geomorphic responses to particular conditions have been accomplished. This research forms the basis for development of the conservative estimates of relationships between unique annual discharges, series of annual discharges, and channel responses presented above (Graf 1983; Nials et al. 1989; Schumm 1977). It should be reiterated that most of these studies took place after substantial human modification and control of the floodplain environment, beginning with initial Euro-American occupation in the Phoenix area and culminating with construction of Roosevelt Dam on the Salt and other upstream dams on the Verde.

Gila River Streamflow, A.D. 534–1988. Figure 5.11 illustrates reconstructed annual Gila River discharge from A.D. 534 to 1988, indicates the six defined analytic intervals, and presents box plots for each of the intervals. Tabular data for the entire period of reconstruction and for each interval are also provided.

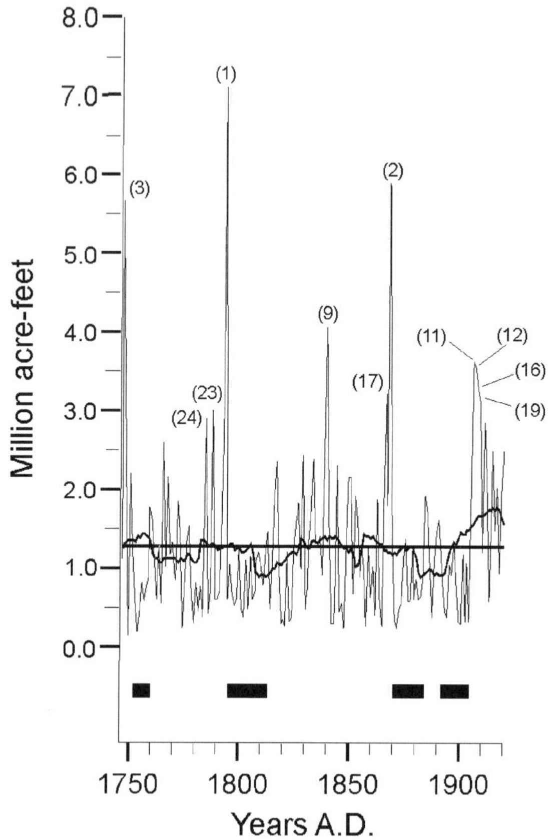

High magnitude annual flows (rank in parentheses)

Rank	Year	Flow	SD above mean	RI
1	1793	7.213954	8.81	1418
2	1868	5.996378	7.03	709
3	1746	5.756900	6.69	473
9	1839	4.143610	4.34	158
11	1906	3.698305	3.69	129
12	1907	3.629400	3.59	118
16	1908	3.361639	3.20	87
17	1866	3.279051	3.08	83
19	1909	3.181350	2.94	75
23	1792	3.067643	2.77	62
24	1787	3.061869	2.65	58

Periods of low flow (bars)

Interval	Length	Mean	SD	CV	Minimum	Maximum
1750 - 1757	8 years	.725814	.286096	.39	.234971	.937327
1794 - 1811	18 years	.871067	.292598	.34	.401384	1.299910
1870 - 1883	14 years	.759100	.324312	.43	.278283	1.433450
1892 - 1904	13 years	.748989	.364935	.49	.320271	1.304717

FIGURE 5.9. Salt River Interval IV: A.D. 1746–1920.

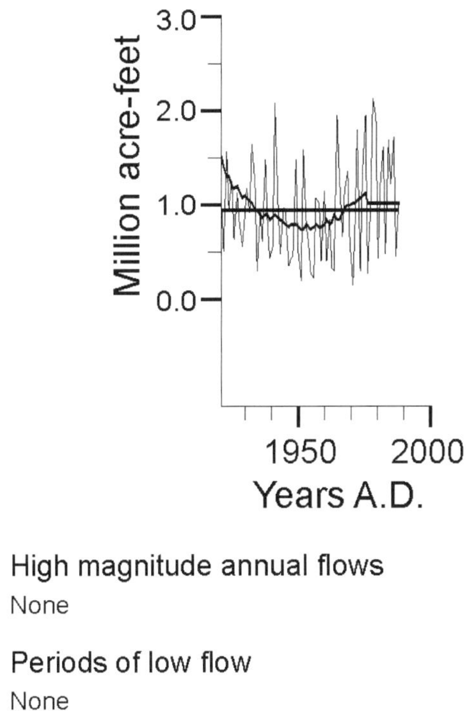

FIGURE 5.10.
Salt River Interval V: A.D. 1921–988.

Gila River Interval I: A.D. 534–748. The first 215 years of the reconstruction exhibit relatively high variability (Figure 5.12). Four of the largest 25 reconstructed annual flows occurred, with two of these being only three years apart (A.D. 728 and 731). Six periods of sustained low flow are identified, accounting for 30.7 percent of the years in the interval. High variability in discharge should have resulted in channel instability, development of a broad active floodplain, and a relatively high probability of avulsion, possibly radical in some instances. Broad, braided channels may have developed and persisted between approximately A.D. 620 and 655 and again between A.D. 715 and the end of interval, with a tendency toward a narrower floodplain and avulsion-prone single-channel conditions in the intervening years.

Gila River Interval II: A.D. 749–1142. Interval II (Figure 5.13) begins with the largest annual discharge in the entire reconstruction (A.D. 749). Numerous other high-magnitude annual flows punctuate the interval, including ranks 2, 3, 4, 8, 9, and 10. Fifteen of the 25 largest annual flows occurred during this interval and 7 of the largest 10. A number of the high-magnitude flows came in clusters: A.D. 820–843, A.D. 897–898, A.D. 983–989, A.D. 1077–1089, and A.D. 1141–1142. Another cluster of high-magnitude but unranked annual flows occurred between A.D. 772 and 784. Eleven periods of sustained low discharge are identified, with three occurring immediately following individual

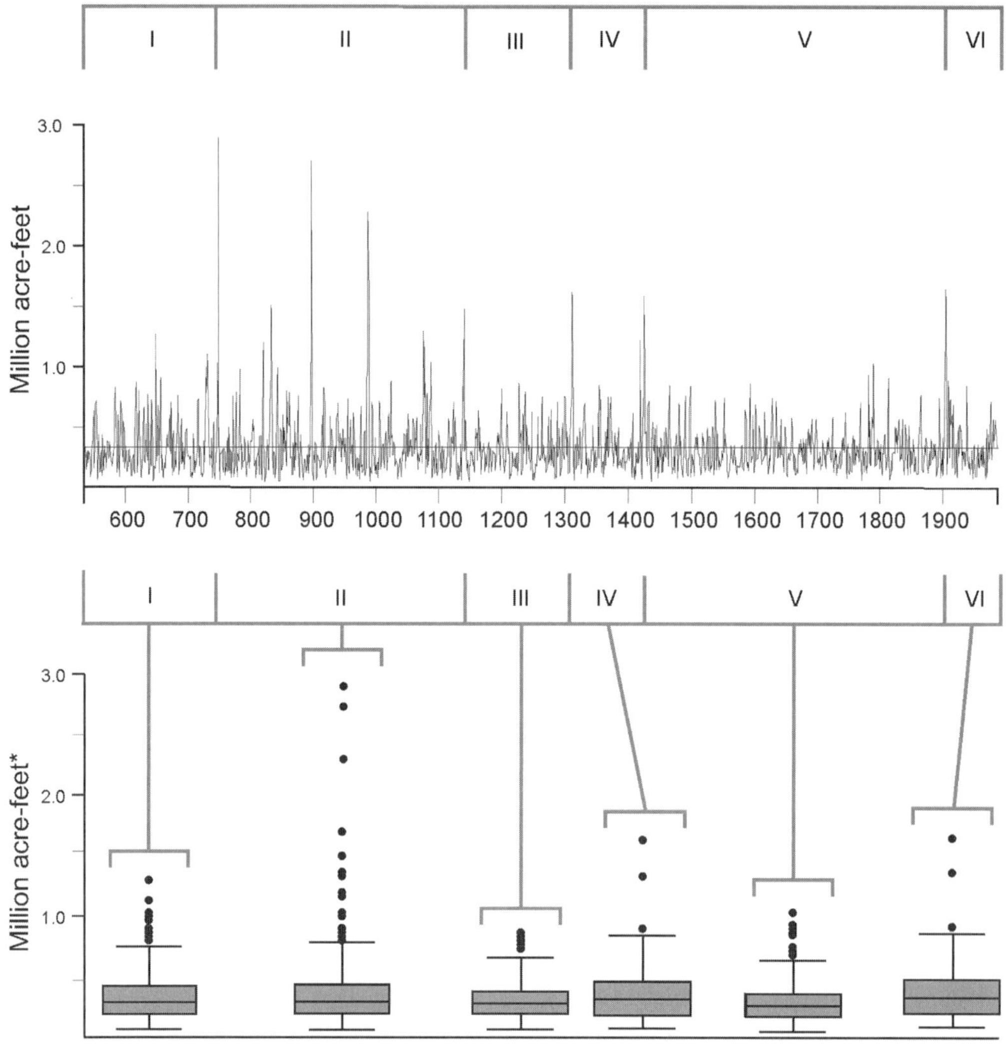

Interval	Dates	N Years	Mean	SD	CV	Minimum	Maximum	> Mean	< Mean	RI > 10	RI > 20
All	534-1988	1455	.326163	.237211	.73	.043114	2.893808	537	918	n/a	n/a
I	534-748	215	.332700	.222836	.67	.048996	1.271704	80	135	31	13
II	749-1142	394	.350871	.316131	.90	.043114	2.893808	155	239	48	29
III	1143-1312	170	.303224	.174948	.58	.052868	.859638	56	114	14	6
IV	1313-1426	114	.339834	.259681	.76	.060790	1.613500	42	72	15	7
V	1427-1905	478	.301965	.163347	.53	.054623	1.025389	163	315	27	13
VI	1906-1988	84	.359798	.258710	.72	.074597	1.540671	39	45	10	4

*Box plots represent medians, quartiles, outliers, and extremes for flow values in each interval.

FIGURE 5.11. Reconstructed Gila River streamflow, A.D. 534–1988, showing identified intervals.

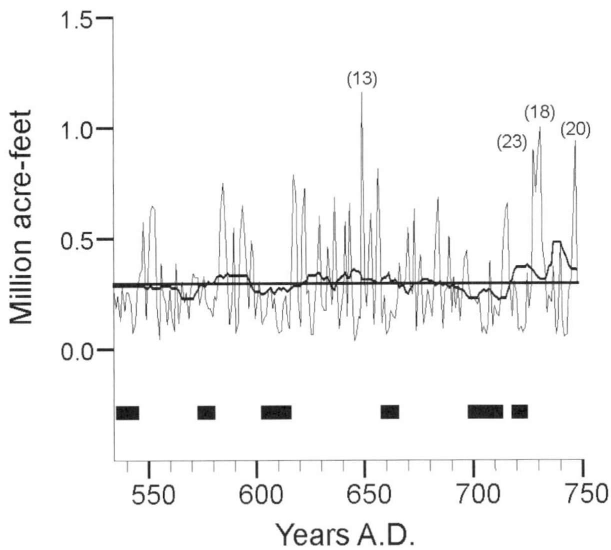

High magnitude annual flows (rank in parentheses)

Rank	Year	Flow	SD above mean	RI
13	649	1.271704	3.99	112
18	731	1.100564	3.26	81
20	747	1.027110	2.95	73
23	728	.995413	2.82	63

Periods of low flow (bars)

Interval	Length	Mean	SD	CV	Minimum	Maximum
534 - 544	11 years	.216917	.071692	.33	.087466	.308115
573 - 581	9 years	.248769	.055990	.23	.173660	.372775
602 - 615	14 years	.187417	.074367	.40	.093415	.315976
658 - 665	8 years	.173531	.062388	.36	.076259	.268540
698 - 713	16 years	.214023	.093316	.44	.078280	.445291
718 - 725	8 years	.200590	.107988	.54	.090854	.383016

FIGURE 5.12.
Gila River Interval I: A.D. 534–748.

or clustered high-magnitude flows (A.D. 751–762, 900–914, and 1129–1138). The eleven low-flow periods account for 30.7 percent of the years in this interval. The combination of frequent high-magnitude annual flows and numerous periods of low flow makes this the most variable of the six analytic intervals for the Gila. High overall variability and the repeated occurrence of sustained low flows immediately following high-magnitude annual flows produced an unstable and variable river bottom. Channel patterns undoubtedly alternated between broadly braided and single-channel conditions, with corresponding variation in the width of the active floodplain. Avulsion would have been more probable during the clusters of higher flow but may have occurred at any time during Interval II.

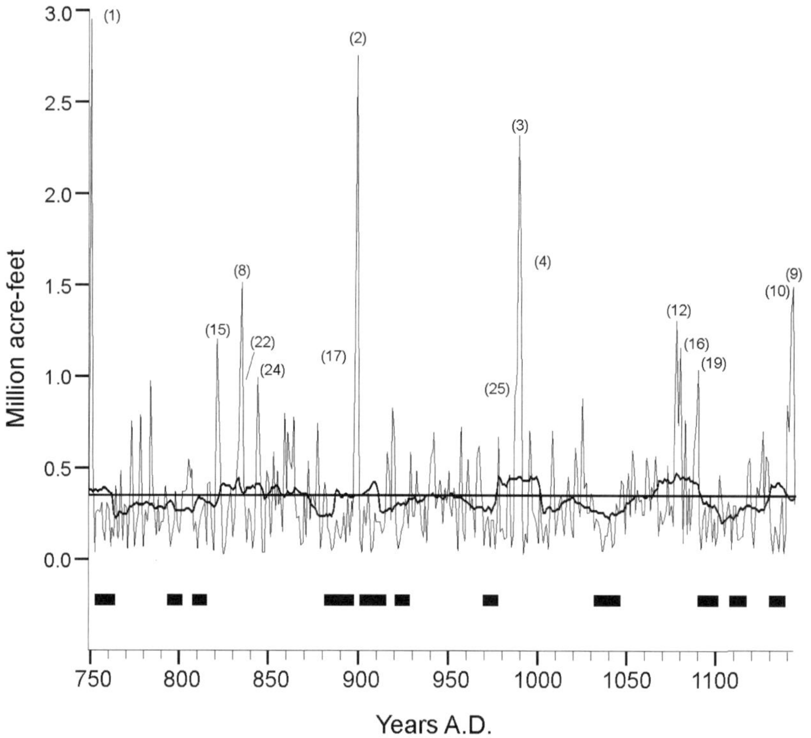

High magnitude annual flows (rank in parentheses)

Rank	Year	Flow	SD above mean	RI
1	749	2.893508	10.82	1456
2	898	2.706683	10.04	728
3	988	2.208618	8.24	485
4	989	1.666660	5.65	364
8	834	1.503529	4.96	182
9	1142	1.480279	4.87	162
10	1141	1.347800	4.31	146
12	1077	1.298731	4.10	121
15	820	1.198715	3.68	97
16	1079	1.158599	3.51	91
17	897	1.148804	3.47	86
19	1089	1.031524	2.97	77
22	833	1.017709	2.92	66
24	843	0.990802	2.82	61
25	983	0.970739	2.72	58

Periods of low flow (bars)

Interval	Length	Mean	SD	CV	Minimum	Maximum
751 - 762	12 years	.211807	.088542	.42	.060469	.320115
793 - 800	8 years	.218544	.085757	.39	.083358	.380580
807 - 814	8 years	.202345	.080384	.40	.079669	.323607
881 - 896	16 years	.201911	.088713	.44	.064196	.358025
900 - 914	15 years	.204023	.092662	.45	.055950	.376903
920 - 927	8 years	.194321	.073889	.38	.082482	.337154
969 - 976	8 years	.180792	.073094	.40	.073479	.256240
1031 - 1045	15 years	.186072	.077435	.42	.066279	.317105
1090 - 1100	11 years	.195836	.087079	.44	.082506	.365380
1107 - 1116	10 years	.218394	.085767	.39	.124220	.321542
1129 - 1138	10 years	.165849	.091680	.55	.057369	.353596

Figure 5.13. Gila River Interval II: A.D. 749–1142.

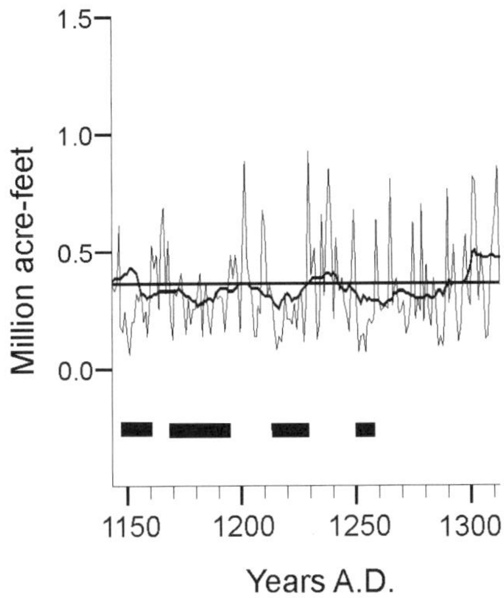

High magnitude annual flows

None

Periods of low flow (bars)

Interval	Length	Mean	SD	CV	Minimum	Maximum
1146 - 1159	8 years	.191022	.069717	.36	.052868	.307393
1168 - 1193	26 years	.243710	.074347	.31	.105799	.374251
1213 - 1228	16 years	.192401	.074303	.39	.068197	.295127
1250 - 1258	9 years	.147405	.060908	.41	.059452	.260296

FIGURE 5.14. Gila River Interval III: A.D. 1143–1312.

Gila River Interval III: A.D. 1143–1312. Interval III (Figure 5.14) exhibits relatively low mean flow and low variability. None of the largest 25 flows occurred during this interval, and four periods of prolonged low discharge have been identified. One has the lowest mean discharge of any low-flow period in the entire reconstruction (A.D. 1250–1258), and low-flow periods account for 34.7 percent of the years in this interval. As seen in Figure 5.14, a further subdivision could be made at approximately A.D. 1229, with less variability before this date and more after. During Interval III, the Gila River flow regime was similar to that of the Salt River at this time. Braided conditions and a broad, active floodplain probably obtained at the beginning of the interval, owing to the high-magnitude flows in A.D. 1141 and 1142. In general, however, the tendency would have been toward development of single-channel equilibrium conditions, especially before approximately A.D. 1229. After this, the probability of avulsion would have increased.

Gila River Interval IV: A.D. 1313–1426. Interval IV (Figure 5.15) witnessed three of the largest 25 high-magnitude annual discharges, beginning with the sixth-ranked flow and ending with the seventh. Several clusters of relatively high, but unranked, flows also occurred: A.D. 1329–1335, 1353–1359, and 1370–1375. Three periods of sustained low discharge are identified, the first of which follows a cluster of high annual flows and the last of which immediately precedes another cluster of high annual flows; these three low-flow periods

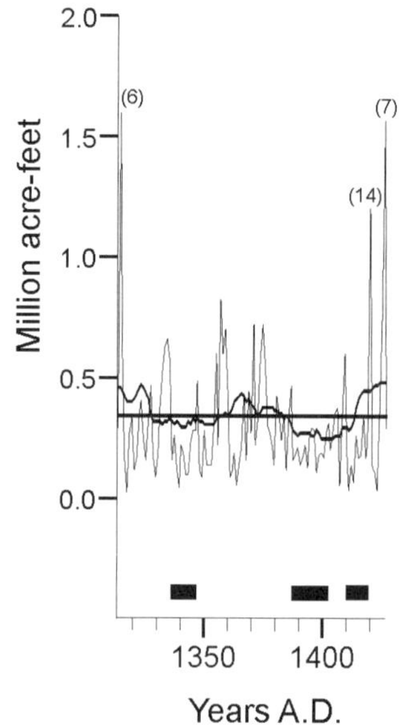

High magnitude annual flows (rank in parentheses)

Rank	Year	Mean	SD above mean	RI
6	1313	1.613500	5.43	243
7	1426	1.583305	5.30	208
14	1420	1.219984	3.77	104

Periods of low flow (bars)

Interval	Length	Mean	SD	CV	Minimum	Maximum
1335-1345	11 years	.210435	.076169	.36	.080445	.311231
1387-1401	15 years	.221211	.052375	.24	.139874	.325204
1410-1419	10 years	.203269	.089210	.44	.066939	.374844

FIGURE 5.15. Gila River Interval IV: A.D. 1313–1426.

account for 31.6 percent of the years in this interval. This period is similar in most respects to the intervals that precede and follow it (see below), but it is bracketed by the high-magnitude flows at the beginning and end of the interval and is further distinguished by higher variability. The several clusters of relatively high flows may have had a greater cumulative influence than their individual magnitudes would suggest, possibly leading to broad channels and braided conditions near the middle of the interval. A long period of relatively low flow in the latter half of Interval IV may have initiated gradual development of a narrower active floodplain and single-channel conditions. Near the end of Interval IV, two high-magnitude flows within less than a decade may have eroded riverbanks and caused a return to a broad, braided channel. Channel avulsion could have occurred at almost any time, but especially during the three years with ranked annual flows or as a result of clusters of unranked but relatively high-magnitude annual events.

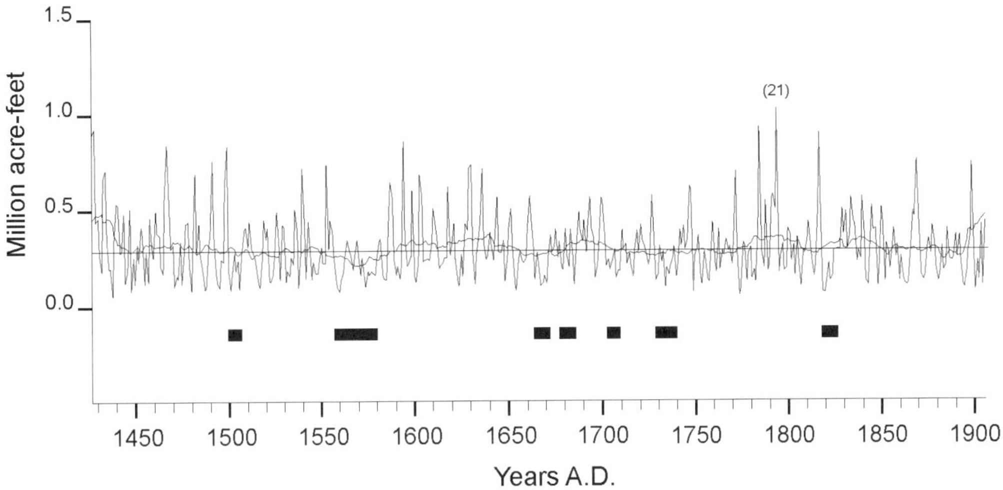

High magnitude annual flows (rank in parentheses)

Rank	Year	Flow	SD above mean	RI
21	1793	1.025389	2.95	69

Periods of low flow (bars)

Interval	Length	Mean	SD	CV	Minimum	Maximum
1500-1507	8 years	.191778	.069288	.36	.087570	.284377
1557-1580	24 years	.206296	.073529	.36	.076717	.346203
1663-1671	9 years	.191036	.071984	.38	.108162	.301773
1676-1686	11 years	.268859	.087445	.32	.112351	.406495
1702-1709	8 years	.211961	.041966	.20	.149216	.295627
1727-1740	14 years	.214060	.064113	.30	.100202	.311250
1818-1827	10 years	.194331	.194331	.46	.071408	.295977

FIGURE 5.16. Gila River Interval V: A.D. 1427–1905.

Gila River Interval V: A.D. 1427–1905. Interval V (Figure 5.16) is characterized by stability and is quite similar to Interval II on the Salt in its length and low variability. Only one of the largest 25 annual flows occurred during this interval (rank = 21), and seven periods of sustained low flow are identified; however, the low-flow periods account for only 17.8 percent of the years in this interval. Interval V has the lowest mean flow and the lowest variability of the six Gila River intervals. Single-channel conditions and a relatively narrow active floodplain would have prevailed throughout most of the interval, and avulsion within that floodplain probably would have been the dominant response to variations in discharge. The single ranked high-magnitude flow of A.D. 1793, preceded by a decade of mostly higher flows, may have initiated development of a broad, braided valley bottom. However, historical records indicate single-channel conditions from the early 1800s to the end of this interval in the Safford Basin (Burkham 1972).

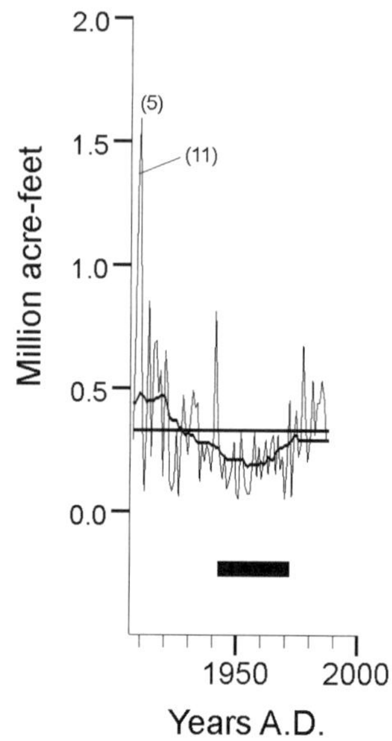

High magnitude annual flows (rank in parentheses)

Rank	Year	Flow	SD above mean	RI
5	1908	1.640671	5.54	291
11	1907	1.347273	4.30	132

Periods of low flow (bars)

Interval	Length	Mean	SD	CV	Minimum	Maximum
1943-1972	30 years	.213700	.086647	.40	.074597	.356189

FIGURE 5.17. Gila River Interval VI: A.D. 1906–1988.

Gila River Interval VI: A.D. 1906–1988. The second year of this final interval (Figure 5.17) was marked by the highest annual discharge in over 700 years (A.D. 1907, rank = 5), and another ranked flow occurred in the following year (rank = 11). By contrast, a 30-year period of sustained low discharge is the longest such interval in the entire reconstruction and accounts for 35.7 percent of the years in this short interval. A cluster of high-magnitude flows between 1905 and 1917 eroded much of the floodplain in the Safford Basin (Burkham 1972), and severe bank erosion formed a wide, shallow, braided channel. A narrower, active floodplain and a narrow, more sinuous channel were reestablished only after approximately 50 years (Burkham 1972). This gradual shift back to single-channel conditions took place during the 30-year period of low flow identified above.

Comparisons

Low-frequency variation in the two flow regimes is highlighted by the analytic intervals defined and discussed above. Interval-specific summaries of flow patterns and predicted geomorphic consequences for both drainages are presented in Table 5.4. Contrasts between the two drainage systems are further highlighted by comparison of several other aspects of the reconstructions.

A simple correlation of streamflow for the 1,417 years shared by the two reconstructions reveals that they have only about 20 percent of their variance in common (Pearson's $r = .454$, sig.: > 0.01). Differences in the absolute amount of discharge are obviously substantial and parallel those described for the modern period: the overall mean flow for the Gila is only 28 percent that of the Salt, and on average the Salt watershed is markedly more productive than that of the Gila—40.2 acre-feet versus 15.9 acre-feet per square kilometer per year. Thus the mean annual flow of the Salt (1,164,030 acre-feet) would represent the fifteenth largest annual flow in the entire Gila reconstruction, with a recurrence interval of over 90 years.

As shown in Figure 5.18, Gila flow was much more variable than that of the Salt. Salt flow is characterized by relatively low variability overall, with only two periods of relatively high variability (corresponding to Intervals I and IV, defined above). In contrast, the Gila is characterized by high variability overall, with only one lengthy period of relatively low variability (corresponding to Interval V, defined above). Finally, prolonged periods of low flow are much more common in the Gila (32) than in the Salt (21) and account for a much greater proportion of the Gila reconstruction—27.3 percent for the Gila (387 of 1,455 years) versus 19.1 percent for the Salt (272 of 1,417 years). These long-term differences are accounted for in large part by the contrasting climate patterns that determine runoff in the two drainage basins.

IMPLICATIONS FOR ABORIGINAL IRRIGATION SYSTEMS

An earlier reconstruction of Salt River streamflow for the period A.D. 740–1370 formed the basis for inferences concerning relationships between variation in streamflow, channel morphology, the operation of canal systems, and more general issues of culture history and process in the Salt River Valley (Graybill 1989; Graybill and Nials 1989; Nials et al. 1989). The extended Salt River reconstruction and the long-term Gila River reconstruction presented here provide greatly enhanced temporal contexts in which to evaluate the influence of both high- and low-frequency variation (Dean 1988a) in streamflow on aboriginal irrigation systems along the two rivers. Extension of both reconstructions back into the sixth century A.D. provides a basis for examination of the early phases of Hohokam canal system development and expansion, and extension of them into the modern period provides important new perspectives on this aspect of environmental variation during the late prehistoric, Protohistoric, and early Historic periods. Having reconstructions for both major drainages introduces a significant geographic dimension for comparison and contrast as well.

The Gila River reconstruction applies most directly to the reach between Safford and the mouth of the San Carlos River. Owing to modern irrigation

Salt River Intervals

Interval	Dates	Characteristics
V	A.D. 1921–1988	Low discharge variability (CV=.47) No high magnitude annual flows No prolonged periods of low flow Trend toward single-channel conditions throughout
IV	A.D. 1746–1921	High discharge variability (CV=.79) Frequent extreme high annual flows 6 prolonged periods of low flow (30.0% of years) Major disruptions of channel form common, resulting in wide, braided channels High potential for radical avulsion
III	A.D. 1381–1746	Moderate discharge variability (CV=.54) Extreme high annual flows infrequent and widely spaced, but the interval begins with the largest annual flow in 480 years 4 prolonged periods of low flow (22.7% of years) Occasional major disruptions of channel form, producing wide, braided channels Trend toward single-channel conditions between high flows Avulsion common, potnentially radical
II	A.D. 900–1381	Low discharge variability (CV=.46) No extreme high annual flows, large floods probably infrequent 9 prolonged periods of low flow (18.1% of years) Trend toward single-channel conditions throughout, with no major disruptions of channel form Avulsion within relatively narrow active floodplain common, radical avulsion minimized
I	A.D. 572–900	High discharge variability (CV=.63) Numerous high annual flows, numerous extreme annual flows 2 prolonged periods of low flow (13.1% of years) High channel variability Wide, braided channel dominates Avulsion common, sometimes radical

Gila River Intervals

Interval	Dates	Characteristics
VI	A.D. 1906–1988	High discharge variability (CV=.76) Extreme high annual flows start the interval, causing major changes in channel form Prolonged low flow results in trend toward single channel conditions
V	A.D. 1427–1906	Low discharge variability (CV=.53) Few high annual flows, only one extreme high annual flow 7 prolonged periods of low flow (17.8% of years) Single channel form probably dominated throughout Avulsion common within narrow active floodplain
IV	A.D. 1313–1427	High discharge variability (CV=.76) Extreme high annual flows at beginning and end of interval 3 prolonged periods of low flow (31.6% of years) Wide, braided channel form at beginning and end of interval, with trend toward single-channel form in between
III	A.D. 1143–1313	Moderate discharge variability (CV=.58) No high annual flows 4 prolonged periods of low flow (34.7% of years) Wide, braided channel form, broad, active floodplain at beginning of interval Trend toward single-channel form until about A.D. 1229, with a trend back toward braided conditions thereafter
II	A.D. 749–1143	High discharge variability (CV=.90) Numerous high annual flows, extreme annual flows common 11 prolonged periods of low flow (30.7% of years) Major disruptions of channel form frequent Wide, braided channel form dominates, but alternates with single channel form; corresponding variation in width of active floodplain Avulsion common, potentially radical
I	A.D. 534–749	Moderate discharge variability (CV=.67) Numerous high annual flows, extreme high annual flows common toward end of interval 6 prolonged periods of low flow (30.7% of years) Wide, braided channel form toward end of interval Avulsion common, potentially radical

TABLE 5.4. Flow characteristics and geomorphic effects for Salt and Gila river analytic intervals.

and related land modification, almost nothing is known of the extent, structure, and history of prehistoric canal systems that drew water directly from the Gila in the Safford Basin. Given the configuration of the channel along this reach, archaeological evidence of a substantial population throughout the last millennium of the prehistoric period (Bronitsky and Merritt 1985; Dean et al.

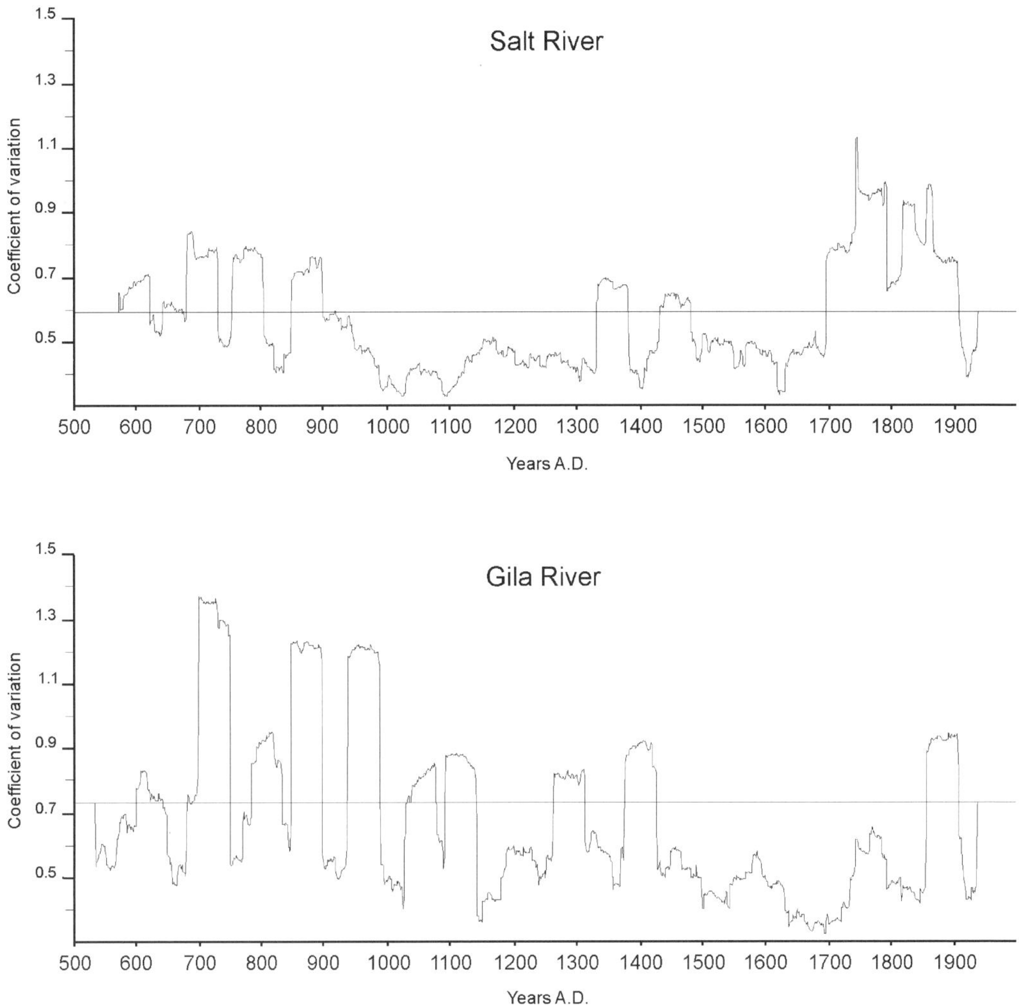

FIGURE 5.18. Coefficients of variation for Salt and Gila river streamflow (50-year running means).

1994; Gregory 1995b; Neely et al. 1997), and minimal direct evidence (Clark et al. 1999), it is a virtual certainty that large-scale irrigation agriculture was practiced in this area. The Gila River reconstruction presented here will provide a useful framework for analysis and interpretation of prehistoric Safford Basin canals that may be discovered and investigated in the future.

The emphasis here is on the well-known Hohokam irrigation systems in the Salt River Valley and along the middle Gila. For purposes of this discussion, it is assumed that variation in flow below the Buttes mirrored that represented in the Gila River reconstruction and that the amount of flow available for irrigation along this reach would not have been greater than the reconstructed values for the Safford Basin. As argued above, it is probable that flow at the Buttes actually would have been less than that at Safford, owing to losses from infiltration, evaporation, and prehistoric diversion of water for irrigation above the Buttes.

Irrigation Technology

Available evidence indicates that all aboriginal populations in the region possessed identical technologies with which to exploit available water resources (Nials and Gregory 1989). The intakes of the principal canal systems along both drainages were positioned at or immediately below bedrock masses or reefs; these optimal intake locations were used by the earliest documented Hohokam systems and thereafter throughout the prehistoric sequence, and by historic Pima canals as well (Gregory 1991; Gregory and Huckleberry 1994; Gregory and Nials 1985; Howard and Huckleberry 1991). These features of the natural landscape ensured a more reliable supply of water during periods of low flow and have the additional advantage of defining zones of relatively greater channel stability under most flow conditions.

All documented systems represent gravity-fed canals that lacked permanent intake structures (Nials and Gregory 1989). Drop structures and canals constructed specifically for drainage are thus far unknown, and augmentation of the amount of irrigable acreage at any given time could be accomplished in only three ways: construction of new main canals, downstream extension of existing main canals, or parallel branching of existing main canals (Nials and Gregory 1989). Refurbishing and repair of intake structures undoubtedly occurred annually, and more substantial repairs or even complete reconstruction of intakes and the headward portions of canals were required on a fairly regular basis, perhaps between 5 and 10 years on average (Nials et al. 1989).

At least one observer indicates that historic Pima canals along the Gila were overbuilt (Southworth 1919), that is, they were constructed to accommodate a much greater amount of water than they routinely carried. This may have been the case with their prehistoric Hohokam counterparts as well (see Crown 1987:Table 1). Provided water could be effectively channeled in downstream segments, overbuilt canals could have more readily accommodated higher-than-normal flows and would have mitigated some potentially serious effects of flood events: silting in or outright destruction of the headward portions through erosion. If demonstrated, such overbuilding may represent one of the few direct technological responses to the highly variable flow of the two rivers available to these irrigation farmers, and is an aspect of Hohokam irrigation technology that deserves attention in future research.

In general, aboriginal canal systems along both drainages had to accommodate the unpredictable flow characteristics of arid-region rivers: high overall variability and non-equilibrium channel conditions punctuated by potentially destructive and unpredictable, channel-changing, high-flow events. Graf (1988b:240) has observed that engineering applications in modern arid-region floodplains need to account for the erodible banks and mobile beds characteristic of such streams. So it was with aboriginal irrigation farmers along the Salt and Gila. However, lacking dimension lumber, concrete, draft animals or machinery, and other advantages possessed by modern societies—including the ability to build massive upstream dams to control flow—these early farmers had to rely primarily on labor-intensive, organizational means rather than technological ones in dealing with the vagaries of streamflow and its effects.

Given a uniform irrigation technology, similar variations in discharge patterns and channel morphology would have had predictably similar effects on canal systems in both drainages. Table 5.5 identifies several general flow conditions, probable geomorphic and other responses, effects on canal systems, and potential human responses.

Drainage System Differences

Discharge patterns and channel morphology influence irrigation systems by virtue of the amount of water available, seasonal and longer-term variation in that availability, and the variable morphology of the streambed from which the water must be diverted. Patterns of discharge over the annual cycle and over the longer term constitute the basic parameters of water availability. Combined with resulting variation in channel morphology, these patterns define engineering and maintenance problems that must be solved using available technology. While facing common problems inherent in drawing water for irrigation from arid-lands rivers, aboriginal populations along the Salt and Gila applied their similar knowledge of irrigation technology to fundamentally different natural systems. Differences in the physiography of the two river valleys, in sediment load, and in the variable amounts and characteristics of flow created different sets of problems for these irrigation farmers.

The physiography of the valleys through which the rivers flowed both imposed constraints and defined opportunities for irrigation agriculture. The influence of valley topography on the shape of aboriginal irrigation systems in the two drainages has been previously discussed (Crown 1987:157–158; Gregory 1991:178–179; Gregory and Nials 1985; Howard and Huckleberry 1991; Nials et al. 1989). Systems along the Gila were more constrained by well-developed terraces, and as a result main canals were strongly linear, paralleling the river rather closely in most cases (Crown 1987; Gregory and Huckleberry 1994; Midvale 1965). In the Salt River Valley, topography allowed Hohokam main canals to extend as far as eight miles from the river and in variable directions. In some cases, this resulted in a stacked configuration of canals paralleling the river at some remove, while in other instances the principal orientation of main canals was perpendicular rather than parallel to the river (Gregory and Nials 1985; Howard and Huckleberry 1991; Midvale 1968; Turney 1929). In both cases, the arrangements ultimately placed most of the length of individual Salt River canals beyond the margins of the geologic floodplain and thus beyond the reach of even a maximum probable flood (Nials et al. 1989).

The greater amount of suspended load carried by the Gila would have had direct implications for routine labor requirements as well as responses necessitated by flood events. Assuming an equal shape, size, length, and gradient, Gila canals would have required more frequent cleaning than those along the Salt. During flood events, infilling of the headward portions of canals by

TABLE 5.5. Flow characteristics, possible geomorphic and related effects, and effects on irrigation systems.

FLOW CHARACTERISTICS	PROBABLE GEOMORPHIC AND RELATED EFFECTS	EFFECTS ON IRRIGATION SYSTEMS	POTENTIAL OR NECESSARY RESPONSES
Prolonged near-average flow conditions	Relative channel stability; trend through time toward single-channel condition with minimal active floodplain	1. Sets upper limits on potentially irrigable acreage 2. Routine deterioration of canal systems (including variable destruction of intake structures)	1. Routine maintenance activities (cleaning, bank repair, and frequent repair of intake structures)
Prolonged low-flow conditions	Landscape dessication; general reduction in landscape vegetation, possible localized increase in channel-marginal and channel-bottom vegetation; increased potential for avulsion or erosion at subsequent high-flow event (see below)	1. Reduction in irrigable acreage 2. Possible failure of downstream systems 3. Increased dependence on systems with most effective hydraulic damming conditions	1. Increase capture efficiency (possibly with the effect of reducing available water for downstream systems) 2. Increase dependence on non-irrigation foodstuffs
High-magnitude discharges (floods)[a]	Channel avulsion (within active floodplain), possibly radical; development of braided conditions	1. Stranded intakes 2. Deposition in canal segments closest to river, with possible complete infilling (destruction of intakes implied) 3. Possible local or major bank destruction in downstream canal segments 4. Possible loss of all or part of one or more irrigation seasons	1. Relocate intake locations in upstream locations and construct new intake structures 2. Clean flood deposits from canals 3. Engineer and excavate new segments to replace silted-in segments; possible relocation of intakes 4. Abandon systems
	Lateral erosion and/or downcutting[b]	1. Stranded intakes 2. Destruction of canal segments closest to river 3. Reduction of irrigable acreage 4. Possible loss of all or part of one or more irrigation seasons	1. Construct new intakes in upstream locations 2. Engineer new segments to connect with isolated canals 3. Abandon systems
Localized flooding from tributary drainages	Creation of fans on and within floodplain; possible avulsion of master river channel to position away from tributary mouth	1. Washout of canal segments built across tributary drainages 2. Possible loss of all or part of one or more irrigation seasons	1. Repair affected segments 2. Abandon systems or portions of systems

[a] Multiple, closely spaced moderate or larger flows may produce the same effects as a single unusually large flood.

[b] Recent geomorphic studies of the Salt River channel (Graf 1981, 1983) demonstrate that earlier suggestions of significant channel deepening (Nials et al. 1989) were in error. Although downcutting can probably be eliminated as a factor in the operation of prehistoric canal systems on the Salt and Gila rivers, avulsion and/or lateral erosion may have caused similar effects on canals and dictated the same responses for proper canal operation.

sediments would have been more severe and extensive; as a result, the need to either clean out or reroute headward segments was likely a more frequent occurrence along the Gila.

Differences in the distribution of flow over the annual period produced differences in risk factors associated with irrigation farming along the two drainages. Assuming an irrigation season falling between May or June and September for both rivers, irrigation along the Gila was a much riskier undertaking. In the Salt, destructive floods are most likely to occur between November and April, providing a relatively predictable window of time during which necessary repairs to intakes and canals could be made before planting. On the Gila, in addition to winter floods there is the possibility of destructive flows *during* the growing season, in July, August, or September. Thus while repairs necessitated by winter floods were probably a regular occurrence in both drainages, along the Gila there was also the chance that labor for large-scale maintenance and repair would have to be quickly marshaled during the critical growing season. Floods during this season could also result in shifts of channel position and configuration, potentially making it even more difficult to repair the systems in time to save crops that were already growing. Thus there may have been substantial long-term differences in the timing and magnitude of labor requirements and in the reliability of foodstuffs derived from irrigation farming.

Variability in discharge over the longer term has been shown to be markedly greater for the Gila than for the Salt, especially during the prehistoric period (see Table 5.4 and Figure 5.18). Before A.D. 1450, the coefficient of variability for Gila flow is .78, whereas this value is only .55 for the Salt. The number of years falling during prolonged periods of low flow is over 30 percent for the Gila during this same period, but for the Salt only about 16 percent of the years fall in low-flow periods. Furthermore, Gila low-flow periods were frequently followed immediately by very high annual flows or clusters of them, greatly increasing the potential effects of floods on channel morphology (A.D. 573–581, 602–615, 698–713, 718–725, 807–814, 881–896, 1129–1138, 1213–1228, 1410–1419). On the Salt, similar juxtaposition of prolonged low-flow periods and high annual discharges is relatively rare. The greater unpredictability of available water from year to year, combined with associated higher variability in channel morphology, defined a less predictable environment for prehistoric irrigation agriculture along the Gila.

Although the Gila River drainage basin is much larger than that of the Salt, the river itself is much smaller than the Salt in the reach above the Salt-Gila confluence. As described above, the Gila watershed produces proportionally much less discharge than does the Salt, and prolonged periods of low flow have been much more frequent in the Gila. These substantial differences in the absolute amounts of water available for prehistoric irrigation have perhaps been underappreciated by archaeologists. A simple example serves to illustrate the scale of the relationship between Gila River flow and the amount of water that may have been diverted prehistorically for irrigation.

Crown (1987:Table 1) estimated a total of 6,247 irrigable hectares (12,783 acres) under the main canals of the Classic period Escalante and Casa Grande systems (Gregory and Nials 1985). These two systems are located on opposite sides of the river, and both had their intakes in the area immediately below the Buttes (Crown 1987:Figure 2). Assuming a relatively conservative irrigation requirement of 5 acre-feet per year, 63,915 acre-feet of water would be required to irrigate all the land under just these two main canals (Nials and Gregory 1989; Nials et al. 1989). Analytic intervals III and IV defined above correspond to the Hohokam Classic period and provide convenient referents for comparison. The amount of water needed to irrigate the land under just these two canals represents 21 percent of the *annual* mean flow for Interval III and 19 percent of the *annual* mean flow for Interval IV (see Figures 5.14 and 5.15). Approximately one-third of the years in these two intervals fall within prolonged periods of low flow (95 of 284 years), and the amount of water needed to irrigate the estimated acreage represents between 26 percent and 43 percent of the *annual* mean flow during these periods. Furthermore, 7 years within this 284-year period had annual flows *less* than 64,000 acre-feet, and 41 years (14.4 percent) had annual flows less than 128,000 acre-feet. Thus about 15 percent of the time, these two canals would have had to capture between 50 percent and 100 percent of the total annual flow in order to irrigate the maximum acreage, and in a few years even 100 percent capture of total annual flow would not have been sufficient.

These calculations use the maximum amount of potentially irrigable land for these two systems, and it is probably unlikely that all of it was actually irrigated. However, neither do the calculations take into account the actual amount of water available during the irrigation season. This amount would have been considerably less than the annual totals and may have been further reduced by upstream diversion of water for irrigation. Furthermore, although the Casa Grande system was by far the largest of those along the middle Gila (Crown 1987; Gregory and Huckleberry 1994), there were at least seven other Classic period systems downstream from the two considered here (Gregory and Nials 1985). All nine of the Classic period Gila systems appear to have had Preclassic antecedents in the same locations (Gregory and Huckleberry 1994; Gregory and Nials 1985).

Taken together, these data suggest that the amount of available water was likely a limiting factor in Hohokam irrigation farming along the Gila and that there were times when irrigation water may not have been available to downstream canal systems. Thus water availability could have influenced development and expansion of canal systems, lack of water may have affected productivity during some intervals, and competition for water between upstream and downstream systems may well have been a factor in social relations among Gila River settlements. Given the substantially greater amounts of Salt River flow and lower variation in discharge, available water was probably seldom, if ever, a problem for Salt River Valley irrigation systems (see Nials et al. 1989).

Differences in the amounts and predictability of water available for irrigation may have been a factor in the balance of subsistence technologies relied on by prehistoric populations living along the two drainages. Non-canal water-control facilities and other agriculture-related features are common along the Gila River. Rockpiles, check dams, gridded gardens, catchment basins, and canals designed to control or divert tributary runoff occur frequently along both sides of the river from Safford to the Salt-Gila confluence, and extensive rockpile fields in particular are associated with all the principal Hohokam canal systems along the middle Gila (Cantley 1991; Crown 1987; Debowski et al. 1976; Gregory and Huckleberry 1994; Gregory 1995b; Lindauer 1984; Neely 1997; Neely and Crary 1998; Woosley 1980). Such features are relatively rare in the Salt River Valley, where a much greater land area could be brought under canals and where sufficient water to irrigate that land was normally available. The need to recognize and analyze the fact of diversity in agricultural technologies employed by riverine Hohokam populations has been previously noted (e.g., Crown 1987). Variable agricultural strategies undoubtedly reflect in part the degree to which streamflow was predictable and sufficient, and thus the degree to which canal irrigation could be depended on by the respective populations.

Temporal Comparisons and Contrasts

The reconstructions presented here also provide a basis for specific temporal comparisons and contrasts, introducing a historical perspective and interpretation of this important aspect of environmental variation. Combined with consistent long-term differences in the two drainages described above, temporal variation in flow characteristics probably created an important dynamic in Hohokam prehistory in the Phoenix Basin. Detailed assessment of the significance of such variation over the entire period of the reconstructions is beyond the scope of this chapter; however, four periods of particular interest are identified and briefly discussed.

The Pioneer Period. Recent discovery of San Pedro phase (1200–800 B.C.) canals in the Tucson Basin (Mabry 1999; Muro 1998a) raises new and fascinating questions about the ultimate origin of Hohokam canal irrigation technology. Although the historical and evolutionary relationships between these early canals and later Hohokam systems are not yet known, available evidence indicates that the enduring locations and characteristic configurations of the principal Hohokam canal systems and the distribution of associated settlements crystallized during the Pioneer period in both the Salt and Gila rivers (Ackerly et al. 1987; Greenwald and Anderson 1994; Gregory 1991; Gregory and Huckleberry 1994; Haury 1976; Henderson 1989a, 1989b; Howard and Huckleberry 1991). The extended Salt River reconstruction and the new

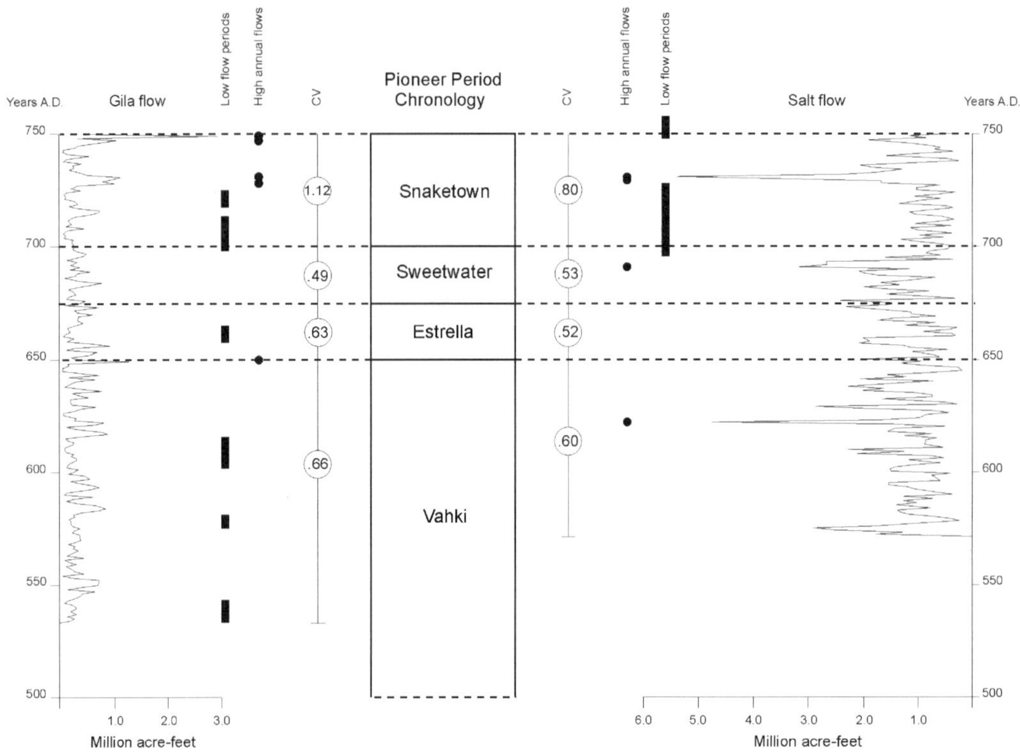

FIGURE 5.19. Salt and Gila river flow during the Pioneer period.

Gila River reconstruction include most of the Pioneer period as currently defined (Cable and Doyel 1987; Mabry 2000b; Wallace et al. 1995), and it is now possible to examine discharge variation during this important interval. Figure 5.19 illustrates characteristics of streamflow in the two drainages in relation to currently recognized Pioneer period phases.

From the beginning of the two reconstructions through the Sweetwater phase, both drainages were characterized by moderate to low variability and conditions that would have been generally favorable for irrigation agriculture. In the Salt, there were no prolonged periods of low flow, and the only two extreme high-magnitude annual discharges (rank of 25 or above) occurred some 70 years apart, one during the late Vahki phase and the other during the Sweetwater phase. In the Gila, five prolonged periods of low flow were more or less evenly spaced over the Vahki and Estrella phases, and none occurred during the Sweetwater phase. Only one high-magnitude annual flow occurred, at the boundary between the Vahki and Sweetwater phases.

There was a marked increase in variability during the Snaketown phase, owing to similar patterns in both drainages. The first three decades of the phase were marked by periods of prolonged low flow, two in the Gila (A.D. 698–713 and 718–725), and in the Salt the longest such period in the entire reconstruction (A.D. 696–727). In both rivers, these low-flow periods were followed closely by extreme high-magnitude annual flows, in the years A.D. 728 and 731 in the Gila and A.D. 730 and 731 in the Salt (see Figures 5.6 and 5.12).

The A.D. 728 and 731 flows in the Gila represent the twenty-third and eighteenth largest in the reconstruction, respectively, and the two intervening years also experienced relatively high-magnitude flows (.78 and .96 million acre-feet, respectively). The A.D. 730 flow is the fourteenth largest in the entire Salt reconstruction, and the A.D. 731 flow the fourth largest. The four following years (A.D. 732–736) also experienced high- magnitude discharges of 2.44, 1.79, 1.86, and 2.07 million acre-feet, and this is the only instance in the Salt during the prehistoric period when a prolonged low-flow period was closely followed by an extreme high-magnitude annual discharge. It is probable that these high flows produced channel-changing events in both drainages, including serious lateral erosion.

Following several years of moderate to low flow, two more extreme high-magnitude annual flows occurred in the Gila at the end of the Snaketown phase, in A.D. 747 and 749. The A.D. 749 flow is the largest in the entire Gila reconstruction, A.D. 747 is the twentieth largest, and the intervening year also produced a relatively high-magnitude flow (.86 million acre-feet). Channel-changing flood events probably occurred in A.D. 749 and may have occurred during the other years as well. In the Salt, the end of the Snaketown phase is marked by the beginning of an eleven-year period of low flow (A.D. 748–758) that extended into the early years of the following Gila Butte phase of the Colonial period.

Snaketown phase-discharge patterns in both the Salt and Gila appear to reflect more widespread climatic patterns. There is a marked decrease in tree-ring growth indices on the Colorado Plateau in the early eighth century, followed at about A.D. 750 by the onset of severe floodplain degradation and sharply dropping water tables (Dean 1988a:Figure 7; Dean and Funkhouser 1995). The strength and geographic scale of these patterns probably account for the strong similarities in the behaviors of the two rivers during the Snaketown phase.

Potential effects of low flow in both drainages during the first three decades of the Snaketown phase are difficult to assess. Based on directly dated canals and the distribution of associated settlements having Snaketown phase components, it is clear that multiple canal systems were present in both drainages by this time. However, there is little basis for quantitative estimates of their size. Depending on the extent of these systems and the average amount of land irrigated in preceding phases, the relatively low amounts of available water may not have been an important factor. Decreased variability in flow actually could have created conditions favorable for expansion of existing systems or even construction of new ones. Conversely, low flow could have resulted in decreased productivity if canal systems were already adjusted to the greater amounts of water available during earlier centuries. It appears that settlements along the Gila and possibly those along the Salt may have experienced population growth during the Snaketown phase (Gregory and Huckleberry 1994:7.1–7.17, Figures 7.1 through 7.4; see also Doelle 1995), perhaps favoring the former hypothesis.

There is no question that these low-flow periods would have magnified the effects of the high-magnitude flows that immediately followed them, and it is highly probable that canal systems would have been severely damaged one or more times between A.D. 728 and 731 in the Gila and between A.D. 730 and 736 in the Salt. Gila River systems were likely adversely affected once again less than a generation later, between A.D. 747 and 749. It has been suggested that similar disruptive patterns of Salt River flow during the subsequent Colonial period may have been a factor in Hohokam settlement in previously unoccupied and marginal areas away from the river as well as some instances of migration out of the Salt River Valley (Gregory 1991:186–187; Nials et al. 1989:66; see also Halbirt and Dosh 1986; Haury 1932; Morris 1970; Stafford 1978). Gila flow patterns between A.D. 750 and A.D. 900 mirror those of the Salt, indicating that similar conditions probably affected Colonial period populations along the Gila as well. More recent research has demonstrated the presence of Snaketown phase Hohokam populations in the Tonto Basin (Crown 1991:154–155, Figure 3; Elson and Gregory 1995; Gregory 1995a), the Lower Verde (Deaver 1997; Hackbarth 1992), and possibly the Safford Basin (Gregory 1995b). These discharge patterns show that the less-than-desirable conditions that characterized the Colonial period in both drainages began to develop in the late Pioneer period. Thus discharge variation may have been a factor in settlement of previously unoccupied and more marginal areas, use of alternative agricultural strategies, and some out-migration from the Phoenix Basin in the late Snaketown phase.

A.D. 900–1150. Beginning in A.D. 900, the Salt entered a nearly five-century period of low variation and few high annual discharges that defined optimal conditions for irrigation agriculture in the Salt River Valley. Although two prolonged periods of low flow immediately followed the extreme high annual discharge of A.D. 899 (see Figure 5.6), low-flow periods were rather evenly spaced thereafter (but see below). By contrast, Gila flow during the interval between A.D. 900 and 1150 continued to be characterized by high variability, including numerous prolonged periods of low flow punctuated by extreme high annual discharges and clusters of them (see Figure 5.13). Clusters of extreme high-magnitude events occurring in the late 900s, in the 1070s and 1080s, and in the 1140s likely created difficult conditions for irrigation agriculture along the Gila. In addition, three prolonged periods of low flow accounted for 60 percent of the years between A.D. 1090 and 1140, perhaps affecting potential productivity of existing canal systems and likely magnifying the geomorphic effects of high flows in A.D. 1141 and 1142. Thus we have a 250-year period during which general conditions for irrigation agriculture appear to have been much more favorable in the Salt than in the Gila.

The latter half of this interval (A.D. 1050–1150) witnessed significant changes in Hohokam culture across the Phoenix Basin and elsewhere: the emergence of platform mounds as an important form of public architecture and a decline in the construction and use of ball courts; a contraction in the Hohokam regional system as defined by the distribution of ball courts and diagnostic buffware ceramics; significant alterations in the composition and internal organization of the principal villages; new forms of residential architecture; new modes of burial and marked changes in the existing cremation complex, including the disappearance of or significant changes in associated material culture—figurines, palettes, and carved stone and ceramic censers, among others; changes in ceramic container technology and applied decorative techniques; shifts in the species used and the forms represented in marine shell artifacts; and the abandonment or marked reduction in scale of important villages (see Doyel 1991a, 2000). Efforts to construct satisfactory explanations for these sweeping changes remain at the intellectual core of Hohokam research.

Clearly, large-scale social processes were at work, and just as clearly, those processes cannot be directly and easily interpreted as responses to paleoenvironmental variation. And although Hohokam culture was rather substantially transformed, it survived, as did most of the principal canal systems along the Salt and Gila. However, it would also be a mistake to ignore the environmental context in which these transformations occurred. Though paleoenvironmental variation cannot explain everything about this period of transition, it is likely that it explains some things. Sorting out what those things are can only bring greater clarity to our efforts to understand the changes overall. Several observations about the A.D. 900–1150 interval are offered in this regard.

Recent studies of buffware ceramic production and distribution have produced evidence that most buffwares found in the Salt River Valley were probably manufactured by specialists in settlements on the Gila. In addition, these studies indicate that movement of buffwares from the Gila into the Salt River Valley reached peak levels during the late Colonial and early Sedentary periods, during the latter half of the A.D. 900–1150 interval (David Abbott, personal communication 1999). Important long-term differences in the two drainages have already been emphasized, as has the corollary that Gila River populations may have relied to a greater degree on non-canal agricultural technologies over time. Given historical differences in the reliability and potential productivity of irrigation agriculture during the A.D. 900–1150 interval, the idea that Gila River populations relied somewhat on exchange relationships to buffer unpredictable agricultural returns is an obvious one.

It has also been suggested that the Salt River population experienced significant growth relative to that of the Gila during this interval (Dean et al. 1994;

Doelle 1995). More generally, if other forces were undermining the legitimacy of socioreligious leaders and the credibility of existing world views, persistent difficult conditions for irrigation agriculture on the Gila would have provided powerful reinforcement for voices advocating change. Thus historical differences in the behavior of the two rivers during the A.D. 900–1150 interval may have strongly influenced demographic, social, and economic relationships between Salt and Gila river Hohokam populations, and research designed to discover, quantify, and explain such variation will likely prove fruitful.

Conversely, there is at least one period during this interval when similarities between the two drainages may have been significant. As noted above, patterns of Gila flow suggest that available water may have been a factor between A.D. 1090–1140, and low flows likely magnified the effects of extreme high annual flows at the end of this period (see Figure 5.13). A closer look at Salt flow shows that the identified A.D. 1034–1042 low-flow period may have been more severe than it initially appears (see Figure 5.6). Although not meeting our criteria for defining prolonged low-flow periods, the mean flow for the 19 years between A.D. 1024 and 1042 was low (.87 million acre-feet) and characterized by low variability (CV = .28). This period may be one exception to the generalization offered above that available water was rarely if ever a problem on the Salt.

Given these patterns, it is of interest that several of the large, important villages that were abandoned or experienced marked reductions in scale at the end of the A.D. 900–1150 interval have at least one feature in common. Snaketown on the Gila, and Cashion, Van Liere, and Villa Buena on the Salt, were served by the farthest downstream canal systems on both drainages at the time. Further, intakes for the canal systems that served the three Salt River sites were the only ones along the Salt not located at optimal locations represented by bedrock reefs. All Salt systems having intakes above these three systems appear to have survived and even flourished during the Classic period. In the case of Snaketown, the associated canal system benefited from the bedrock mass represented by Gila Butte, and there is a strong possibility that other factors were also involved in the abandonment of that important settlement (Wilcox et al. 1981). However, it is also possible that the large Classic period Casa Grande canal system was consolidated during this interval (Gregory and Huckleberry 1995), and increasing upstream diversion of water, combined with prolonged low flows, may have diminished the positive effects of Gila Butte and created difficulties for the people at Snaketown.

The Late Prehistoric Period. The role of environmental variation in the significant changes in southwestern demography that occurred in the fifteenth century has long been a matter of study and debate, including the role of floods in the demise of the Classic period Hohokam. Based on an earlier, shorter Salt River reconstruction (A.D. 740–1370), we argued in favor of the hypothesis that the substantial reduction in population and major changes in organization that marked the end of the Civano phase were precipitated by catastrophic floods and associated geomorphic effects that occurred in the mid to late 1300s (Nials et al. 1989). The long-term reconstructions presented here allow reassessment and refinement of that position as well as a comparison of Gila River behavior during the same interval. Figure 5.20 illustrates long-term trends in Salt and Gila flow (A.D. 1050–1550) in relation to a hypothetical Classic period and Postclassic period chronology proposed below. Shown in this figure are coefficients of variability for arbitrary 50-year intervals, prolonged periods of low flow as defined above, and high-magnitude annual flows defined and discussed below.

The long-term pattern that had defined optimal conditions for irrigation agriculture in the Salt River Valley since about A.D. 900 changed radically in the mid-1300s. The years A.D. 1357, 1358, and 1359 all produced annual flows greater than 2 million acre-feet (2.17, 2.52, and 2.05, respectively). Although none of these years ranks in the top 25 for the entire reconstruction, the flow of A.D. 1357 was the largest in 82 years. More important, this was the first time in 278 years that flows greater than 2 million acre-feet occurred in two consecutive years and the first time in over 500 years that three consecutive years witnessed flows of this magnitude. In addition, these high flows occurred immediately after two prolonged periods of low flow (A.D. 1335–1342 and A.D. 1344–1353), and the two years following the high-magnitude flows witnessed very low flows (A.D. 1360 and 1361, .39 and .65 million acre-feet, respectively).

Less than a generation later (23 years), another cluster of high annual flows occurred, in A.D. 1381, 1382, 1383, and 1384. These years produced flows of 2.10, 4.82, 2.42, and 3.26 million acre-feet, respectively, and the A.D. 1382 flow was the largest annual discharge in 483 years. The A.D. 1382 discharge represents a volume of water nearly double that of the next highest annual flow that occurred during this 483-year period (A.D. 929, 2.6 million acre-feet) and is the sixth largest flow in the entire reconstruction; the A.D. 1384 flow is the eighteenth largest. These four years represent one of only two instances during the prehistoric period in which four consecutive years produced flows greater than 2 million acre-feet, the other having occurred 575 years earlier, in A.D. 803–806. The high-magnitude flows of A.D. 1381–1384 were followed closely by a 10-year interval of low flow (A.D. 1386–1395), the first six years of which experienced extremely low flows (mean flow = .50 million acre-feet).

Annual flows of 2 million acre-feet or more have recurrence intervals of about 10 years or greater over the entire period of the Salt reconstruction. However, the period from A.D. 1000 to A.D. 1356 witnessed only 15 flows of this magnitude. In contrast, 7 flows exceeding 2 million acre-feet occurred in the 28 years between A.D. 1357 and 1384—the span of a single human generation. Furthermore, these high-magnitude flows occurred in two clusters of successive years, the likes of which had not been seen in over five centuries, and the second of these clusters included the largest annual discharge in nearly 500 years.

Between about A.D. 1150 and A.D. 1420, the Gila River also experienced a

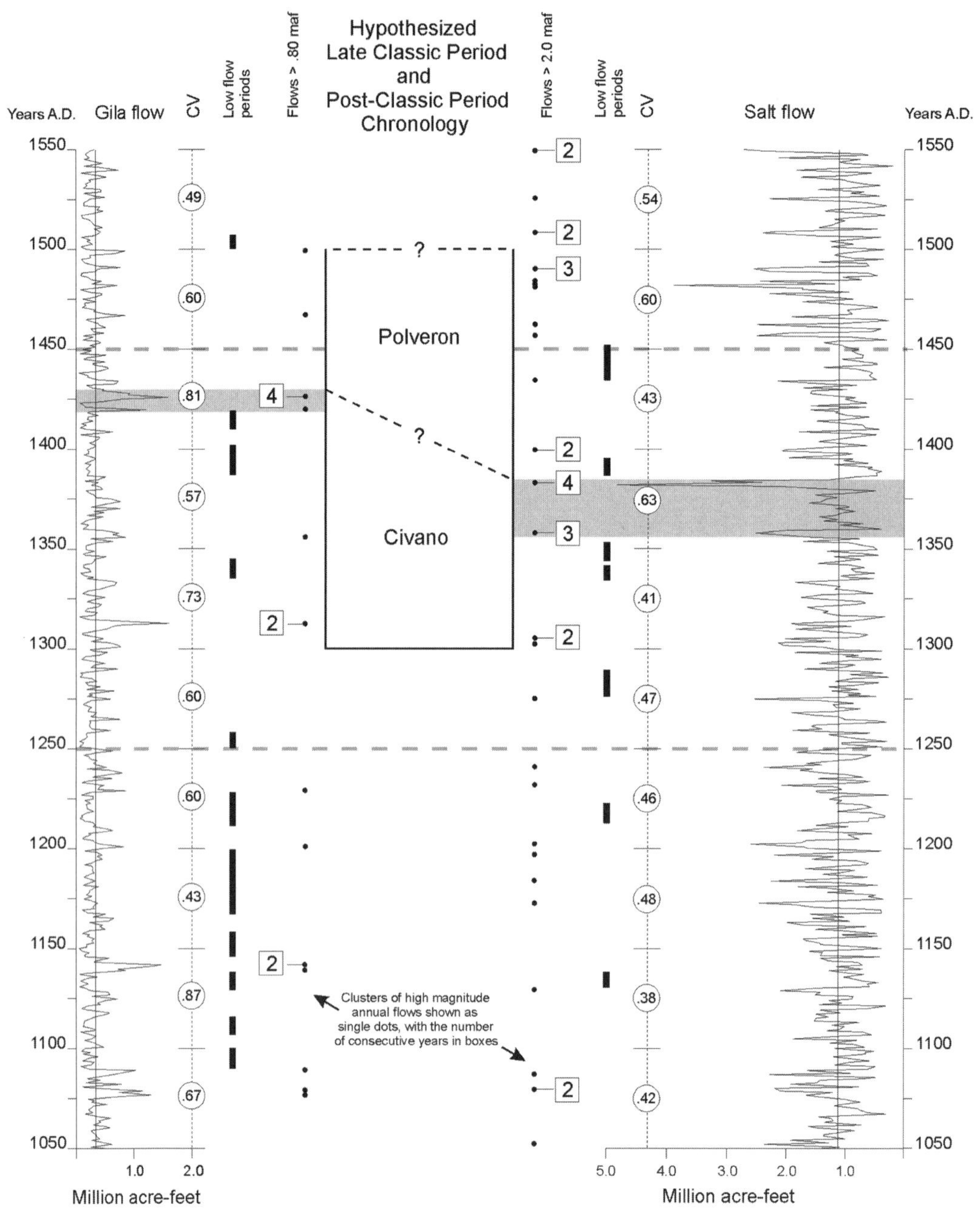

FIGURE 5.20. Long-term trends in Salt and Gila river flow in relation to the Classic and Postclassic periods.

long interval of conditions favorable for irrigation agriculture. During this 270-year period, Gila flow exhibited relatively low variability and few high-magnitude annual flows, and after A.D. 1250 even the prolonged periods of low flow of preceding centuries were infrequent. A cluster of two high-magnitude annual flows occurred in A.D. 1312 and 1313, the first time in 170 years that two consecutive years experienced flows greater than .80 million acre-feet. The flow of A.D. 1313 was the highest annual discharge in 325 years (rank = 6), and it is highly probable that canals suffered damage, possibly severe damage, during these years. However, this two-year cluster represents an isolated occurrence neither preceded nor followed closely by prolonged periods of low flow or other high-magnitude events or clusters of them. As shown in Figure 5.20, the coefficient of variation for the period A.D. 1300–1350 is .73; if the flow of A.D. 1313 is removed, this value drops to .60, similar to that of the preceding century and the following one. More significant events occurred in the second decade of the fifteenth century.

In A.D. 1420, the largest annual flow since A.D. 1313 occurred, representing the fourteenth largest flow in the entire period of reconstruction. This high annual flow was preceded by two prolonged periods of low flow (A.D. 1387–1401 and 1410–1419). Although the eight years between these two low-flow periods do not meet the criteria used here to define periods of low flow, the mean flow for the 33 years between A.D. 1387 and A.D. 1419 was only .25 million acre-feet. Thus geomorphic effects of floods that may have occurred in A.D. 1420 would have been substantially increased. The next three years experienced extremely low flows (.16, .14, and .07 million acre-feet), and one year later a series of four consecutive years with flows greater than .80 million acre-feet occurred. The years A.D. 1425 through 1428 produced flows of .81, 1.58, .90, and .92 million acre-feet, respectively, and the A.D. 1426 flow is the seventh largest flow in the entire reconstruction. This was the first time in 112 years that two consecutive years produced flows of this magnitude, and one of only two instances during the prehistoric period when four consecutive years produced flows greater than .80 million acre-feet, the other having occurred over 400 years earlier, in A.D. 986–989.

Annual flows of .80 million acre-feet or more have recurrence intervals of 30 years or greater over the entire period of the Gila reconstruction. However, between A.D. 1150 and A.D. 1419, only five flows of this magnitude or greater occurred. Between A.D. 1420 and 1428—in less than a generation—there were five annual flows of a magnitude matched only five times in the preceding 270 years, and four of these occurred in consecutive years.

Thus discharge patterns in the two drainages show strong parallels during the late prehistoric period. In both cases, several centuries of relative stability and conditions generally favorable to canal irrigation were interrupted by short-term, radically altered flow patterns. In the Salt, the 29 years between A.D. 1356 and 1384 twice produced discharge patterns that had not been seen in over 500 years; in the Gila, discharge patterns like those over the 9-year span between A.D. 1420 and 1428 had not been witnessed in over 400 years. In both cases, the onset of high flows was preceded by prolonged low-flow periods that probably magnified the effects of subsequent floods, and in the Salt the last set of high-magnitude annual flows was followed immediately by several years of extremely low flow. In both cases, these patterns clearly exceeded the levels of variability to which both human organizations and the physical facilities of the canal systems had become adapted over the preceding four or five centuries.

On the basis of these patterns, we reiterate the hypothesis that catastrophic floods and associated geomorphic events were causal factors in the decline of Salt River Classic period systems. Arguments in support of this hypothesis remain essentially unchanged, and extension of the reconstruction beyond A.D. 1370 has provided additional supporting evidence as described above. The only real difference resulting from the new reconstruction is that a previously identified period of low flow between A.D. 1322 and 1355 does not appear to have been as long or as deep as originally thought (Nials et al. 1989:69–70). Some or even all of the Salt River systems may have recovered from the A.D. 1357–1359 flows during the two decades that followed. However, the flows of A.D. 1381–1384 likely produced severe damage and substantial lateral erosion, which immediately rendered the Civano phase systems inoperable and which—because of the bedrock masses that previously had been such an important asset—effectively prevented any recovery, given the available technology (Nials et al. 1989:69–70, 75). Discharge patterns described above suggest that this hypothesis may now be extended to the middle Gila systems as well.

If these hypotheses are correct, then Civano phase platform mound-canal systems along the Gila River may have endured for more than a generation longer than those along the Salt. The collapse of Civano phase Salt River systems undoubtedly resulted in some attrition as well as out-migration, and one interesting possibility is that some portion of the Salt River population may have sought refuge in the Gila during the late 1380s and thereafter. The subsequent Polverón phase represents a modified adaptation by populations that remained in both drainages after the decline of Civano phase systems (Chenault 1996; Crown and Sires 1984), one in which irrigation agriculture played a much reduced role. One obvious test implication of these hypotheses is that the appearance of Postclassic period Polverón phase manifestations should be later in the Gila than in the Salt.

Recent analyses of radiocarbon and archaeomagnetic dates suggest that the Civano-Polverón transition occurred between A.D. 1300 and 1400 in the Salt River Valley, and the original A.D. 1330–1450 range for the whole phase is strongly supported (Zyniecki 1996; see also Crown and Sires 1984). However, a few dates and yet more recent investigations suggest that the phase may have lasted longer in the Salt than previously recognized, perhaps until A.D. 1500 or even later (Bostwick et al. 1996; Chenault 1994; Zyniecki 1996). The only excavated Polverón components lying in the Gila drainage are situated well away from the river, including the type site (Sires 1984), and no Polverón phase site actually located on the river has yet been excavated. Single-component Polverón phase sites exist along the Gila, and some Classic period sites undoubtedly include later Polverón components (Gregory and Huckleberry 1994:7.49–7.51; see also Wilcox and Shenk 1977 and Wilcox and Sternberg 1981). Given these data, the hypothesis that the Polverón phase began as much as a generation later in the Gila than it did in the Salt is a plausible one.

These interrelated hypotheses concerning the timing of developments in the late Classic period and the Postclassic period in the Phoenix Basin are represented in the proposed chronology shown in Figure 5.20. They may tested by reexamination of existing data and against new data from both drainages that accumulate in the future.

Salt and Gila discharge patterns during the late prehistoric period appear to reflect the influence of large-scale climatic patterns. The manner in which these larger factors affected the two rivers have important implications for interpretations of demographic processes on a macroregional scale. Dean and Funkhouser (1995:92–95, Figure 9) note that between about A.D. 1250 and 1450, long-term climate patterns on the Colorado Plateau break down into a totally aberrant configuration:

> These results specify a major climatic disruption that would have had a major impact on the Colorado Plateau. For two centuries a fairly simple, long-term, stable pattern was interrupted by a complex, chaotic configuration that represents an unprecedented change in conditions to which the human and biotic populations in the region had become accustomed.... Just the change from a persistent stable pattern to an unstable one would have had important adaptive repercussions. (Dean and Funkhouser 1995:94)

As described above, changes in long-term Salt and Gila discharge patterns are also apparent during the A.D. 1250–1450 interval, highlighted in Figure 5.20. In the Salt, there was a gradual shift away from long-term stable patterns that had obtained since about A.D. 900. These changes included some increase in mean flow and, beginning about A.D. 1300, an increase in the occurrence of clusters of high annual flows that culminated in the radical events of the mid to late 1300s. In the Gila, the change was away from a long-term pattern of high variability and prolonged periods of low flow, with conditions more favorable for irrigation agriculture. Aside from the two high-magnitude flows of A.D. 1312 and 1313, these salutary conditions persisted until the radical flows of the 1420s.

Amelioration of conditions for irrigation agriculture along the Gila and continuing favorable conditions on the Salt are of particular interest in that they provide a broadened context in which to understand late-thirteenth- and early-fourteenth-century migrations from the Colorado Plateau into the Safford Basin, the Point of Pines area, the San Pedro drainage, and the Tonto Basin (Clark 1995; Clark et al. 1999; DiPeso 1958; Lindsay 1987; Stark et al. 1995; Woodson 1999). More advantageous conditions in the Safford Basin and environs likely caused this and adjacent areas to be targeted by refugees from the Colorado Plateau. Similarly, migrations from the same general area into the Tonto Basin occurred in the mid- to late 1200s (Clark 1995; Stark et al. 1995), prior to the short-term radical events of the mid- to late 1300s and during a period when the Salt carried abundant water and was characterized by relatively low variability in annual flow. Thus while large-scale climatic changes produced almost immediate effects on the Colorado Plateau, their influence on Salt and Gila discharge developed more gradually and probably contributed to the attractiveness of these areas as *refugias* for hard-pressed Colorado Plateau populations.

With respect to the environment for irrigation agriculture in the Salt River Valley and along the middle Gila, the short-term, radical discharge patterns of the mid- to late 1300s in the Salt and during the 1420s on the Gila define exactly the sort of unprecedented variation in existing long-term, stable patterns identified by Dean and Funkhouser (1995) for the Colorado Plateau. New adjustments to changing conditions are represented by Polverón phase materials, suggesting that paleoenvironmental variation during the late prehistoric period had important adaptive repercussions across the Phoenix Basin as well.

The Protohistoric and Early Historic Periods. When the Spanish arrived in the southern Southwest, they found the Pima irrigating along that segment of the Gila where extensive Classic period Hohokam irrigation systems had operated less than 300 years before. Although arguments continue about the relationship between the Pima and the prehistoric Hohokam, it is clear that the Pima were practicing successful canal-irrigation agriculture at the time of contact and had been for some considerable period of time. Following the unprecedented events of the 1420s, Gila River flow entered the lengthy period of low variability, few high-magnitude annual flows, and relatively few prolonged periods of low flow defined as Interval V above (see Figure 5.16). In most respects, this 481-year period is quite similar to A.D. 900–1350 on the Salt, when conditions were amenable to irrigation agriculture.

By contrast, there were no permanent settlements in the Salt River Valley at this time. At the time of the Spanish entry, the Salt River Valley was a boundary area between the Yuman-speaking Yavapai and Hualapai to the north and northwest, the Maricopa to the southwest, the Pima to the south, and the newly arrived and troublesome Apaches to the northeast. Although a number of historical factors combined to produce this contact-period distribution, it is notable that the latter decades of Salt River Interval III and, most markedly, Interval IV exhibit characteristics extremely unfavorable for canal irrigation (see Figure 5.9). Thus the differential behavior of the two streams may have been an important factor in aboriginal demography and subsistence during the Protohistoric period.

Finally, it may be noted that the first attempts to construct Historic period canals in the Salt River Valley were undertaken during a flow regime extremely unfavorable to irrigation agriculture, resulting in frequent failures due to floods (Davis 1897; Zarbin 1980). Ultimately, the experience of these early Anglo farmers led to the construction of Roosevelt Dam and other control facilities and an unprecedented degree of human control over Salt River discharge. By contrast, the early Historic period on the Gila was characterized by low variability and conditions generally favorable to irrigation agriculture. Hence Anglo irrigation systems in the Safford area and in the nonreservation area above Blackwater along the middle Gila were established early and underwent continual expansion. As early as 1871 and continuously thereafter, the Gila River Pima complained of a lack of water because of upstream diversion by Anglo farmers, and warfare was threatened at least once as a remedy (Southworth 1919:118–120). Once again, the variable discharges of the two streams differentially influenced the historical course of human interaction among the people who drew water for irrigation from them. The reconstructions presented here provide a new context in which to examine history and process along the two drainages during the Protohistoric and early Historic periods, before the advent of modern streamflow records.

SUMMARY AND CONCLUSIONS

The theoretical and methodological bases for tree-ring-based retrodictions of Salt and Gila river streamflow have been reviewed, and long-term reconstructions based on application of these principles have been presented, covering the period A.D. 534–1988 for the Gila and A.D. 572–1988 for the Salt. These reconstructions provide accurate approximations of what is arguably the single most important aspect of the desert environment with respect to canal irrigation as a subsistence strategy. The implications of geomorphic processes peculiar to arid-lands rivers have been reviewed, and several important differences in the two natural systems have been identified and discussed in light of these processes. Low-frequency variation in discharge has been identified by definition of several analytic intervals for each reconstruction, and discharge characteristics of these intervals and their probable effects on channel morphology have been specified.

Given a uniform irrigation technology, long-term variation in the behavior of the two rivers likely influenced the specific adaptations forged by aboriginal irrigation agriculturalists living along them. Greater overall variability in flow made the Gila a less predictable environment for canal irrigation. The greater sediment load of the Gila, coupled with a more variable distribution of flow over the annual period, produced greater risk factors and influenced the timing and magnitude of labor needed to maintain canals and effect necessary repairs. Substantial differences in the amounts of water available for irrigation had a number of effects. There was significantly less water available in the Gila and thus differences in the potential productivity of canal irrigation between the two rivers. Water availability was probably not a limiting factor for Salt River canal systems but may well have been a significant factor for Gila River systems. Some potential implications of these differences for aboriginal irrigation agriculture along the two drainages have been suggested.

Historical differences in the behaviors of the two rivers may have had important implications as well, and several periods of particular interest have been identified and discussed. Perhaps most notably, it has been argued that catastrophic floods and associated geomorphic effects were causal factors in the decline of Classic period Hohokam social systems along both drainages. This interpretation has important implications for current views of the cultural chronology of the Phoenix Basin, and an alternative view has been proposed and illustrated. The probable influence of large-scale climatic variability on Salt and Gila river discharge has also been discussed, and some implications of these large-scale patterns for interpretations of demographic processes on a macroregional scale have been suggested.

The several hypotheses offered here may be tested with existing data and with new data resulting from future investigations. These preliminary perspectives deserve continued attention in future research involving comparison and contrast of aboriginal subsistence, organization, and demography along the two drainages. Long-term Salt and Gila river streamflow reconstructions provide a new context in which to assess and interpret patterned variation in the archaeological record, and constitute a significant addition to the consultable record of paleoenvironmental variation in the American Southwest.

NOTES

One of the anonymous reviewers of this volume recommended that we update the references to reflect relevant studies that have accumulated since we completed this work in 1999. Among these would be Abbott 2003, Ensor et al. 2003, Waters and Ravesloot 2001, and numerous others. We decided not to follow this advice, as it would require substantial additions to what is already a long manuscript. We leave it to the reader to evaluate the implications of more recent studies for the perspectives and conclusions presented here.

Lest some think it odd to name Don Graybill the senior author of this chapter—a decade after his death—a word of explanation is in order. All the authors were involved in the original reconstruction of Salt River streamflow conducted in the context of the Arizona Department of Transportation–sponsored Las Colinas project, focused on excavation of a large Hohokam settlement in metropolitan Phoenix and carried out during 1982 and 1983. Gregory, as project director, engaged the services of Graybill and Nials to pursue studies that would elucidate the relationship between the settlement at Las Colinas, the behavior of the Salt River, and the functioning of the prehistoric Hohokam canal systems that served the inhabitants of Las Colinas specifically and the Salt River Valley more generally. The climate-geomorphology-irrigation complex of relationships that spurred and informed that study was central to the project research design as originally envisioned (Gregory et al. 1985) and became an even more important aspect of the research as the results of the streamflow reconstruction were examined and the implications fully realized (Graybill and Gregory 1989). The findings were reported in a series of chapters in a volume (Graybill et al. 1989) that detailed environmental studies conducted in conjunction with the project (Graybill 1989; Graybill and Gregory 1989; Graybill and Nials 1989; Nials and Gregory 1989; Nials et al. 1989), as well as a number of papers presented at various professional meetings.

The desire to extend the period of time covered by the reconstruction, to duplicate the study in the Gila River drainage, and to relate the reconstruction to geomorphic and climatic events and processes on a larger scale were pursued in subsequent years by Graybill and Nials. During this period, Graybill re-collected numerous tree-ring sites to extend the modern period of calibration and collaborated with David Meko (also of the Laboratory of Tree-Ring Research) to produce reconstructions of Gila River streamflow from A.D. 1663 to 1985 (Graybill and Meko 1995; Meko and Graybill 1993). Funkhouser worked as Don's assistant at the Tree-Ring Laboratory during this time. Don subsequently accomplished a longer-term reconstruction of the Gila and extended the Salt reconstruction. He supplied the results to Gregory and Nials for further analyses, and work was begun on what turned out to be the present chapter. During this period, Funkhouser worked closely with Graybill in establishing the chronologies used and in implementing the calibrations and reconstructions. Don's untimely death interrupted this process. Funkhouser subsequently took up the mantle, refining the two reconstructions and applying new statistical tests of their validity. However, geography and other commitments mitigated against completion of the project until all of us ended up in Tucson once again in 1997, at which point the editors suggested that the work would be an appropriate addition to this volume.

Assignment of lead authorship to Don Graybill is an appropriate recognition of his role and a way for the other authors to acknowledge Don's contribution not only to this paper but to tree-ring-based studies of paleoenvironment in general. As an irreplaceable friend and colleague, he is sorely missed.

6

Subsistence Management Strategies in the Grasshopper Region, East-Central Arizona

J. Jefferson Reid, Donald A. Graybill, and Ann Clair Seiferle-Valencia

A TREE-RING-BASED CLIMATE RECONSTRUCTION FOR THE GRASS-hopper region of east-central Arizona provides the environmental context for explaining rapid changes in subsistence and settlement strategies during the fourteenth century. Canyon Creek Pueblo, a cliff dwelling partially excavated by Emil Haury in 1932, is the centerpiece of a subsistence and settlement model based in traditional techniques of agricultural management under conditions of recurrent climate change. The success of agricultural management in contrast to the failure of wild plant and game management in this case study provides insights into the rapid development and long-term stability of Pueblo village farmers, as well as a more balanced understanding of human ecological relations in prehistory.

Archaeologists recognize that the marginal subsistence characteristics of the American Southwest are largely the products of low and variable precipitation. Documenting that variability, as well as the behavioral responses of prehistoric farmers, is therefore important for understanding prehistoric cultural change. Simple accounts of prehistoric people as pawns of environmental whimsy, however, are as unrealistic as views of early farmers fine-tuning multiple subsistence technologies to the shifting requirements of local environments.

Since 1963, research by the University of Arizona Archaeological Field School at Grasshopper has attempted to reconstruct an accurate picture of variation in human subsistence adjustments to the environmental characteristics of the Grasshopper region in the mountains of central Arizona. Concurrent dendroclimatic studies at the university's Laboratory of Tree-Ring Research have resulted in data and methods that allow quantitative reconstruction of annual precipitation in instrumental units (inches) for critical periods of the Grasshopper occupation.

Growth models for Grasshopper Pueblo and Canyon Creek Pueblo frame a well-defined prehistoric example of managing a mixed subsistence system in which the productivity of major components—farming and hunting mule deer—varied directly with precipitation. Our analysis contrasts a rather remarkable ability to anticipate and respond quickly to alterations in the direct relationship between precipitation and agricultural productivity with an apparent inability to manage the more complicated relationship between precipitation and deer productivity.

THE REGIONAL SETTING

The natural environment of the Grasshopper region includes agricultural land situated between 5,000 to 6,100 feet in elevation with an estimated 120 to 140 frost-free days per year and about 19 inches of mean annual precipitation. Modern vegetation includes western yellow pine at the higher elevations and mixed-evergreen woodlands of pinyon pine, juniper, and oak along with chaparral and grasslands at the intermediate and lower elevations (Holbrook and Graves 1982).

Corn, beans, squash, and cotton (Bohrer 1982:97–98; Welch 1996) were cultivated with dry-farming methods marked in the archaeological record by simple soil- and water-conservation features, such as check dams and contour terraces. The close proximity of the eleven largest pueblos (35–500 rooms) to agricultural soils (Tuggle et al. 1984; Welch 1996) confirms this developing agricultural orientation (Figure 6.1). The principal buffering response to agricultural shortfalls was to expand the land under cultivation, which often required residential moves. There is no evidence of food shortages being buffered by interregional exchange or technological improvements to the agricultural technology, such as irrigation (Tuggle et al. 1984).

TREE-RING DATA AND PRECIPITATION RECONSTRUCTION

Tree-ring reconstruction of quantitative climatic and hydrologic variability in instrumental units for the past two millennia is feasible in several regions of the American Southwest (Graybill 1989; Rose et al. 1981). This capability is made possible in part by the development of long, regional tree-ring chronologies based on wood from archaeological contexts (Dean and Robinson 1978). These chronologies furnish a basis for extending the record of a climatic variable that may be monitored by living trees in the same region. Both the living trees and the archaeological samples are selected for analytical purposes when they evidence high year-to-year ring-width variability. In the semiarid Southwest, this ring-width variability is usually an indication that the trees are strong responders to and recorders of variation in annual precipitation. Determination of the strength and quality of the precipitation signal in the living trees with respect to instrumented precipitation values during the twentieth century is accomplished with common parametric and nonparametric statistical procedures. This correlation results in an equation, or transfer function,

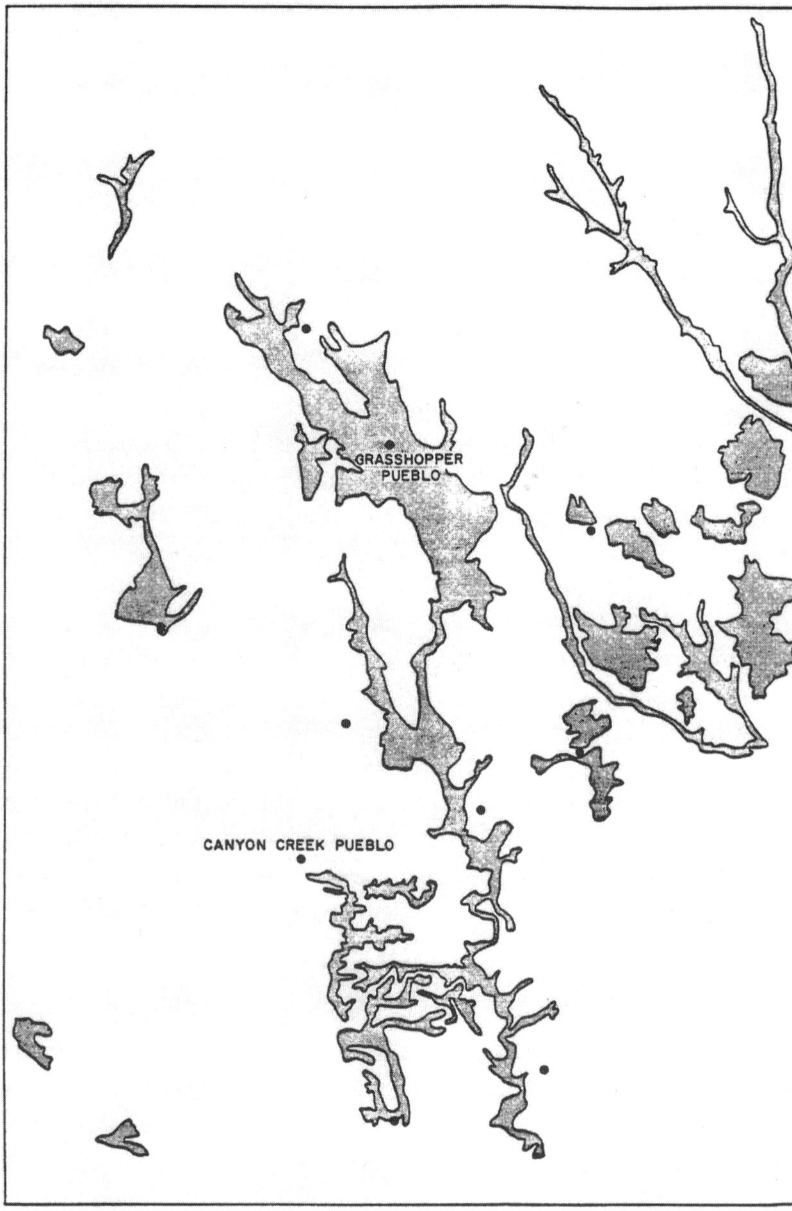

FIGURE 6.1.
Site distribution and agricultural soils.

which is then used to reconstruct precipitation for periods when only tree-ring chronology data are available.

In the Grasshopper reconstruction, two sets of drought-sensitive ring-width data from ponderosa pine (*Pinus ponderosa*) are used in an average regional chronology for calibration with instrumented climatic data (Graybill and Rose 1992). One set is from trees in the vicinity of Show Low, Arizona, and the other is from trees growing near Grasshopper Pueblo. Ring-width series for each of the chronologies are processed according to standard techniques

(Stokes and Smiley 1968), indexed to remove some nonclimatic signals, and then averaged together to form tree-ring index chronologies (Fritts 1976; Graybill 1982). The respective dates of the Show Low and Grasshopper chronologies are A.D. 1597–1986 and A.D. 1643–1985. Tree-ring chronologies based on wood from archaeological sites in or near the Grasshopper region are averaged to provide a record from A.D. 1039 to 1376 (Central Mountains North and South, Grasshopper Ruin; Dean and Robinson 1978, 1982). Both the living-tree chronologies and the archaeological chronology are further transformed to remove statistically significant autocorrelation (Box and Jenkins 1976), a necessity for the subsequent regression analysis (Graybill 1989; Wonnacott and Wonnacott 1981).

Monthly precipitation totals are obtained for surrounding climate recording stations in order to develop a regional instrumental series (1912–1985). The stations and their period of record used in the final average include Fort Apache–Whiteriver (1912–1985), Payson (1912–1985), Pinedale (1912–1968), and Show Low (1969–1985) (electronic media, National Climate Data Center, Asheville, North Carolina; Green and Sellers 1964). A few missing values are estimated by the ratio method (McDonald 1957), and all individual precipitation series appear to be homogeneous with respect to one another (Kohler 1949). The latter fact implies that each record is free of major nonclimatic aberrations, such as those that might be induced by change in station location or instrumentation.

Precipitation for an annual period representing August of the year prior to tree growth through July of the year of tree-ring formation is selected as the primary analytical unit of climate. This choice is due, in part, to the fact that trees integrate climatic information over those months through biological mechanisms (Fritts 1976). Empirical correlation analysis of the transformed tree-ring average and the regional precipitation average support this decision. The final, averaged climatic time series does not evidence significant autocorrelation and is not subjected to the type of transformation used for the tree-ring data.

Precipitation during the August–July period is relevant with respect to the productivity of field crops, native plants, and animal populations. For example, fluctuations in deer-herd size are largely the result of variation in fawn survival, which is positively related to doe survival and precipitation during the preceding year and during the critical summer production season (Brown 1981; Short 1979). Additionally, this annual period monitors the fall–winter time when soil-moisture reserves that are crucial for germinating crops such as maize develop, as well as the early-summer precipitation component that is important for crop maturation.

Development of an equation for retrodiction of annual precipitation values from tree rings involves the use of linear-regression procedures and associated statistics (Draper and Smith 1981). Stability of the climate–tree growth relationship during the twentieth century is evaluated by first using half the

TABLE 6.1. Calibration and verification statistics.

Calibration periods	1914–1949	1950–1985
r^2	0.59 (0.000)	0.53 (0.0000)
F	51.14 (0.000)	40.12 (0.000)
Verification periods	1950–1985	1914–1949
t-test	0.74 (0.533)	0.91 (0.630)
r^2	0.54 (0.000)	0.60 (0.000)
Final calibration	1913–1985	
r^2	0.56 (0.000)	
F	92.44 (0.000)	
Final calibration equation $RP_t = (8.891 * TR_t) + 19.373$, where RP_t is reconstructed annual precipitation in inches for year t and TR_t is the tree-ring value for year t		

Note: Values in parentheses are significance levels associated with each statistic.

available data to establish such a relationship. This calibration procedure uses the average of the transformed Show Low–Grasshopper chronologies as the independent variable and the regional average of August–July precipitation as the dependent variable. An equation derived from this regression is used to predict annual precipitation for the remaining portion of the data. Statistical evaluations, or verification of the quality of the prediction, are then made. These procedures are performed a second time by reversing the roles of the two halves of the data in calibration and verification. A summary of these exercises in Table 6.1 indicates that the relationships between tree growth and precipitation for the twentieth century are reasonably consistent in terms of the amount of precipitation variability explained during each calibration period (R). For the verification periods, the question is how well the distribution of precipitation values had been predicted or reconstructed. This question is first evaluated with the nonparametric Wilcoxon matched-pairs ranked-signs test (W). The null hypothesis is that the population of values is not identical. Parametric statistics are used to determine whether the mean of the reconstructed values was significantly different from that of the actual data (t-test) and whether the covariation between the reconstructed and actual data is different from 0.0 (r^2). An alpha level of .05 is used in all cases, and the test results indicate that the predictions are acceptable in all respects. Given the quality of these results, a final calibration using all data from 1913–1985 is made, resulting in a transfer function (calibration equation) for reconstruction purposes (Table 6.1).

The adequacy of the transfer function for reconstruction can be seen in Figure 6.2, which shows the actual and reconstructed precipitation values during the twentieth century. Some of the higher actual values are not fully

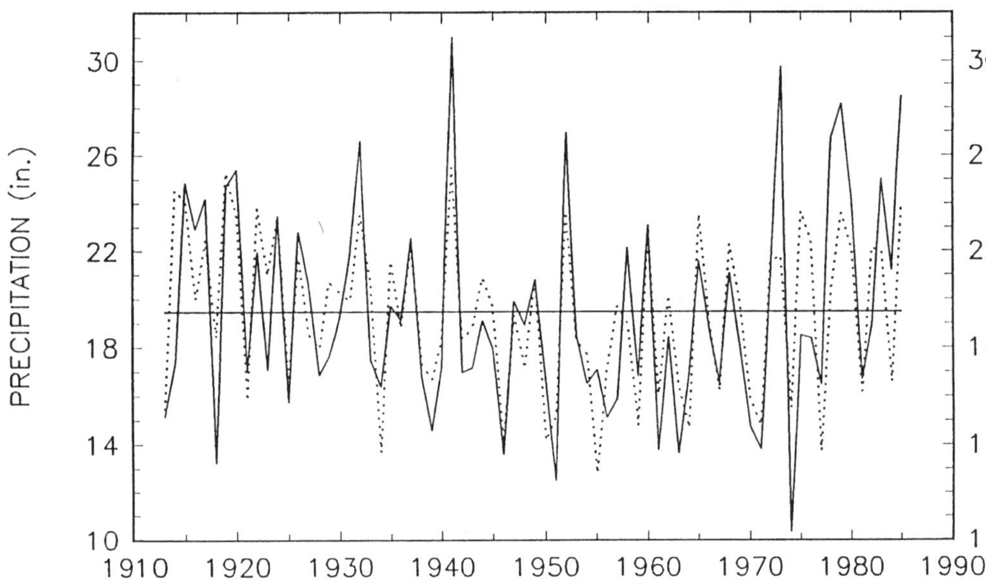

FIGURE 6.2. Actual *(solid)* and reconstructed *(dot)* annual precipitation for the Grasshopper Region, 1913–1985.

reconstructed, whereas those in the lower range appear to be well represented. This pattern reflects a common response of southwestern conifers to precipitation: under exceedingly wet conditions the trees may not have a strong uniform response and may not need to use all available moisture, whereas under dry conditions they tend to have a more uniform and direct growth response (Fritts 1976; Graybill 1989).

Ten-year, non-overlapping averages of the full precipitation reconstruction for the period of A.D. 1040–1376 are provided in Figure 6.3. This allows perspective on the lack of overall trends in precipitation throughout this time and on the relatively unique nature of shorter-term changes during the last century of this record, corresponding with activity at Grasshopper. Notably, the sustained high precipitation from about 1300 to the mid-1330s is the longest in the record. Variation in the annual values of precipitation during the Grasshopper developments is shown in Figure 6.4.

PRECIPITATION AND SUBSISTENCE MANAGEMENT

The relationship among precipitation, demography, and subsistence management is framed within the four periods of community growth at Grasshopper Pueblo (Reid 1978, 1989; Reid and Shimada 1982; Reid and Whittlesey 1999). These periods are Establishment (A.D. 1275–1300), Aggregation (A.D. 1300–1330), Dispersion (A.D. 1330–ca. 1355), and Abandonment (A.D. 1355–ca. 1400). The original periodization model (Reid 1973) is supported by recent analyses of tree-ring dates (Graves 1991) and architecture (Riggs 2001).

The Establishment period (A.D. 1275–1300) was characterized by a decrease in precipitation throughout the American Southwest known as the

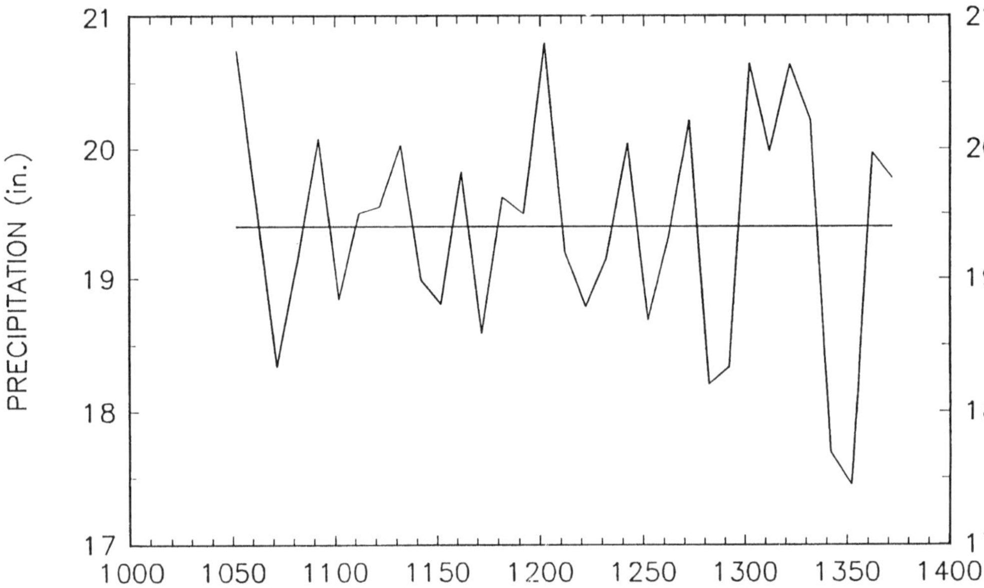

FIGURE 6.3. Ten-year, non-overlapping averages of reconstructed annual precipitation for the Grasshopper Region, 1047–1376.

Great Drought (Dean and Robinson 1978, 1982). In the Grasshopper region, this period was marked by relatively low precipitation and high variability in precipitation. A long-term, annual mean precipitation of 19 inches, however, meant that even in relatively dry years there was more precipitation than in surrounding desert regions. Additionally, springs, seeps, and a high water table were common.

The population settled in small, masonry pueblos of as many as 20 rooms, which were loosely clustered near agricultural soils, and we infer that these habitual inhabitants of the mountains were mobile hunter-gatherer-gardeners (Reid et al. 1996; Welch 1996). Evidence suggests that initial habitation was part time but that residential stability increased at the height of the drought in the A.D. 1280s, a time of rapid population increase due to immigration (Ezzo and Price 2002). Chodistaas and Grasshopper Spring pueblos were inhabited at this time.

Throughout this period and into the next, garden produce was heavily supplemented by hunting and gathering (Ezzo 1993). In addition, the quantity of animal bone (McKusick 1982; J. W. Olsen 1982, 1990; S. J. Olsen 1982) and the skilled manufacture of large quantities of arrow points (Lorentzen 1993; Whittaker 1984) confirm the local importance of hunting mule deer and the continuation of the traditional hunting emphasis in the Arizona mountains, where the value of hunting was higher than among contemporaneous populations in the deserts or on the Colorado Plateau.

The Aggregation period (A.D. 1300–1330) coincided with a precipitation regime that was above the long-term mean and sometimes exceeded that value

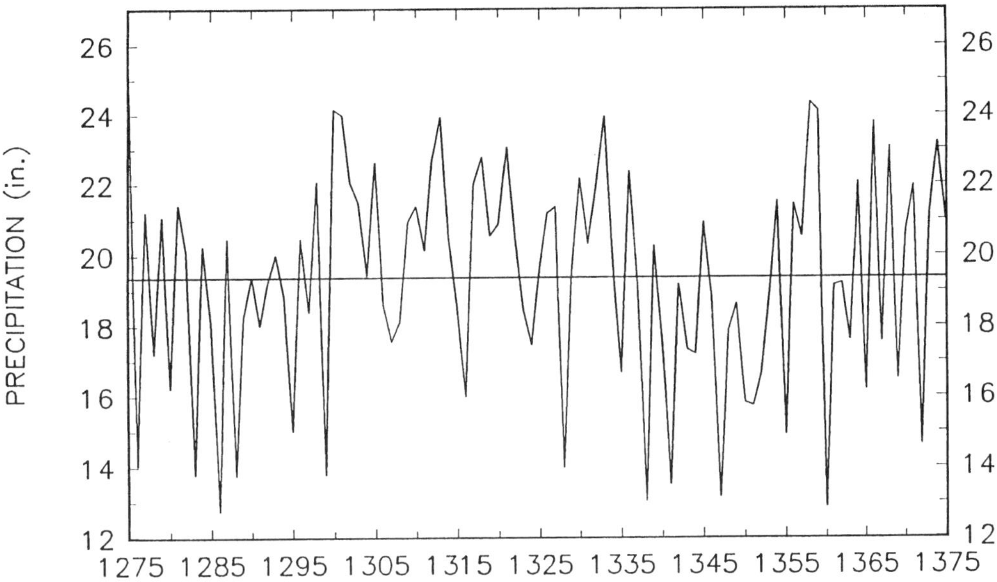

FIGURE 6.4. Reconstructed annual precipitation for the Grasshopper region, 1275–1375.

by as much as five inches. Soil-moisture levels would have been at their highest through this period, and dry farming would have been relatively successful.

Rapid growth at Grasshopper Pueblo was the result of local population aggregation and immigration of people from beyond the region followed by normal population growth; continued immigration and growth contributed to the fissioning of satellite communities (see Figure 6.1) (Longacre 1975, 1976; Reid 1989). Toward the end of this period, Canyon Creek Pueblo was founded (Graves 1982, 1983; Haury 1934).

The Dispersion period, beginning around A.D. 1330, and the Abandonment period, ending by A.D. 1400, overlapped during a time of low precipitation in the mid-1330s and high annual variability, followed by a prolonged dry period until A.D. 1355. At this time the population made maximum demands on subsistence resources and management skills.

New settlements arose in part from the dispersion of households from Grasshopper into smaller patches of more marginal agricultural lands (see Figure 6.1) (Tuggle et al. 1984; Welch 1996). Defensible locations of small pueblos and extensive use of cliffs for unobtrusive storage suggest conditions of social and economic instability (Tuggle and Reid 2001). Increased physiological stress among subadults at Grasshopper Pueblo during the Dispersion–Abandonment period (Hinkes 1983) supports an inference of periodic subsistence shortfalls.

The most frequent buffering response to agricultural shortfalls was to expand the land under cultivation, which often required movement of people to new villages. There is no archaeological evidence of buffering mechanisms involving technological improvements to the agricultural system such as water diversion (a possibility demonstrated by historic Apache irrigation) or interregional trade, which was most extensive during the Aggregation period's optimum agricultural productivity.

Processes reflecting agricultural management are best documented in the founding of Canyon Creek Pueblo (Graves 1982, 1983; Haury 1934). Stockpiling of timbers prior to room construction suggests an expectation of movement to that locality (Graves 1982:111–116) in anticipation of the need to expand and diversify cultivated land in response to agricultural shortfalls. It is not inconsistent to imagine that the prehistoric farmers, with memories of the Great Drought, were not lulled by above-normal precipitation and sought to buffer an anticipated crash by founding satellite communities. Initial construction at Canyon Creek in the late A.D. 1320s coincided with peak population at Grasshopper Pueblo and preceded by only a few years the dry spell of the mid-1330s. The founding of Canyon Creek Pueblo suggests a short response time for these farmers to exercise a mobility option for subsistence management and an ability to monitor annual variation in precipitation rather closely. Again, one imagines that the condition being most closely monitored was actual agricultural productivity.

Although other management options could have operated with short-term effectiveness, there is no evidence for increasing labor per unit of land or for interregional trade in either high-value commodities or subsistence goods. The Canyon Creek case suggests the existence of a traditional, intergenerational management strategy of precipitation awareness and farming adjustments that could be translated rather quickly into a residential-mobility response.

In contrast to the management of agriculture, in the case of mule deer the people of Grasshopper appear to have been unable to manage the more complex relationship between precipitation and deer productivity. Compared with the Aggregation period, mule deer procurement during the Dispersion and Abandonment periods was marked by increased harvesting of immature animals (Olsen 1990) and more thorough reduction of bone for its food content (Olsen 1990:169, Table 6). Although increased reliance on deer may have adequately buffered previously smaller populations in the mountains against crop shortfalls, under the conditions that prevailed at Grasshopper this practice would have accelerated the reduction in deer productivity during a time when decreased precipitation was already adversely affecting local herds (see Olsen 1990:172–182 for a qualitative summary of Grasshopper fauna).

We conclude that traditional mechanisms of effectively monitoring agricultural productivity, probably incorporated as ceremonial knowledge, came together with game-management ineffectiveness under conditions of population aggregation during a time of low and erratic annual precipitation. The result in the Grasshopper region in the mid-1300s was to force the development of a Pueblo subsistence system totally dependent on agriculture (Welch 1996). A dry-farming system could not be sustained in the Grasshopper region, with the result that the traditional management strategy of residential mobility compelled the inhabitants to pick up and move to localities with a more predictable water supply.

THOUGHTS ON DEAN'S "MODEL OF ANASAZI BEHAVIORAL ADAPTATION"

Dean's (1988b) model thrives on geomorphological, palynological, and dendroclimatic data sets, sets that are usually marked by a broad range of accurate, well-integrated reconstructions. Integration of demographic, environmental, and behavioral components provides a framework for analysis of cultural change. The fundamental assumption of the model is that cultural processes and change, which are loosely qualitative characteristics, can be inferred from environmental and, to a lesser extent, demographic data, both of which are quantitative characteristics. The transition between these two types of data is rough, and analysis can become an issue of quantifying qualitative characteristics—the clichéd technique of comparing apples with oranges. As a result, many of the cultural inferences derived from this model are conceptual, hard to justify, and prone to criticism on the grounds of subjectivity. The process-frequency model would benefit greatly from a more rigid mathematical definition. This is not to say that the model should be reduced to a series of formulae. Rather, new-found mathematical precision and the intuition implicit in the process-frequency model should be elegantly integrated to provide a model that is satisfying to anthropologists and can stand up to criticism from the so-called hard sciences.

The process-frequency model uses frequency as the primary unit of analysis. From a more mathematical viewpoint, frequency is not necessarily the most important quality of a cyclic trend. Something like a sinusoid wave, for example, can vary only a minimal to moderate amount in frequency before it is classified as a polynomial trend. The beauty of sinusoidal behavior is that it is, to a certain extent, predictable: a mathematician knows that a true sine wave will repeat itself every 360 degrees; variance from that pattern alters the classification of the wave. The characteristic that allows for more variability in cyclic trends is amplitude. It is suggested here that the process-frequency model be more precisely defined in terms of sinusoidal trends and that those trends then be analyzed in terms of amplitude. This avoids the random mathematical definition now employed by the model; as it stands, the model is defined by frequencies denoted by generations, which are the assumed correlative units of cultural data transmission. Amplitude analysis would suggest that high-frequency process (HFP) trends can occur over long periods of time (greater than one generation) and, conversely, low-frequency processes (LFP) can occur in shorter time periods. These interpretations mean that the material remains found in sites, when analyzed in a frequency-over-time model, can illustrate cultural data transmission more accurately.

The behavior of trends becomes increasingly self-illustrative; without the restrictions on frequency, trends illustrate the nature of cultural changes rather than the frequency of such changes. Mathematically, this is comparable to slope analysis, in which the change in slope is an important means of comparing two functions that may not necessarily occur within the same parameters. Anthropologically, amplitude analysis would provide a means of comparing cultural trends to one another. Frequency analysis currently employed by the model does an excellent job of such comparisons among environmental data but stops short of convincing cultural analysis. The use of frequency and amplitude analysis together in this model would improve it holistically and would strengthen it by adopting mathematical criteria used in the analysis of cyclic trends; mathematicians rely on amplitude as well as frequency to describe and define the quality of trends. If frequency and amplitude were used in the analysis of cultural data sets, the occurrence and quality of cultural changes could be investigated more thoroughly. Although the general defining parameter of one generation leads to more restricted cognitive interpretation, a more precise definition would lead to more freedom in interpretation and would allow researchers to fine-tune the model to particular data sets.

Secondarily, the model's capacity for cognitive interpretation needs to be improved. For example, a generation is defined as circa 25 years (Dean 1988b). In cognitive terms, this definition is imprecise; exactly what does "circa 25 years" mean to an individual? Certainly life is not experienced in circa years. The definition is not an accurate reflection of the way culture and cultural changes are perceived by individuals. The model runs on the concept that people pay little heed to LFP, virtually ignoring long-term change because it is so slow. At any given time, however, certain individuals may be living who remember when a trend actually began. All trends, whether LFP or HFP, have to start somewhere; the origins of trends are inevitably recorded in the memories of the individuals who witnessed them. The model fails to reconcile with this point, and thus the parameters of generations create a static cognitive model.

The model as it stands cannot adapt to the reality of human experience. Rather vague definitions, such as "circa 25 years," and difficulty in the modeling of cognition have prevented such an adaptation. Again, increasing mathematical precision might improve this point. Once the underbrush of imprecision has been cleared away, the more defined model would allow for more latitude and precision in cognitive interpretation. Additionally, mathematical improvements would reduce the vagueness of the model, allowing researchers to see more precisely what is actually going on. With more accurate representations of data behavior, more accurate interpretations could be made.

Overall, this model exceeds in some areas—dendrochronological analysis, for example—and needs improvement in others. This discrepancy is probably due, in part, to a lack of data with which to work. Trees laying down growth bands every year provide excellent data to manipulate and interpret. Cultural and behavioral data are never so precise or predictable, and they prove more difficult to analyze. The process-frequency model is well adapted for dealing with complete data but relies too heavily on fuzzy concepts when it comes to interpreting human behavior. A newly improved, mathematically justifiable model would go a long way toward tackling the challenges of interpreting cultural change and cognition.

NOTE

Donald Graybill's research on reconstructing annual precipitation for the Grasshopper region and Reid's on subsistence management strategies first went public at the 1984 Society for American Archaeology meeting as a joint paper with slides. Don and Reid had been discussing the issues and analytical problems for several years before that meeting. This lengthy incubation period is noted to illustrate a significant and timeless quality of Don Graybill's research in pursuit of the past. Unlike so many of his archaeological contemporaries, Don was not compelled to rush research into publication. He adamantly refused to publish until he felt the data and analytical techniques supported the conclusions. And although this could be excruciatingly frustrating for those of us who worked closely with him, his reasons were solid. On the one hand, Don could take his time because he knew that no one had his high-resolution climatic data. Who else was going to reconstruct annual precipitation, temperature, and streamflow? And this unique position within the southwestern archaeological community kept his services in high demand and his research time spread entirely too thinly across the academic landscape. On the other hand, Don was consumed by the protocols of hard science and stubbornly resisted any pressure to make allowances for the soft, mushy scientific pretensions of processual archaeology. Where many archaeologists found quick salvation in factor and cluster analyses, Don saw flagrant violations of the analytical assumptions required of these multivariate routines. He often commented that he had never seen a nontrivial normal distribution of archaeological data.

After 20 years, it cannot be claimed that Don would be ready to publish, but even he would probably concede that the climatic data are not likely to improve after the University of Arizona Archaeological Field School at Grasshopper closed in 1992. He would agree, however, that the current reconstruction is more accurate than the earlier one, and to Don Graybill accuracy was everything.

7

Settlement History and Environmental Variability in the Upper Little Colorado River Valley, Arizona

David E. Doyel

PROCESSES OF AGRICULTURAL DEVELOPMENT, SEDENTISM, AND INtensification are dominant research themes in southwestern archaeology. By A.D. 1050, the Chaco, Hohokam, and Mimbres regional systems were undergoing spatial and formal elaboration, with great houses, great kivas, and engineered roads at Chaco in the north; great kivas, large towns, and elaborate ceremonialism among the Mimbres in southwestern New Mexico; and ball courts, platform mounds, and massive intervillage irrigation networks among the Hohokam in southern Arizona (Doyel and Lekson 1992). Central to the study of these evolving systems are questions about subsistence, demography, technology, social organization, and related issues of resource production and allocation, social integration, decision-making structures, and the nature of ecological and ideological systems.

These issues should also frame research in lesser-known regions of the Southwest where basic questions of *who, where, when,* and *why* remain to be more fully addressed. Using a broad brush, I present a summary of settlement and land-use patterns in the Upper Little Colorado River Valley (ULCRV) between Springerville (Round Valley) and St. Johns in east-central Arizona (Figure 7.1). Research between the mid-1950s and the 1970s led by the Chicago Field Museum produced survey and excavation data and a chronology for the region (Danson 1957; Martin and Plog 1973), and contract archaeology projects have since contributed additional knowledge (e.g., Doyel and Debowski 1980; Schreiber and Sullivan 1984; Van West and Huber 1995; Westfall 1981). The ruling hypothesis maintains that the late prehistoric economies were fueled by agricultural intensification, but most research was conducted prior to systematic investigations of land-use practices and sampling for subsistence remains. Recent trends include documenting specific practices of production, preparation, consumption, and disposal in culturally prescribed ways (Fish and Donaldson 1991; Gumerman 2001; Netting 1975, 1993; Steward 1955), the

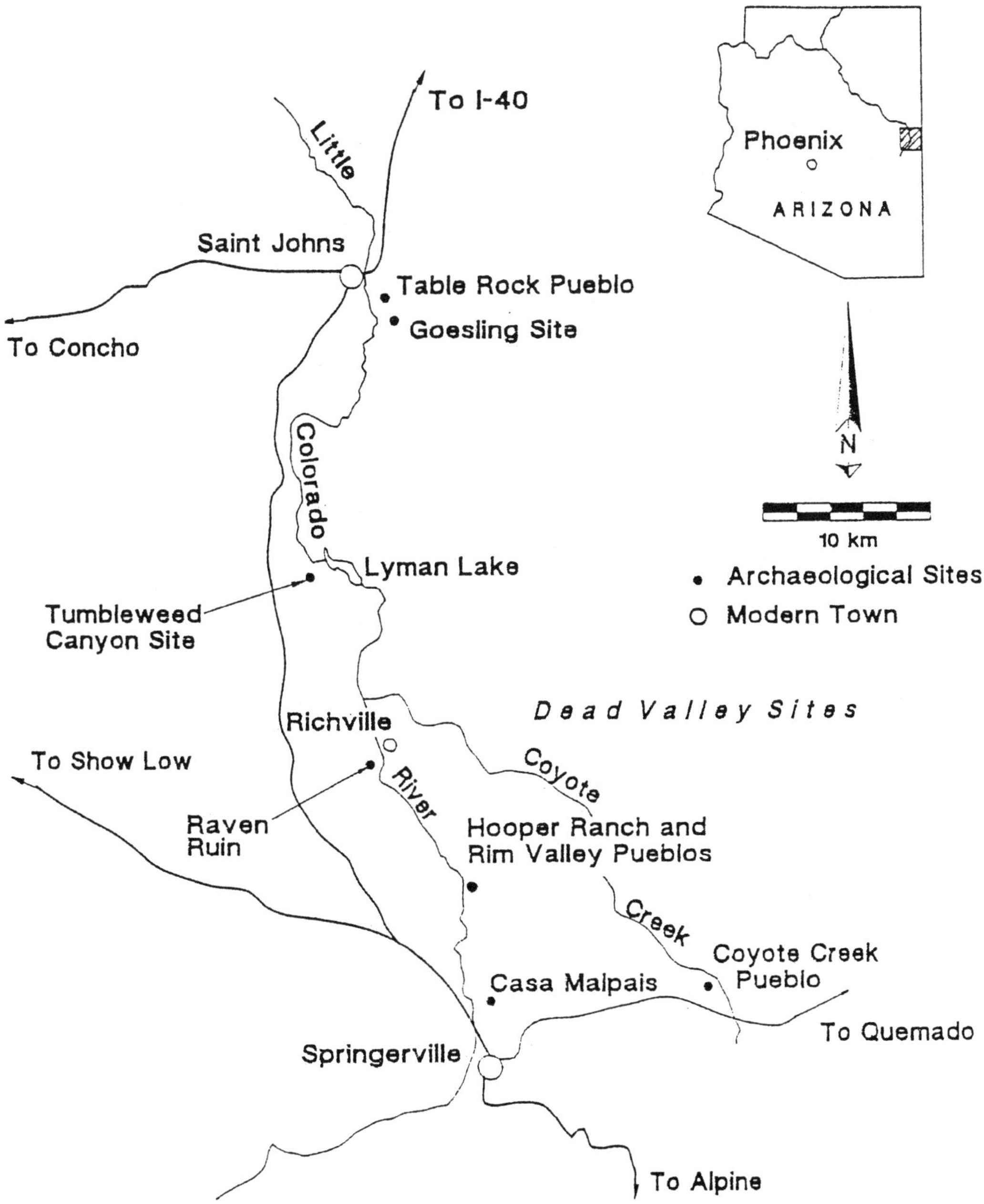

FIGURE 7.1. The Upper Little Colorado River Valley and archaeological sites.

explanation of which should be a goal of research, not an afterthought (Martin 1970).

Previous research indicates that multiple agricultural strategies and settlement patterns adapted to diverse environmental contexts existed during the prehistoric occupation of the ULCRV. A central topic addressed here is the value of and potential for small-site phenomena to yield data sensitive to issues of food production, land-use practices, their temporal and spatial expressions, and the integration of small-site data in the formation of regional settlement patterns. A traditional challenge has been functional identification of small sites, especially those that may lack features (Ward 1978). Indeed, Sullivan (1992) claims that agriculture is often incorrectly inferred at small sites without sufficient evidence. In this chapter, analysis of excavated small sites within their environmental context provides a strong case for a specific type of agriculture—sand-dune dry farming—between A.D. 900 and 1150 in a nonriverine locality known as Dead Valley. The influence of climate, primarily rainfall, on agricultural practices, settlement patterns, and variability in cultural patterns is addressed. The results are then placed in a broader temporal and spatial framework to explore variability in the settlement history of the ULCRV. But first, a brief review is presented of agricultural practices in the northern Southwest.

AGRICULTURAL DIVERSITY IN THE NORTHERN SOUTHWEST

Water was the critical variable for successful agriculture in the arid Southwest. Specifically, the frequency, amount, and distribution of moisture were key factors. Modern corn requires 45–60 cm (18–24 in) of water to mature, but native maize can be grown on the Hopi mesas with 25–33 cm (10–13 in) (Vivian 1992). With an average precipitation of 20 to 30 cm across the Colorado Plateau, insufficient moisture could have affected local production (Vivian 1992:46). Diverse techniques used by the Hopi to maximize available moisture included *ak chin*, floodplain, sand-dune, seep, spring-side, and terrace gardens (Hack 1942). Situated at an elevation of about 7,000 feet, the Hopi mesas require a frost-free growing season of 120 to 130 days to grow corn (Hack 1942:22). Higher elevations, especially those approaching 7,500 feet, have shorter growing seasons that are likely marginal for maize agriculture under average conditions.

The practices recorded at Hopi and other locales have great time depth in the Southwest (e.g., Brandt 1995; Doyel 1979, 1993b; Fish and Fish 1984; Travis 1990; Woosley 1980). For example, field systems at Wupatki in the Lower Little Colorado area included floodwater, rockpile, bordered-terrace, and composite fields and bordered and terrace gardens (Travis 1990:10). Associated feature types included rock piles, pads, and alignments; terrace walls; check dams; cinder berms; and upright stones. To the east, farming practices along the Middle Little Colorado River Valley included sand-dune, floodwater, and diversion systems (Lange 1996). Systems in Chaco Canyon included (1) *ak chin* fields along the edges of drainages, (2) sheet runoff to gardens on terraces, (3) stone terraces to hold soil and water that drained over the cliff edges, (4) dry-farming fields along the canyon edges, and (5) canal-fed fields in the canyon bottoms. Reservoirs, dams, and gridded fields were also present, and check dams, linear borders, and mulched gardens were added later, indicating continued diversification (Vivian 1974, 1992).

Northwest of the Hopi mesas, upland forested areas near the Grand Canyon farmed before A.D. 1300 were abandoned in favor of lower-elevation alluvial settings (Sullivan 1992). In the White Mountains in central Arizona, alluvial and groundwater systems used in the chaparral and transition zones were later replaced by open meadows and other field situations, some of which were associated with water-control features, suggesting a stronger commitment to agriculture (Donaldson 1993; Sullivan 1982; Tuggle et al. 1984; Welch 1991).

These examples indicate that agriculture in the northern Southwest was not a monolithic and unchanging form of land use. Time-tested, locally adapted practices and technologies were developed within diverse contexts. Redundancy in form and function was present owing to basic engineering principles, but significant variation existed in emphasis, intensity of use, technology, and system scale due to environmental and organizational factors.

AGRICULTURE AND SETTLEMENT HISTORY IN THE UPPER LITTLE COLORADO RIVER VALLEY

Aboriginal agricultural practices have not been a focus of study in the ULCRV. Of the strategies described above, what types of systems have been identified? Can a developmental sequence of agricultural types and practices be predicted and/or documented? Were agricultural and land-use patterns similar to those in neighboring areas and/or were local adaptations present in this area, located in the rain shadow of the White Mountains? The evidence hints at a diversity of practices that varied by time, place, and circumstance.

Archaeological reconnaissance survey was undertaken in the "Triangle" (Figure 7.2) by Rinaldo and Longacre (Longacre 1964) and by Beeson (1966) in a slightly different but overlapping area. Longacre describes the Late Basketmaker (A.D. 500–700) and Pueblo I periods (A.D. 700–900) as a time of "established village farming," with small clusters of pithouses located along drainages in diverse settings. Few sites in the ULCRV dating to this period are known, and information on agricultural practices is scarce (Doyel 1980d; Duff 1999). The locations of Pueblo I (PI) sites along floodplains imply use of groundwater and floodwater systems, and dune fields in nonriverine areas were also farmed (Beeson 1966; Longacre 1964).

An increase in the number of sites in the Pueblo II period (A.D. 900–1150) indicates a growing population (Danson 1957; Duff 2002; Longacre 1964). Many small pueblos of 8 to 15 rooms were present along the river; Rim Valley Pueblo, with 25 rooms, was also occupied in this period (Duff 2002). Beeson (1966) notes that between PI and PII the number of "living units" increased from 45 to 112 in the Richville area and from 132 to 300 in his larger survey area. Between A.D. 1050 and 1150, Chaco-era communities were established in the

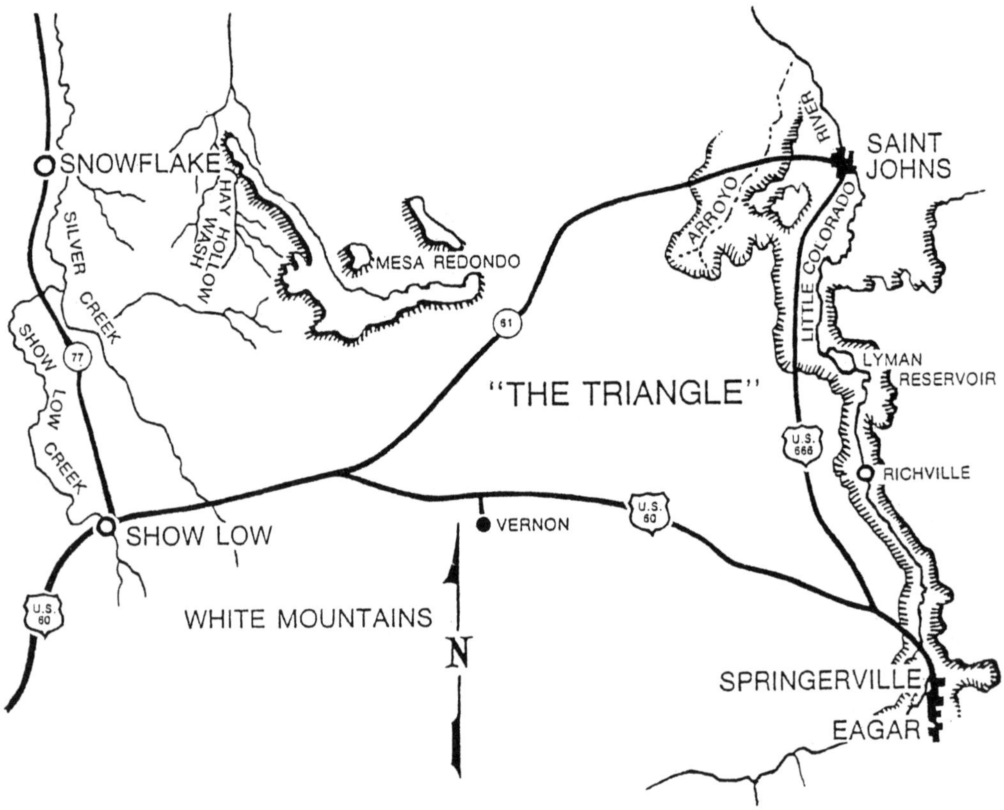

FIGURE 7.2. The Triangle survey area (from Longacre 1964).

ULCRV near St. Johns (Kintigh 1996:Figure 9.1) along an east-west line pointing to Quemado (e.g., Cox Ranch) (Duff 2002).

In the Pueblo III period (A.D. 1150–1275), settlements contained room blocks of 3 to 50 rooms (Duff 1999). The large Coyote Creek Pueblo, situated in the uplands to the northeast, resembles a Tularosa phase village (DeGarmo 1975; Duff 1999). Survey data from the mountain zone south of Round Valley recorded sites dating between A.D. 1050 and 1300 with up to 10 rooms that contained small round and square kivas (Wood 1979:68).

In the Pueblo IV period (A.D. 1275–1400), earlier settlements were abandoned and new sites were established along the Little Colorado River in areas favorable for irrigation (Duff 1999, 2002; Kintigh 1996). Population aggregated into sites of 25 or more rooms containing room blocks, plazas, and kivas (Beeson 1966; Kintigh 1996:140). Whether hierarchical organization existed among these settlements is debated (Duff 2000, 2002; Hantman 1989; Lightfoot 1978). In the late Pueblo IV period, the presence of settlement voids between St. Johns, Rattlesnake Point, and Hooper Ranch, combined with variability in material culture and site structure, suggests occupation by groups of differing historical backgrounds (Duff 2002).

In summary, population more than doubled between the Pueblo I and Pueblo IV periods in the ULCRV (Beeson 1966; Duff 1999). Longacre (1964) estimates a population peak between A.D. 1200 and 1400. Duff (2002) partitions the Pueblo IV period into early (e.g., Casa Malpais) and late (e.g., Table Rock, Rattlesnake Point) subdivisions, which indicates that not all PIV sites were occupied for the entire period and suggests that population levels were lower than previously reported. Kintigh (1996:138–139) states that the available data are insufficient to support reliable estimates of relative population size.

Environmental instability affected agricultural practices and associated settlement patterns in the region. After A.D. 1130, a reduction in average precipitation, an increase in erosion, and a slight lowering of water tables in the northern Southwest resulted in a discontinuous distribution of population as groups relocated or reorganized to adjust to environmental changes (Dean et al. 1994). The subsequent Pueblo IV period witnessed a severe drop in precipitation, even lower water tables, and arroyo cutting unfavorable to agricultural economies, all contributing to regional depopulation between A.D. 1400 and 1450 (Dean et al. 1994:150; Dean and Van West 2002).

In summary, this review suggests that (1) settlement of the area by farmers had considerable time depth; (2) population increased through time; (3) diverse agricultural practices were employed; (4) minimal systematic data are available on land-use and agricultural practices and changes through time; (5) environmental instability characterized the late prehistoric occupation; and (6) regional populations reflected considerable cultural diversity. Kintigh suggests (1996:136) that although settlement patterns "are not well understood," the area was not "filled up," and that the changes in settlement and subsistence patterns do not appear to have been density-driven.

AGRICULTURE AND LAND USE IN DEAD VALLEY

Dead Valley is located 7 km east of the Little Colorado River north of Richville at an elevation of 2,075 m (7,000 ft) (Figure 7.3). Pinyon and juniper are dominant along with an understory of yucca, saltbush, snakeweed, and grama grass. Intensive survey of this small locality (5 km^2) recorded numerous sites, most of which were surface artifact scatters, some associated with subsurface features (Wilson 1976). The excavation (data recovery) program included 15 sites, the research plan for which focused on (1) cultural identification, (2) temporal ordering, (3) site activity analysis, (4) ecological analysis, and (5) settlement and subsistence reconstruction (Doyel 1981; Doyel and Debowski 1980). Lacking clear indications of why the sites were located where they were, we asked the following questions: when did the occupation occur? what was the range of site variation? what resources were used? could agriculture have been practiced, and if so, what strategies were employed?

The Dead Valley Sites

Excavation determined that some sites were, in fact, surface scatters, but substantial features were present at other sites. One site, AZ Q:12:10 (ASM), had a deep pithouse equipped with a stone-lined ventilator shaft; also present was a slightly later component containing ramada features with post supports and

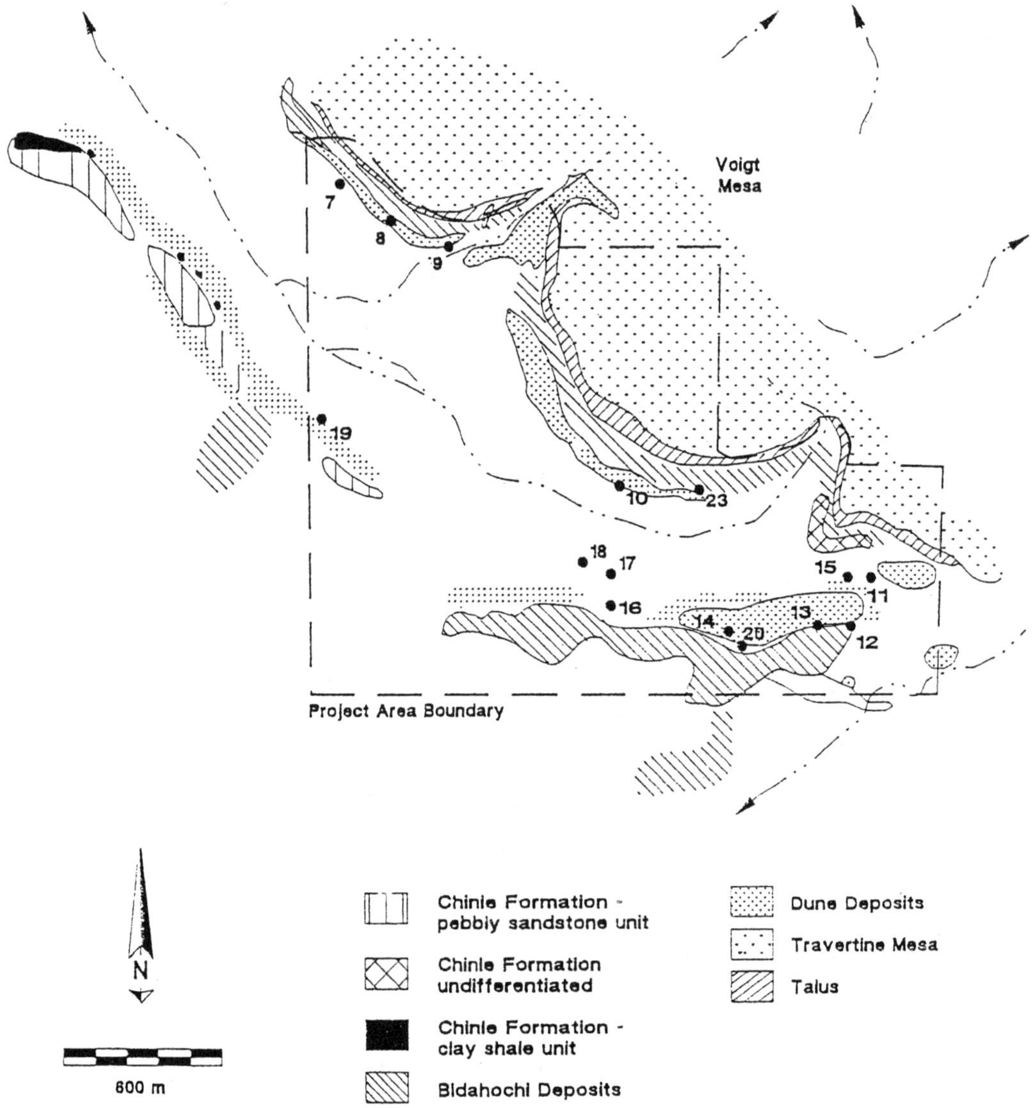

FIGURE 7.3. Dead Valley: site locations and surface geology.

low stone walls. A two-room masonry structure partially enclosed by a walled patio was present at AZ Q:12:9 (ASM), where the smallest room had a mealing bin (Figure 7.4). The largest site, Dead Valley Pueblo, AZ Q:12:16 (ASM), had two contiguous room spaces, a larger semi-enclosed area, and a small detached kiva (Figure 7.5). Tree-ring samples from the kiva, which had burned, produced dates between A.D. 1093 and 1130(vv), with six of nine between A.D. 1102 and 1130.

Ceramic wares common to the Cibola area dominated. Average frequencies included gray paste corrugated ware (37 percent), Cibola White Ware (31 percent), brown paste corrugated and plain wares (30 percent), and White Mountain (Cibola) Red Ware (2 percent) (Table 7.1). Decorated types included Reserve Black-on-white (50 percent), Puerco Black-on-white (40 percent),

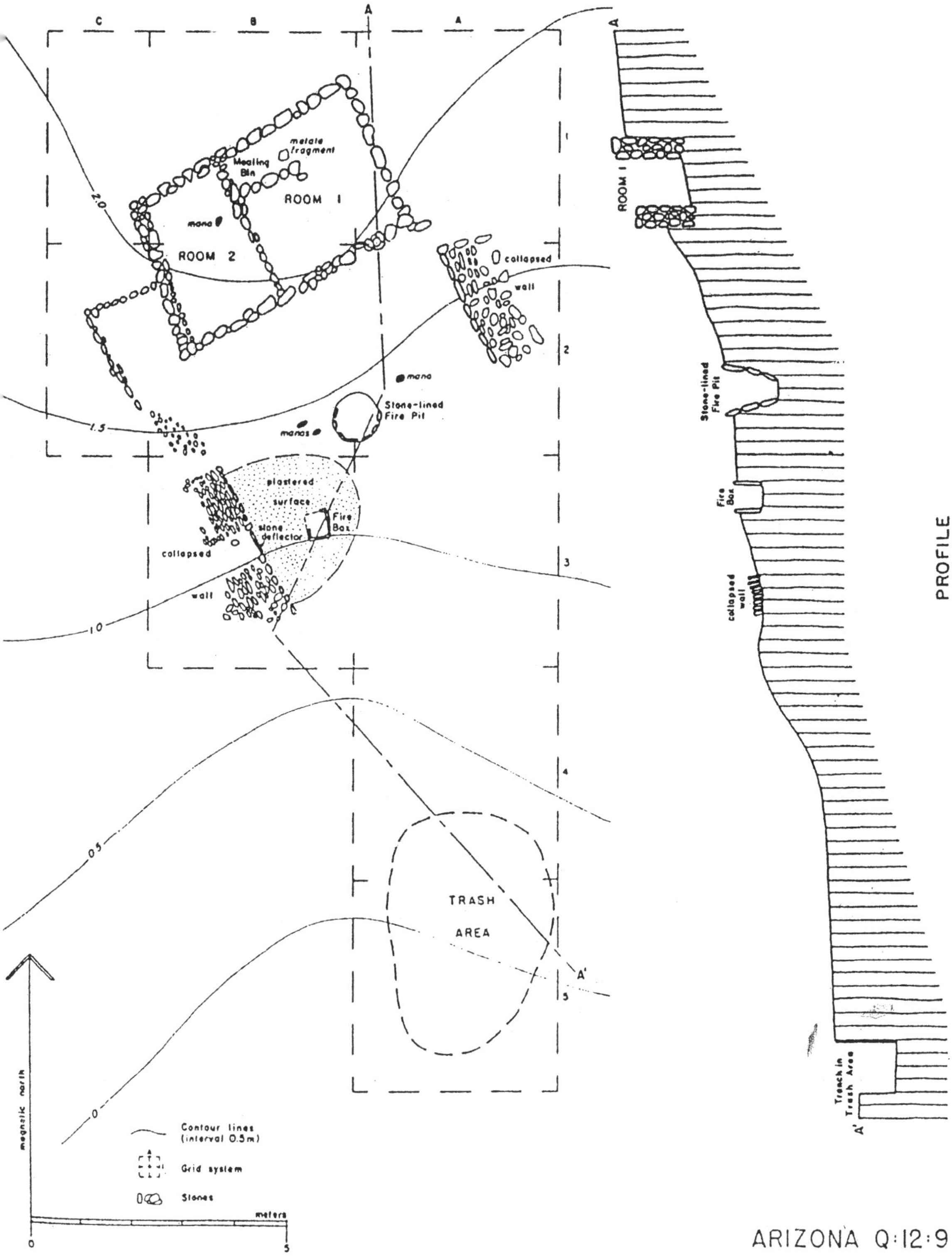

FIGURE 7.4. Site AZ Q:12:9 (ASM).

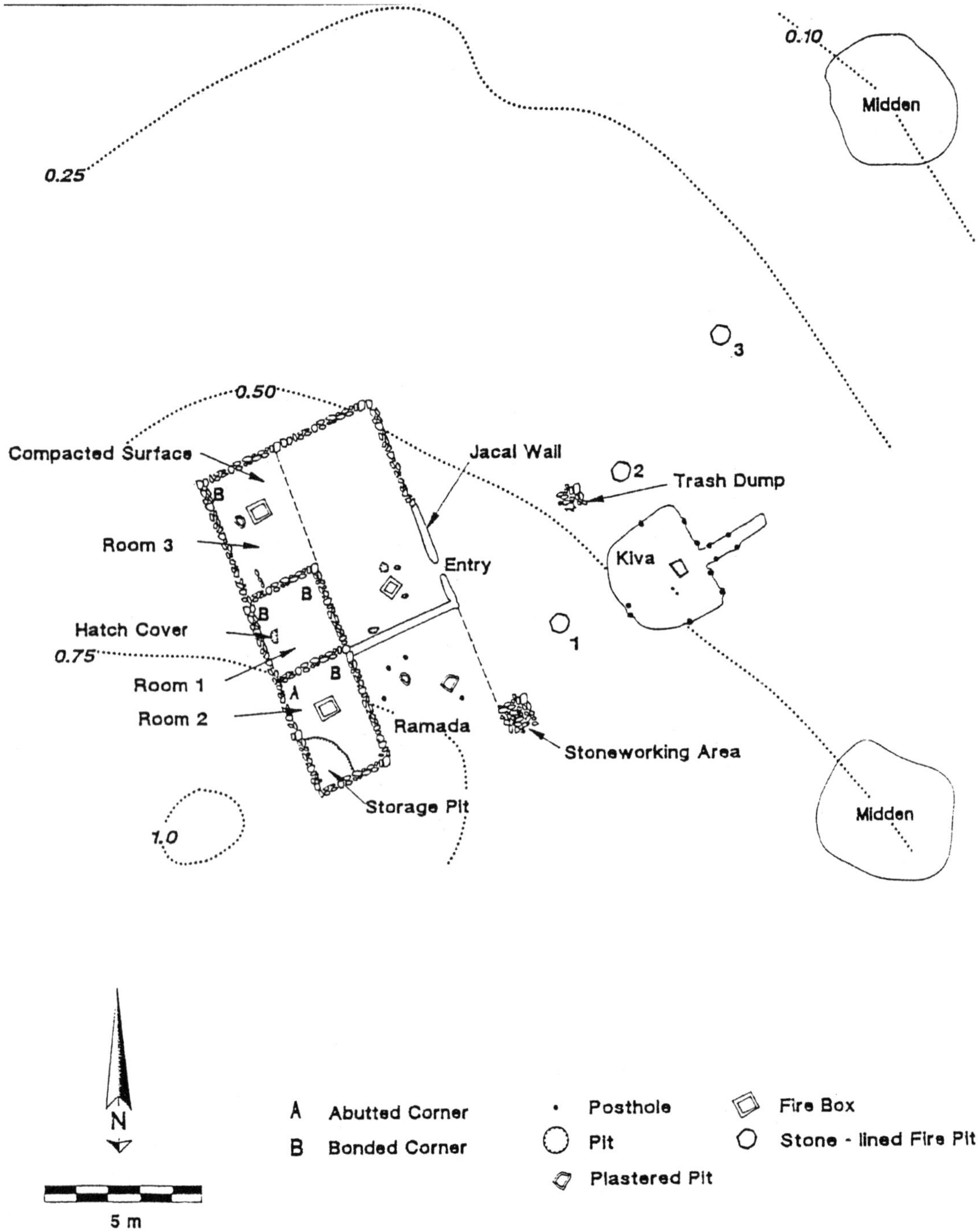

FIGURE 7.5. Site AZ Q:12:16 (ASM), Dead Valley Pueblo.

TABLE 7.1. Pottery Sherds Recovered from the Dead Valley Sites (in Percent with Total N).

ASM SITE NO.	SITE TYPE	WHITEWARE	REDWARE	GRAYWARE	BROWNWARE	TOTAL SHERDS
AZ Q:12:13	Pithouse (Basketmaker era)	0	0	20	80	203
AZ Q:12:15	Artifact scatter	23	2	45	30	287
AZ Q:12:17	Artifact scatter	50	1	40	9	631
AZ Q:12:19	Artifact scatter	25	1	39	35	310
AZ Q:12:23	Lithic artifact scatter	0	0	0	0	0
AZ Q:12:14	Artifact scatter/ fire box	34	2	23	41	197
AZ Q:12:7	Windbreak/fire boxes	32	4	61	13	3,093
AZ Q:12:8	Windbreak	26	1	13	60	89
AZ Q:12:10	Pithouse/ramada	34	1	42	23	4,377
AZ Q:12:11	Ramada	28	4	8	60	660
AZ Q:12:12	Fire hearth	24	1	44	31	364
AZ Q:12:18	Windbreak/fire boxes	36	2	49	13	569
AZ Q:12:20	Ramada	31	1	32	36	2,801
AZ Q:12:9	2 rooms and walled patio	40	0	41	10	2,028
AZ Q:12:16	3 rooms with kiva	35	1	40	24	19,571
Average		31	2	37	31	35,180

Red Mesa and transitional Reserve-Tularosa Black-on-white (4 percent each), and small amounts of Wingate and Puerco Black-on-red (2 percent) (Doyel 1980c, 1984).

Gray paste corrugated and neck-banded were the dominant utility wares, with slightly globular, medium-size jars a common form (Doyel 1980b). Brown paste wares of the Reserve series (Rinaldo and Bluhm 1956) were dominant at several sites that included plain, brown smudged, and corrugated wares with incised, pattern-incised, and punched-surface texturing. These utility types occurred as bowls and medium to large jars, many of which were well made. Use of different tempering materials, vessel shapes, and finishing techniques indicated different production technologies, if not production areas, for these wares (Doyel 1980b, 1980c, 1980d; Rugge and Doyel 1980).

Higher frequencies of plain gray and brown wares, lower frequencies of redware, and a few late Red Mesa–style designs on whiteware sherds suggested that several sites, such as AZ Q:12:9 (ASM), may date between A.D. 950 and 1050. Several pioneering sites were possibly present between A.D. 950 and

1050, but most of the sites dated to a 75-year period between A.D. 1050 and 1130 (Doyel 1980d, Doyel 1984; Graybill and Doyel 1980).

About 50,000 artifacts were recovered, half of which were from Dead Valley Pueblo (Table 7.1). Of this total, sherds made up 70 percent ($N = 35{,}180$); chipped stone, 29 percent ($N = 14{,}688$); and ground stone, 0.3 percent ($N = 189$). The large chipped stone assemblages present included 38 projectile points, 18 from one site, suggesting hunting and/or the production of points (Jeter 1980). Ground stone tools were common and included a dominance of two-handed manos and basin and trough metates (McPherson and Doyel 1980). A few ornamental shell and other small artifacts were present (Debowski 1980). Overall, the assemblages appear to resemble those from small habitation sites. Few human burials were encountered, as would be expected at these short-term sites occupied by small social groups.

Notable intersite variation was observed in the artifact distributions. Although expressing a bimodal distribution, the average size of four artifact scatter sites that lacked features was much larger than eight sites that had fire boxes and ephemeral structures, 1,450 versus 295 m^2, respectively (Table 7.2). The sites with features, however, had average artifact counts five times greater than the sites without features, 2,244 versus 434, respectively. Ceramics associated with the artifact scatter sites were often limited to corrugated jars and bowl forms, and whiteware jars were uncommon, whereas the feature sites contained greater variation. A wider range of activities is inferred for the feature sites, including food processing and preparation, storage, and manufacturing.

Agriculture and Land-Use Practices

Environmental variables examined included available moisture, length of growing season, and soil type. The combination of elevation and climate produce a short growing season averaging 130 frost-free days between May and October. With an average annual precipitation of 29 cm (Sellers and Hill 1974), the area is not blessed with an abundance of moisture. Variation in climate and/or precipitation could have rendered agriculture precarious in some years. Surface water was likely absent, with a possible exception of seep springs emanating from the fractured limestone on Voight Mesa, which towers above the northwestern edge of the valley (Rugge 1980). Little water would have been available for pot irrigation. Surface water was available at Coyote Creek, 3 km to the south. Historically, agriculture has not been practiced in or near Dead Valley but is successful in the Richville and Round Valley areas, where irrigation is employed.

Architecture and site structure provided insights into land-use practices. Four artifact scatter sites lacking architectural features were present. Eight other sites contained low, coursed masonry walls (ca. 50–70 cm high) or ephemeral ramada structures. Most of these masonry features consisted of

TABLE 7.2. Sites Excavated and Recorded by the Dead Valley Project (DVAP)

ASM SITE NO.	SITE TYPE	SITE SIZE (M^2)	ARTIFACT COUNT	TIME PERIOD (A.D.)
	DEAD VALLEY SITES			
AZ Q:12:13	Pithouse	75	381	600–800
AZ Q:12:15	Artifact scatter	3,200	339	1050–1150
AZ Q:12:17	Artifact scatter	2,000	812	1050–1150
AZ Q:12:19	Artifact scatter	100	478	1050–1100
AZ Q:12:23	Lithic artifact scatter	500	106	P II?
AZ Q:12:14	Artifact scatter/fire box	1,000	366	1050–1150
AZ Q:12:7	Windbreak/fire boxes	50	3,864	1100–1150
AZ Q:12:8	Windbreak	50	115	1050–1100
AZ Q:12:10	Pithouse/ramadas	400	6,817	950–1100
AZ Q:12:11	Ramada	600	998	1100–1150
AZ Q:12:12	Fire hearth	500	524	1000–1100?
AZ Q:12:18	Windbreak/fire boxes	60	1,070	1050–1100
AZ Q:12:20	Ramada	200	4,197	1050–1100
AZ Q:12:9	Two rooms and walled patio	150	2,607	950–1025
AZ Q:12:16	Three rooms and kiva	600	27,542	1100–1150
	SURVEY SITES			
AZ Q:11:62	8 rooms	100+	NA	1100–1150
AZ Q:11:63	1–2 rooms	300	NA	1100–1150
AZ Q:11:64	2–3 rooms	1,000	NA	1050–1150
AZ Q:11:65	1 room?	425	NA	1050–1150
AZ Q:12:24	None visible	250	NA	950–1150?
AZ Q:12:25	Slabs present	150	NA	1050–1150
AZ Q:12:26	None visible	1,000	NA	Lithic only
AZ Q:12:21	1 room	1,500	NA	1050–1150
AZ Q:12:22	Slabs and fire boxes	4,000	NA	1000–1100

single walls that served as windbreaks. Artifact concentrations, charcoal-stained soil, and, in cases, small trash areas were associated with these features. Stone-lined fire boxes or fire pits were present. Multiple occupations were common, as indicated by superimposed use surfaces and fire boxes (Figure 7.6). Ramada features were areas containing low stone walls and/or ephemeral use surfaces associated with ash, charcoal, postholes, and tabular stone, the latter likely used to anchor roofs made of brush (Figure 7.7). This architecture was not designed to provide year-round shelter at 7,000 feet on the Colorado

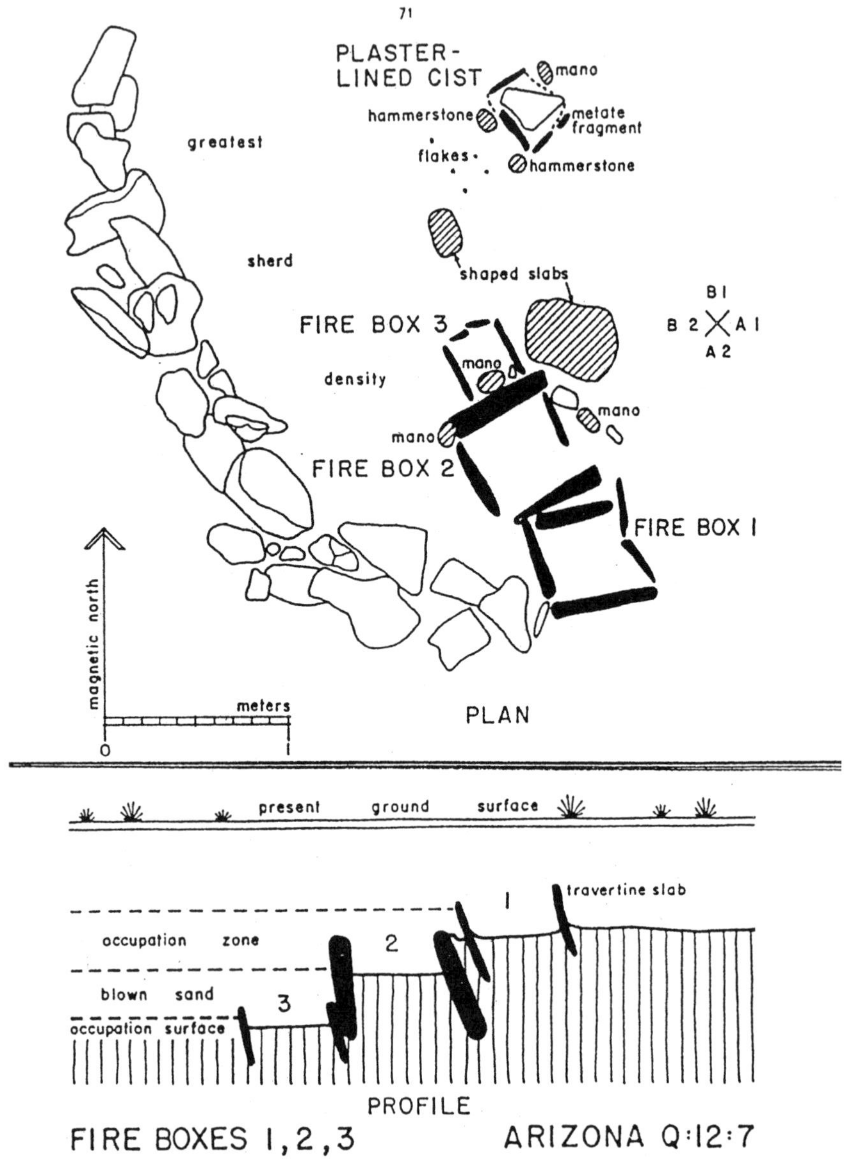

FIGURE 7.6.
Site AZ Q:12:7 (ASM).

Plateau but is similar to field houses used historically by native peoples in the growing and harvest seasons (Doyel 1993b; Mindeleff 1989; Ward 1978) (Figure 7.8).

Field survey identified the presence of extensive sand dunes and dune soils across Dead Valley. Strong winds passing through the area deposit large volumes of sand along the mesas and ridges. Machine trenching revealed that some dunes were over 3 m deep and other areas contained 20 to 30 cm of deposits. Dune soils are known for their capacity to retain moisture, but all dune soils are not of equal value relative to successful farming. Native peoples, including Navajo and Pueblo, select dune fields that contain under-

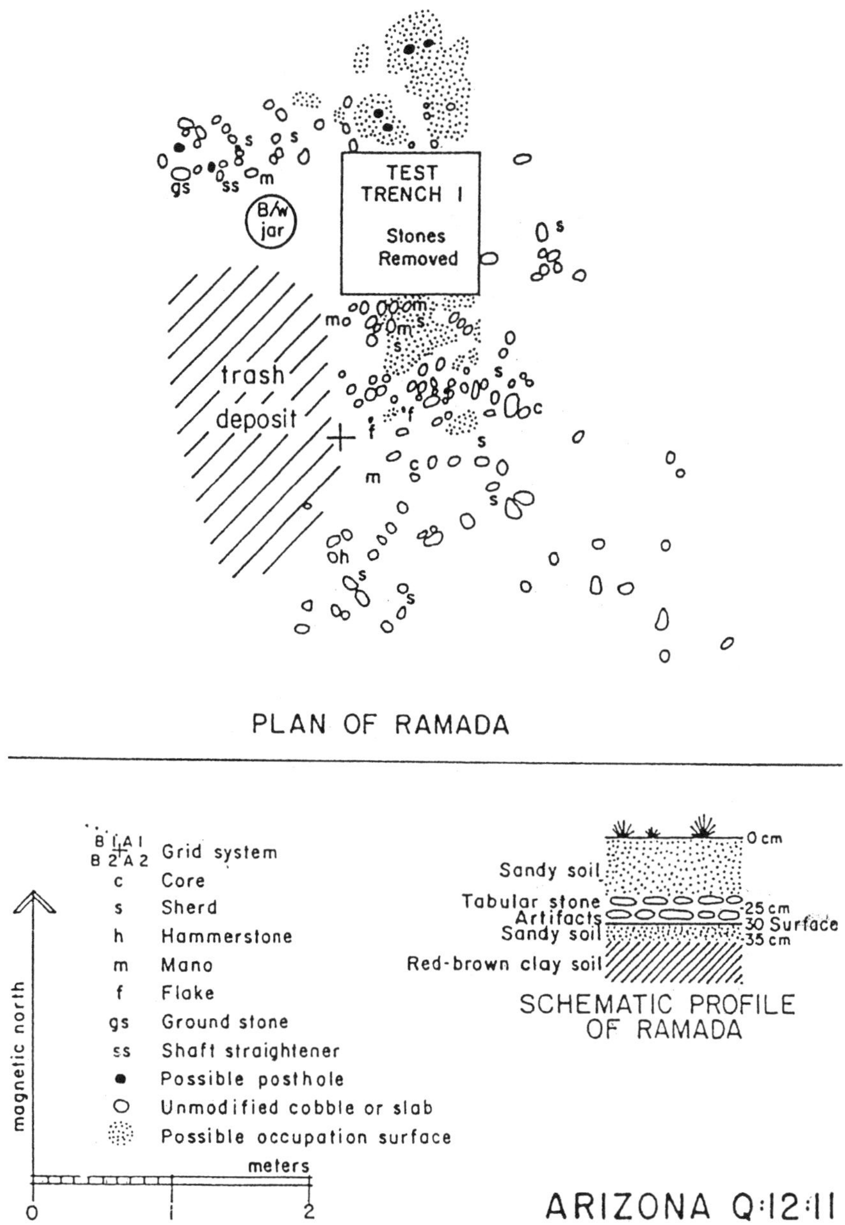

FIGURE 7.7. Site AZ Q:12:11 (ASM).

lying, relatively impervious, clay substrates that reduce moisture loss and serve as moisture reservoirs (Bradfield 1971; Hack 1942). In Dead Valley, the sandy to loamy dune soils overlie substrates consisting of clay loam and silt loam (Rugge 1980).

Multiple lines of evidence strongly support the practice of sand-dune farming. Site location was correlated with dunes, and sites were absent in the valley bottom where dune soils were absent (Figure 7.3). Some sites were located on top of or near dune deposits. Isolated grinding tools were found, and large stones and slabs found on dunes likely served as deflectors to protect

FIGURE 7.8.
A field house in the pueblo area (from Mindeleff 1989).

plants from wind damage, a practice recorded among the Pueblo people (Hack 1942).

Excavated data, most of it from Dead Valley Pueblo, support the inference of agricultural functions. High percentages of pollen and plant remains from the *Chenopodium* and *Amaranthus* (cheno-am) groups (e.g., goosefoot, pigweed) were present in contexts where historically they occur in low frequencies (Kelso 1980; Reichardt and Nabhan 1980). The higher frequencies in prehistoric contexts are likely attributable to local disturbances due to agriculture, as these plants thrive in disturbed soils. Kelso notes (1980:369): "The soil disturbance accompanying the introduction of agriculture definitely resulted

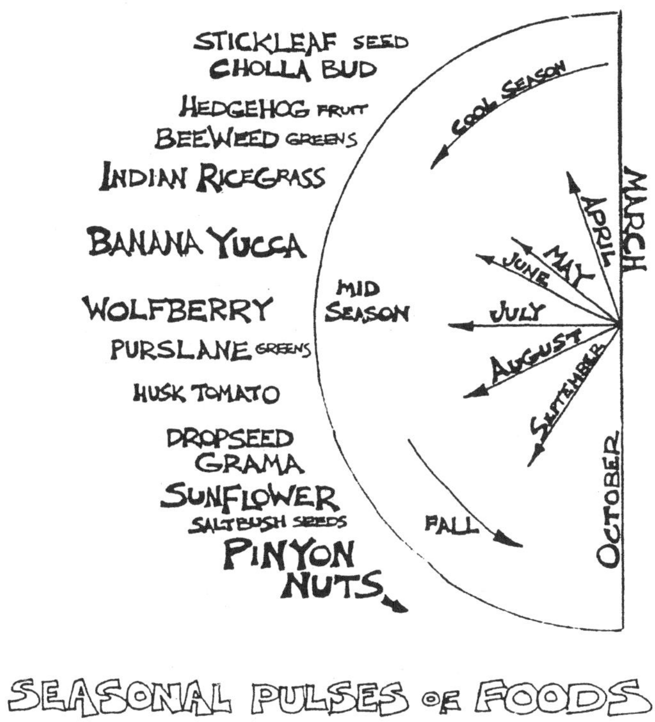

FIGURE 7.9. Availability of native food plants (from Reichardt and Nabhan 1980).

in the proliferation of plants producing cheno-am-type pollen, and, with a suggestion of human selectivity, apparently reduced the available habitat to the formerly dominant wind-pollinated members of the Compositae family during Pueblo II times. Local populations of grass were apparently reduced as well."

Pollen samples ($N = 78$) indicated that maize pollen was present on grinding implements, in storage pits, and on room floors; use of squash, sunflower, Rocky Mountain bee weed, and grasses was also indicated (Kelso 1980). Corn remains and cobs, most of which were eight-rowed, were found in half the nonsterile macrobotanical samples ($N = 46$) and accounted for over 60 percent of the recovered seeds. Native plants represented in the macrobotanical remains included goosefoot, cheno-ams, banana yucca, Indian rice grass, pinyon nuts, and prickly pear cactus (Minnis 1980). Reichardt and Nabhan (1980) inventoried more than a dozen useful native plants in the area that mature between spring and fall, many of which were likely used (Figure 7.9). Most of the 216 nonhuman bones recovered were unidentifiable but included rat, squirrel, prairie dog, and rabbit, suggesting opportunistic and garden-related hunting rather than hunts for large game (Bayham 1980). The relatively low presence of faunal remains further supports the inference of agricultural functions for the sites.

As a test of this developing land-use model, archaeological survey was undertaken in a 10-km^2 area divided into six judgmental sample units that surrounded Dead Valley (Figure 7.10). Of the nine additional sites located, the largest had about eight rooms; some had one or two rooms, and some had windbreak features similar to those excavated in Dead Valley (Table 7.2) (Doyel 1980a:401–410). Sites were not present in areas that lacked dune soils, but dune soils were present in areas that lacked sites. Ground stone artifacts were commonly associated. Based on the associated pottery, most of the sites dated between A.D. 1050 and 1150. These observations led to the prediction that relatively higher site densities would also be present in nearby areas that contain dune soils in close proximity to surface water, especially west along Coyote Creek toward Richville.

In total, at least 20 of the 24 sites excavated or surveyed dated to A.D. 900 and 1150, and most of them dated between A.D. 1050 and 1150. Five sites had architectural features that *may* suggest extended occupation, but most were occupied seasonally. These data may prove useful to projects limited to survey without benefit of excavation; for example, it became clear that wall alignments visible on the surface may not represent full room spaces.

The small habitation sites, such as AZ Q:12:9 and AZ Q:12:16 (ASM), may have been occupied by independent, extended family units or possibly by lineage-segment groups. The field-house sites were likely used by household or extended family groups tethered to larger communities located elsewhere. Some of the artifact scatter sites may represent task-specific loci associated with residents at AZ Q:12:9 and AZ Q:12:16 (ASM), but this possibility does not account for the density and content of all the small sites in the valley.

In summary, the length of the growing season, available moisture, and soil type provided suitable conditions for sand-dune agriculture in Dead Valley. In addition, site content and location and the associated artifact and archaeobotanical assemblages strongly support the inference that agriculture was the primary form of land use. Repeated use of these sites was characteristic, indicating specialized short-term occupations. Archaeological survey revealed an absence of any large sites and documented a wider spatial distribution for the findings in Dead Valley. Since long-term habitation sites were rare, groups using the valley must have traveled from elsewhere, which indicates seasonal mobility.

DEAD VALLEY IN REGIONAL CONTEXT

Regional perspectives can often strengthen the interpretive potential of site- and locality-specific studies and can identify factors that may affect site structure and content. For example, detailed analysis of the pottery yielded rather surprising results. It was estimated that about 900 pottery containers were represented in the sherd collection, about 350 of which were from Dead Valley Pueblo, with an average of 43 vessels for the remaining sites (Doyel 1980b). Elsewhere in the region, excavated field-house sites have much lower artifact counts (Gregory 1975). The small sites reported by Gregory in Hay Hallow

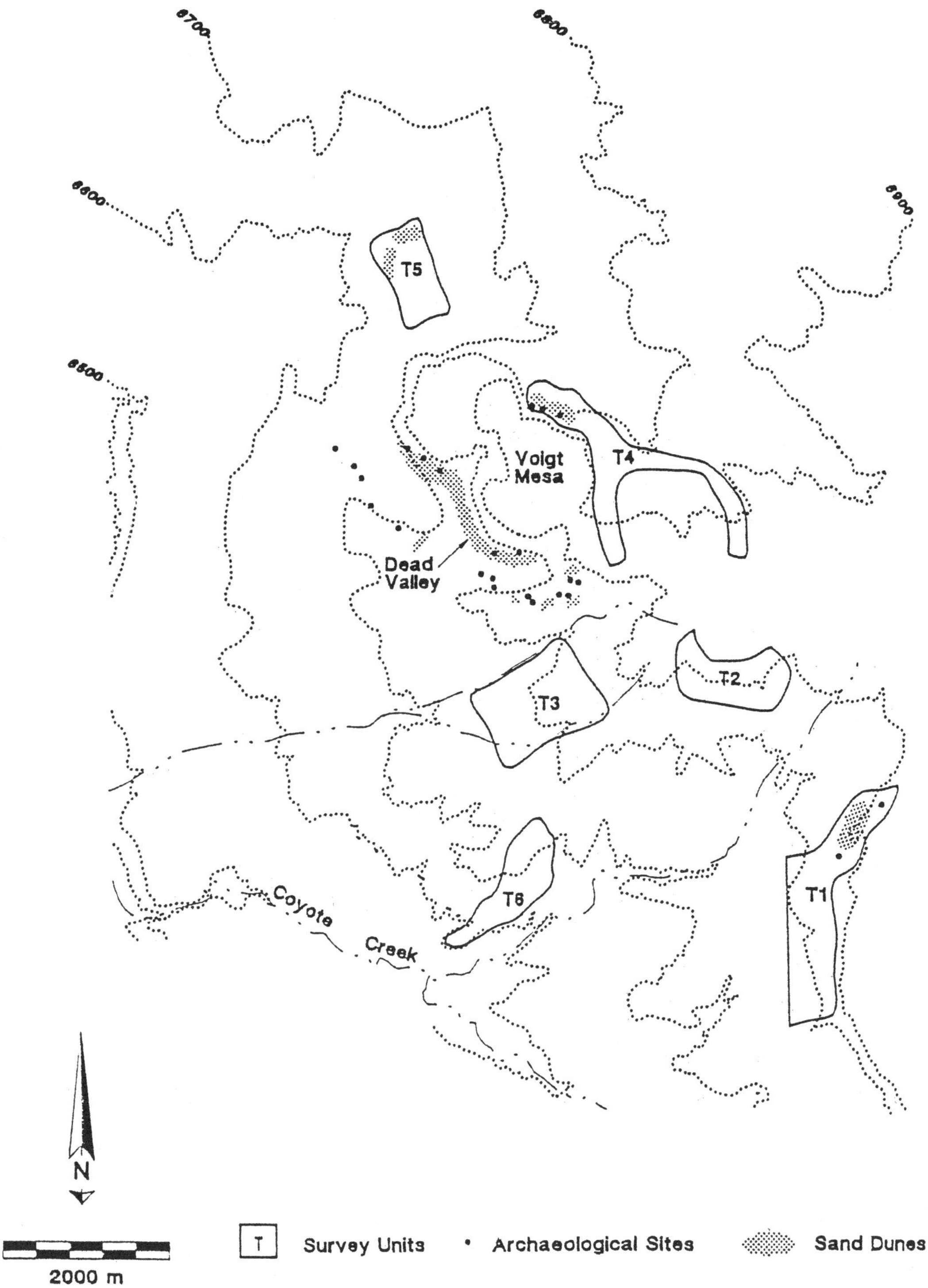

FIGURE 7.10. Archaeological survey adjacent to the DVAP area.

Valley are located only short distances from primary habitation sites. Since habitation sites were rare in Dead Valley, groups had to travel from some distance, resulting in more extended occupations and a wider range of activities and producing larger numbers of artifacts. The assemblages indicate that the Dead Valley sites essentially functioned as temporary, fair-weather residences. This finding underscores the value of using regional frameworks to interpret site structure and settlement systems.

Regional perspectives can also help us interpret environmental factors affecting site distributions. For example, a particular site may conform to a regional pattern of locating sites near arable land in association with specific hydrologic conditions. In this case, the fact that the Dead Valley sites were used only between A.D. 900 and 1150 is more interesting in light of the long settlement history of the region and along the Upper Little Colorado River, the latter located less than a half-day walk to the west. These facts prompt questions: why was Dead Valley used intensively in the Pueblo II period and not in other times, and what factor(s) can account for this pattern?

Several interpretations could account for this finding. An absence of earlier sites, the intensive use documented for the Pueblo II period, and the relatively sudden abandonment could have been a response to short-term population-resource imbalances, real population increases, climatic conditions, use of a particular organizational strategy, or some combination of these factors. Long-term use of fields may have strained near-river environments, perhaps stimulating the expansion of existing farming and gathering practices into more marginal areas. Or a regime of higher rainfall could have provided opportunities to expand the existing repertoire of practices into areas such as Dead Valley. These practices, especially sand-dune agriculture, were more dependent on direct rainfall—true dry farming—than on groundwater or alluvial sources.

Based on existing climatic models (Dean 1988a; Dean and Robinson 1977; Graybill 1989), Van West (1994b) partitions the 1,000-year period from A.D. 572 to 1545 into 33 periods of various lengths. The 43-year period between A.D. 1045 and 1087 is characterized by a "consistent, moderate-to-wet" regime that would have provided favorable conditions for dune agriculture. The following period, A.D. 1088–1151, was less favorable due to a "variable, moderate-to-dry" regime accompanied by high spatial variability. Plog (1974:40) notes a strong emphasis on agriculture in the Hay Hollow Valley between A.D. 900 and 1100. Dean (1992) suggests that environmental conditions favorable to agriculture existed on the Colorado Plateau between A.D. 900 and 1130. The severity of a drought between A.D. 1130 and 1180 has been verified by more recent research in the Four Corners region (Dean and Van West 2002; Van West and Dean 2000).

This reconstruction corresponds with the occupation of the Dead Valley sites. Since most of the use of the dune fields occurred over a period of three or possibly four generations (75 to 100 years), this adaptive strategy was not a simple adjustment to high-frequency environmental processes occurring over short periods (i.e., year to year, or less than one generation) (Dean 1988a). Use of Dead Valley appears to have been a response to variability in long-range amplitude of the precipitation regime, namely, predictably higher levels of usable moisture more conducive to a diversity of agricultural practices. Knowledge of diverse practices likely existed within regional populations before this favorable change in precipitation patterns. The change was impermanent, however, and when precipitation returned to "average" conditions, agricultural practices that depended on direct rainfall became less viable.

Bradfield (1971:18) notes that sand-dune farming was not a reliable long-term strategy, as fields made on sand dunes are "a wasting asset." The productivity of dune fields decreases rapidly because there is no means of renewing the soil, in contrast to floodplain soils, which were frequently enriched with nutrients by floodwaters. Decreasing soil fertility, possibly combined with less favorable climatic conditions, may have hastened the abandonment of Dead Valley.

Strictly environmental explanations, however, are insufficient to account fully for the settlement history of the region. Before the Pueblo III period, agriculture was constrained by the natural distribution of water. By the PIII period, technological solutions that included canals, dams, terraces, and other features were used to manipulate the distribution of water (Duff 2002; Plog and Garrett 1972). Large villages with public architecture including plazas and great kivas likely functioned as ritual and economic centers. Agricultural intensification along with more integrated social systems and labor allocation may have rendered the distant dune fields obsolete. Longacre (1964:210) suggests that production in the late periods was sufficient to sustain the population with less widespread land use. In the long view, the remote dune fields were perceived as high-value resources in the Pueblo II period, much as suitable slopes for gridded gardens and terraces, and flatlands for irrigation, were critical resources in other periods (Pueblo III and IV).

CULTURAL DYNAMICS IN THE UPPER LITTLE COLORADO RIVER VALLEY

In the larger view, the scale of the occupation should also be noted. The ULCRV area was not a population magnet, for the occupation was not as substantial as elsewhere. The largest sites contain 300 rooms, small in comparison with contemporary developments nearby (e.g., Quemado, Zuni) where villages and towns of 500 to 1,500 rooms were present (Danson 1957; Kintigh 1996:Figure 9.1; LeBlanc 1989). As Kintigh (1996:136) observes, the area does not appear to have been "filled up." The area was lower in elevation and hotter than neighboring areas. It has been suggested that the potential for flooding could have rendered intensive agriculture precarious along the river floodplain (Lightfoot and Plog 1984). The growing season at Springerville averages

127 days, compared with 159 days at St. Johns (Plog et al. 1978:190) and 150 to 180 days at Homol'ovi (Lange 1996:242). Future research could investigate the possible marginality of the ULCRV relative to its ability to sustain agrarian economies and large populations. Additionally, situated on the southern prehispanic Pueblo frontier and the eastern edge of the Mogollon region, the occupants were socially isolated relative to goods and services present in the respective core areas.

Although not a focus of this effort, cultural affiliation is a necessary consideration when we wish to develop more robust models, as it speaks to historical trajectories and prior probabilities. Who were the agrarian people(s) of this area: prehispanic Pueblo, Mogollon? Or was the region occupied by a lesser-known tradition, by "people-in-between," derived from the Archaic period populations, perhaps supplemented by other groups (Martin et al. 1962)? Paul Martin (Martin and Plog 1973) associates the ULCRV with prehispanic Pueblo but notes that the area was transitional, tending toward prehispanic Pueblo north of St. Johns and toward Mogollon to the south, a view supported by recent research (Duff 2002). Wood (1979:68) suggests that the pueblo sites in the mountain zone near the southern end of the valley reflected a "conservative Mogollon cultural pattern." Was the late settlement history of the area characterized by "two traditions mixing with strange results?" (Longacre in Martin et al. 1962:160). Duff notes (2002) that the area was within the Chacoan interaction sphere and further describes the post-Chaco era villages as "relatively autonomous entities with little regional coherence," which could suggest that the agrarian folk were derived from multiple source areas.

DEAD VALLEY PUEBLO AGAIN

Dead Valley Pueblo provides some insight into the question of affiliation for one period in time. The site plan, with its room block, kiva, and trash area aligned on an east-west axis, is traditional prehispanic Pueblo (Reed 1956). The surface rooms had roof entryways and hatch covers. In contrast, the detached, semi-subterranean kiva, about 8 m^2 in area, had a D-shape plan and an enclosed ramp entry. A stone-lined fire box was centered in the floor next to two small holes that likely supported a vertical loom. The heavily plastered walls and floor included wall sockets that held perimeter-support beams, from which most of the tree-ring samples were obtained. Similar architectural plans with ramp entries and recessed roof-beam sockets have been reported from the Mogollon area beginning in earlier periods (Haury 1936). The kiva was intensely burned and had a thick layer of trash deposited immediately above the burned floor, which may suggest a closure procedure similar to those reported for later sites in the ULCRV (Duff 1999).

Three well-used, slab-lined roasting pits present are almost identical to features illustrated from near Quemado (Smith 1950); these features were also found at sites AZ Q:16:10 and AZ Q:16:11 (ASM) in Dead Valley. Interestingly, both prehispanic Pueblo (full groove) and Mogollon (three-quarters groove) ax types were present, and the projectile point assemblage also contained types reportedly associated with both cultures (Jeter 1980).

Petrographic analysis of pottery revealed a consistent background of volcanic and other nonplastic materials in the clays and/or tempers of sherds from the site, suggesting that most vessels were produced in the transition zone, which includes Dead Valley, the Little Colorado River between Lyman Lake and Nutrioso, and Coyote Creek (Rugge and Doyel 1980). The presence of pottery clay, many polishing stones, and sherd scrapers suggests that pottery was produced at Dead Valley Pueblo, an inference further supported by the recovery of 35,000 potsherds from this relatively small site. Some vessels of Puerco-style whiteware and a few gray paste utility wares were produced in area(s) to the north or northeast.

The ceramic ware groups present in Dead Valley provide a high-order contrast with sites in the Hay Hollow Valley to the west, with sites in the mountain zone south of Round Valley, and with Reserve phase sites to the south and east (Doyel 1980c:188; Longacre 1964; Martin and Rinaldo 1950; Martin et al. 1961; Wood 1979). Major differences between these areas and the Dead Valley sites include much higher frequencies of gray paste utility and Cibola decorated wares in the latter. These data are inconsistent with the observation that the occupation of the upper portion of the valley was more Mogollon than prehispanic Pueblo. The consistent co-occurrence of high frequencies of gray and brown paste wares indicates a high degree of interaction, the meaning of which should be pursued in future work.

Overall, the pottery wares from Dead Valley are closely related to what McGregor (1941:291) refers to as the "western Zuni" area, or to what Gladwin (1957:205) calls the "Cibola Branch." Gladwin's conception begins in the ninth century and integrates the cultural developments in the region between Zuni and the upper Gila River. Within these models, the Dead Valley sites may represent the southwesternmost extension of the Cibola (Puerco-Zuni) ceramic ware group in the PII period. The presence of Chaco-era and Chaco-inspired populations has recently been documented, with great-house communities located 25 km due north near St. Johns and 40–50 km east-northeast at Cox Ranch (Duff 2002; Kintigh 1996). It is possible that the use of Dead Valley may have resulted from range expansion from either of these directions rather than from the LCRV to the west as initially proposed (Doyel 1980c, 1984). The pottery types present at the Chaco-era Cox Ranch site, which include white and red wares, corrugated grayware, and high frequencies of brown paste plain and corrugated wares, are similar to the Dead Valley site assemblages. These new data may suggest new possibilities.

Further support for Cibola Branch prehispanic Pueblo affiliation is the focus on sand-dune dry farming within a pinyon-juniper environment, an adaptation more common to prehispanic Pueblo than to Mogollon (Martin and Plog 1973). From a more general perspective, Binford (1972:203–205) notes that major cultural alignments and processes identifiable in the archaeological record include those of tradition, adaptive area, and interaction sphere, each of which can vary independently. As noted elsewhere (Doyel 1976:35), although underused, this scheme will prove useful in deciphering the complex records of regional cultural expressions in the ULCRV and across the Southwest. In the present case, the occupants of Dead Valley were allied with the Cibola prehispanic Pueblo tradition and were engaged in agricultural practices common among prehispanic Pueblo groups, but they also maintained active relationships with Mogollon potters who produced the brownwares common on the sites.

SUMMARY AND DISCUSSION

A primary goal of this chapter was to review the settlement history of the Upper Little Colorado River Valley between Springerville and St. Johns. Available information suggests the presence of diverse agricultural practices and land-use patterns. Additional research is needed to investigate in greater detail the relationships among environmental, cultural, and behavioral parameters. Dead Valley provides a strong case for inferring land-use practices at sites with chronologically unmixed assemblages through use of an integrated approach to the study of settlement patterns, site structure, material culture, and biological remains. As Dead Valley shows dramatically, small sites can produce important data sensitive to land-use practices and settlement patterns, data that can contribute substantially to a broader understanding of settlement history. The study further suggests that more robust models of agricultural diversity and regional settlement patterns can be expected as other tributary drainages and upland areas are inspected. The sites may also assist in defining the nature of the "cultural understory" of the region and determining how the occupants of such sites related to the larger communities.

Dead Valley may also serve as a proxy for regional settlement trends insofar as the widespread distribution of Pueblo II sites and their short-term use may be attributable to rapid depletion of dune soils and/or a change in climate. Evaluation of the observations presented here will require more field studies of land-use practices and long-term change to complement the current emphasis on large late sites and regional modeling. Greater effort is needed to document the extent, scale, and diversity of agricultural practices in the region. Development of meaningful data sets for controlled comparisons will require measures of how various populations used the resources available to them in culturally prescribed ways. To this end, the ULCRV appears well suited for applications of analytical techniques such as Graphic Information Systems (GIS) to model and simulate resource availability and agricultural potential.

As a subsistence strategy, agriculture was an integral part of a larger adaptive system that involved population size, environment, social organization, and technology. In the Upper Little Colorado River Valley, agricultural practices were altered as the relationships among these variables shifted in response to environmental variation, sociocultural dynamics, and adaptive requirements. Challenges for the future include development of more refined methods to document past agricultural and land-use practices and the types of systems that were productive and sustainable within definable conditions. The degree to which land-use practices may have been influenced by divergent cultural histories is another interesting direction for future research.

Archaeology can provide empirical data to measure the relationships among these variables and can identify choices made as human populations adapted to the challenges of life in the Southwest. Identification of these behaviors and processes will produce more integrated and interesting models of what happened in the past.

ACKNOWLEDGMENTS

The Dead Valley Archaeological Project (DVAP) was funded by the Tucson Electric Power Company (TEP). Although conducted 25 years ago, the project remains unique in its contributions to understanding the settlement history of the region. Our colleagues, cited in the text, were instrumental in making the project productive and memorable. Discussions with Bruce Donaldson, Andrew Duff, David Gregory, Fred Plog, Alan Sullivan, and Carla Van West were helpful. Thanks to Jeff Dean and Andrew Duff for reading an earlier version. This chapter is dedicated to Don Graybill and to our friendship along the way. Don's comment on this study would likely have approximated, "Okay, Mr. Science! Now let's do something *really* interesting!"

8

Subsistence Stress and Food Storage at Kiet Siel, Northeastern Arizona

Jeffrey S. Dean

HUMAN BEHAVIORAL ADAPTATION TO SUBSISTENCE STRESS IS A topic of perennial interest in anthropology. Responses to such stresses are major stimuli of sociocultural change and, consequently, important factors in the evolution of cultural systems. Archaeologists have focused on this issue for two principal reasons. First, archaeology controls a time depth sufficient to isolate and study genuine cases of long-term culture change. Second, the data produced by archaeological research are capable of elucidating many independent and dependent variables involved in adaptive culture change. This chapter examines a prehistoric instance of behavioral adaptation to environmentally induced stress on a relatively simple agricultural subsistence system in the Southwest. The analysis illuminates aspects of the general adaptive process and archaeology's role in understanding the process. The study is focused on food storage, a component of sociocultural systems that is particularly sensitive to perturbations in food supply, at Kiet Siel, a late-thirteenth-century "cliff dwelling" at Navajo National Monument in northeastern Arizona.

THE PROBLEM

Food storage has long been recognized as an important technique for buffering the effects of variability in food procurement (Burns 1983:152–166; Clarke 1978:135; Cordell and Plog 1979; Dean 1988a; Jorde 1977; Plog et al. 1988; Powell 1983:116). When crop yields vary little from year to year, predictability is high and food storage capacity changes little through time. In contrast, when year-to-year variability in food production is high, the storage of food surpluses produced in favorable years is a viable means of compensating for the reduced yields of leaner years. In theory, therefore, storage should vary directly with the degree of variability in the productive system. That is, storage capacity should increase when year-to-year variability in production is high and decrease when yields are relatively stable from one year to the next. Given the limited "shelf life" and vulnerability to animal pests of most native foods, storage is most useful as a response to rapid fluctuations in production (Dean 1988b), that is, variability that has a periodicity of around five years or less. In contrast, storage responses alone may not be sufficient to offset sudden long-term reductions in crop yields.

Assuming that storage behavior responds primarily to annual variability in food production, investigation of this aspect of adaptive behavior should be focused on the independent variables that control the high-frequency aspects of crop production. Such variables are relatively easy to identify in the records of high-frequency paleoenvironmental variability on the Colorado Plateau (Dean 1988a, 1996a; Dean et al. 1985; Dean and Funkhouser 1995; Dean and Van West 2002; Euler et al. 1979; Van West and Dean 2000). Reconstructed variability in factors such as annual precipitation and the temporal structure of precipitation can be used to "predict" storage responses to these variables, which should be evident in the archaeological record (Burns 1983; Plog et al. 1988; Van West 1994a).

Any attempt to assess predicted storage responses to environmental perturbations must satisfy two conditions. First, it must be possible to characterize with reasonable accuracy variability in the independent and dependent factors involved, in this case crop yields and storage behavior. Second, a plausible relationship between the two variables must be established for the specific test case. Only rarely is it possible to reconstruct prehistoric crop yields directly (Burns 1983; Dean et al. 2000; Van West 1994a). Usually surrogate measures of production, such as dendroclimatic reconstructions, must be used. Storage capacity is a valid measure of the dependent variable, but it can be determined only when excavation or preservation is complete enough to provide representative structure area or volume figures. Given satisfactory control of these variables, they must be related to each other in such a way that the reaction of one to changes in the other can be perceived and evaluated. One way to achieve this goal is to compare two cases that differ in terms of the high-frequency characteristics of crop production. The case with higher variability also should possess greater storage capacity. A weakness of this approach lies in the difficulty of controlling intervening variables such as between-case differences in agricultural techniques, social organization, and exchange relationships. A more satisfactory test would be provided by a situation in which a single society experienced significantly increased or decreased variation in production. Such changes should stimulate, respectively, increases or decreases in storage capacity.

Conceptually, this hypothesis can be evaluated against archaeological data at a wide range of geographical scales of analysis. Unfortunately, regional efforts founder on the paucity of analogous quantitative data on prehistoric storage behavior across large areas (Plog et al. 1988). Because of this problem, computer modeling has been used to indirectly examine the role of storage in human adaptation to environment-driven fluctuations in food production.

Burns (1983) used dendroclimatically reconstructed annual crop yields to isolate potential effects of different durations of food storage on prehistoric societies in the Mesa Verde area of southwestern Colorado. Closer to Kiet Siel, the Artificial Anasazi project incorporated ethnographically based household storage capacity and duration limits into an agent-based model of prehistoric subsistence-settlement behavior in Long House Valley near the mouth of Tsegi Canyon (Axtell et al. 2002; Dean et al. 2000; Gumerman et al. 2003). Although these large-scale studies illuminate the potential role of storage in human responses to changing agricultural production, neither incorporated archaeological data on actual storage behavior.

Although we lack suitable data for regional evaluations of the expected relationships between variability in food production and storage capacity, a test at the locality level is possible using the thirteenth-century site of Kiet Siel (Dean 1969:85–150) in northeastern Arizona. A fortunate combination of circumstances at Kiet Siel satisfies the conditions for an adequate examination of storage responses to subsistence stress. High-quality, low- and high-frequency paleoenvironmental reconstructions (Cooley 1962; Dean 1996a; Dean et al. 1985; Dean and Funkhouser 1995; Dean and Robinson 1977; Euler et al. 1979; Hack 1942, 1945; Karlstrom 1983; Karlstrom and Karlstrom 1986; Karlstrom 1988) document the replacement of a stable supply of agricultural water (alluvial groundwater) by a much more variable and less dependable source (precipitation), which substantially increased annual variability in crop yields. The degree of preservation of this cliff dwelling allows the accurate measurement of room area and volume. Detailed architectural data coupled with a suite of 140 tree-ring dates provide a clear picture of the development of the site through time, which in turn permits both the documentation of changes in storage behavior at the site and control of population variation through time.

ENVIRONMENTAL CHANGE AND FOOD PRODUCTION AT KIET SIEL

Abundant plant remains preserved in the dry rockshelter show that subsistence at Kiet Siel was based on agriculture; maize was the primary crop, supplemented with beans and cucurbits. Hunting and the gathering of wild plant foods were far less important than farming, but quantitative estimates of the contribution of nonagricultural food to the diet of the villagers cannot be made. In any case, the preserved vegetal remains indicate that the granaries of Kiet Siel were devoted mainly to agricultural products rather than to gathered foods. Although the granaries probably housed the full variety of agricultural produce, few of them have permanent subfacilities, and access to all such chambers is restricted to a single doorway or roof hatch. These facts suggest that waste space and space devoted to access were minimal and therefore that granary capacity is probably a good measure of the food storage capability of Kiet Siel.

Because most of the input into the storage system at Kiet Siel took the form of agricultural produce, variations in expected agricultural productivity should be the major cause of changes in storage capacity. Therefore, changes in annual crop yield that could have increased variability in storage input must be identified. Crop yield is a function of innumerable variables, including agricultural technology, crops grown, field location, soils, hydrology, pest infestation, disease, and weather, among many others (Bradfield 1971; Forde 1931; Hack 1942; Muenchrath and Salvador 1995). It is manifestly impossible to estimate, much less measure, the single and combined effects of all these factors in a prehistoric context; there are simply too many unknowns. It is possible, however, to identify variables that under specified conditions were limiting to subsistence agriculture in Tsegi Canyon during the time Kiet Siel was occupied. Some factors—such as soil types, crops grown, field location, and farming technology—can be treated as constants for the 50-year period under consideration. There is no tangible evidence for changes in other possible factors—such as disease, animal damage, and human predation—and these too can be treated as constants. Two remaining variables that limited crop production changed significantly during the late thirteenth century. These variables are climate and hydrologic conditions in the Laguna Creek drainage.

U.S. Weather Service records (Desert Research Institute Western Regional Climate Center [http://www.wrcc.dri.edu]) for the Betatakin meteorological station on the rim of Tsegi Canyon indicate that climate alone could not have supported prehistoric crop production at Kiet Siel. Although the average yearly frost-free period of 155 days is ample for maize agriculture, the mean annual rainfall of 300 mm is not sufficient for successful corn production in the absence of natural or technological means of augmenting the water supply (Hack 1942:20). Furthermore, annual precipitation is too variable, ranging from 91 mm (2002) to 514 mm (1982), to have produced crops consistently from one year to the next. Because long-term subsistence farming could not have been sustained solely by rainfall, some other factor or factors must have intervened to compensate for the inadequate precipitation of the area.

The deficiencies in precipitation were offset by hydrologic conditions in the Laguna Creek drainage, which, at the beginning of the Tsegi phase, were the principal limiting factor in the Kiet Siel agricultural system. Around 1250, the alluvial sediments extended virtually unbroken across the floor of the canyon (Hack 1945). Under such conditions, the alluvial water table would have been high enough to intersect the alluvial surface, perhaps support a flowing stream, and contribute substantially to crop production. The Tsegi phase people could have relied solely on groundwater for farming, or they could have supplemented it with water diverted from surface runoff. The water storage capacity of the alluvial system, which was supplied by precipitation and by seepage from the massive sandstones in the canyon walls, compensated for the general deficiency of and high year-to-year variability in rainfall to produce a stable agricultural situation characterized by only minor fluctuations in

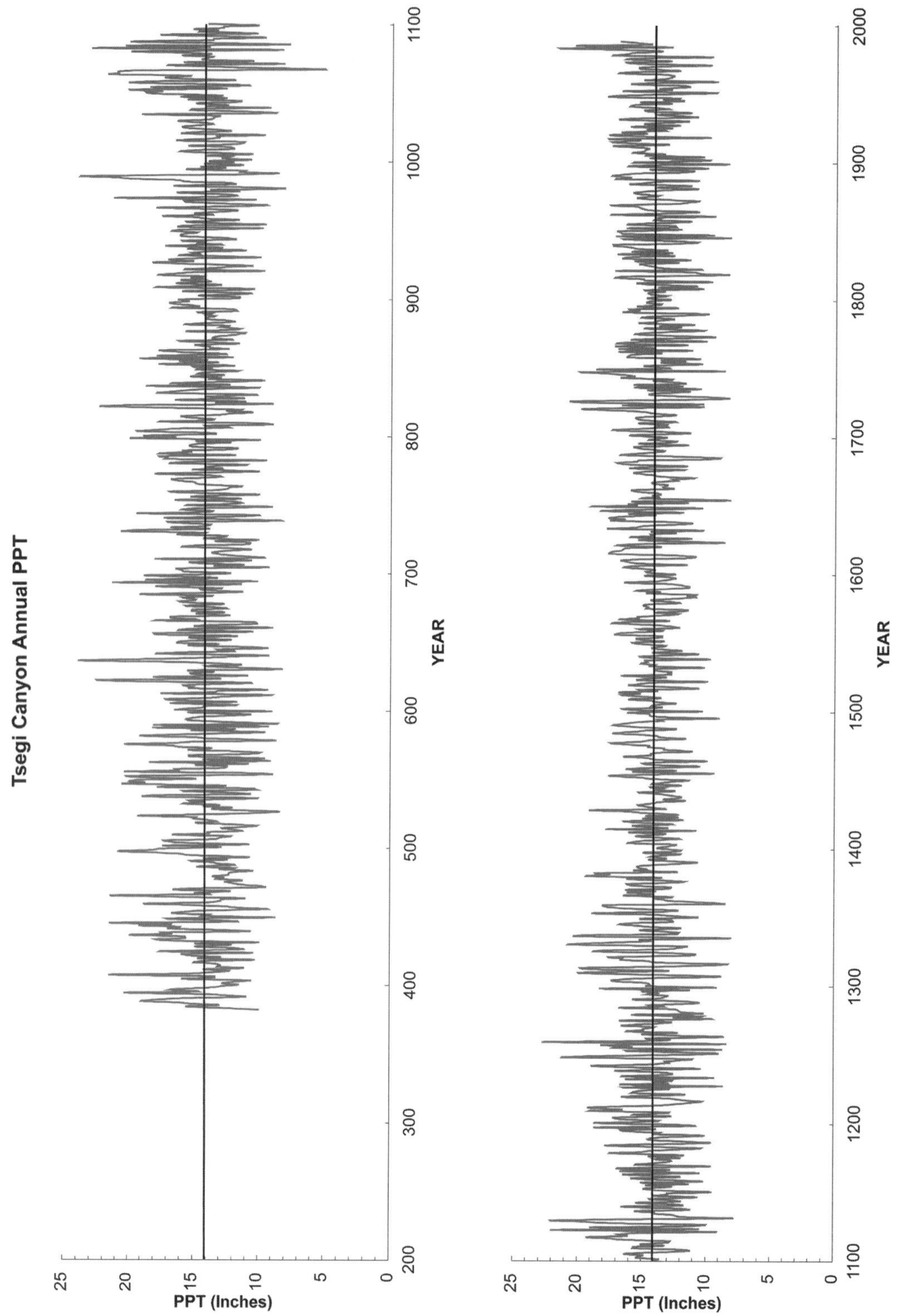

FIGURE 8.1 Dendroclimatic reconstruction of annual tree-year (previous August through current July) precipitation in inches for Tsegi Canyon, northeastern Arizona, A.D. 382–1988. The reconstructed mean of 14 in (356 mm) is higher than the Betatakin weather station mean of 12 in (305 mm) because the former incorporates precipitation data from a number of weather stations, including Betatakin, in the U.S. Weather Service's Northeastern Arizona Climate Division.

annual crop yields. Thus, the Kiet Siel storage systems at the beginning of the Tsegi phase would have experienced little year-to-year variability in input.

In the last half of the thirteenth century, a drastic change in the heretofore stable relationship between arable land and water supply destabilized the prevailing agricultural system by increasing interannual variability in crop production. This crisis was caused by the cutting of a fully integrated system of arroyos into the floodplain alluvium (Cooley 1962; Dean et al. 1985; Euler et al. 1979; Hack 1942, 1945; Karlstrom 1983; Karlstrom and Karlstrom 1986; Karlstrom 1988) and the concurrent subsidence of alluvial groundwater levels. Arroyo cutting would have adversely affected agriculture through the destruction of fields and crops, the lowering of the alluvial water table, and the channeling of surface runoff away from the fields. Although the first factor undoubtedly adversely affected crop production, the second had far more serious long-term consequences for subsistence. As the arroyos cut deeper into the sediments, the water table that had supported farming and the surface water that could have been used for irrigation dropped 15 or more meters below the level of the arable land. Lacking advanced techniques of water transport, Tsegi phase farmers were unable to raise enough water from the arroyo bottom to the level of the terraces where the fields were located; consequently, as entrenchment progressed, it became increasingly difficult for them to get water to their crops. Furthermore, large-scale floodwater farming, such as that practiced by the Hopis (Hack 1942), would have been impossible in the narrow canyons, where sufficient land suitable for floodwater farming simply was not available. Finally, instead of spreading evenly across the land surface, precipitation runoff was concentrated by incised channels and carried rapidly out of the drainage system. These fundamental changes in Tsegi Canyon agricultural systems caused increasingly severe subsistence stress.

The hydrologic changes in the Laguna Creek drainage stemming from channel entrenchment constituted a basic alteration of the way Tsegi phase farming was articulated with the natural environment in which it was practiced. The environmental variable that had been primarily limiting to crop production, the alluvial water table, was replaced as the principal limiting factor by another variable, precipitation. Basically, the Kiet Siel farmers were faced with decreasing access to groundwater and a consequent necessity to rely to a greater extent on rainfall, which exhibited far more variability within and between years than did the supply of groundwater. Furthermore, the increasing dependence on precipitation coincided with an extended period of below-normal rainfall (Figure 8.1). The "Great Drought" of 1276 to 1299 could only have intensified the stress on the Kiet Siel subsistence system. This dire situation was exacerbated by a breakdown in the long-term climate pattern that, beginning around 1250, disrupted the prevailing seasonal distribution of precipitation (Dean 1996a; Dean and Funkhouser 1995).

The shift in emphasis from groundwater to precipitation would have impinged on two important components of the agricultural process: seed germination and crop maturation. Prior to the arroyo cutting, germination would have been ensured directly by the high groundwater levels near Kiet Siel and/or by ditch irrigation from the stream segments or ponds that developed where the water table intersected the ground surface. After the lowering of the water table, germination would have depended more on residual soil moisture resulting from previous winter precipitation, a far less reliable supply of water. Similarly, the depressed water table would have made crop maturation more dependent on summer rainfall and runoff, both of which vary considerably from year to year and from place to place within the year. Even in years of adequate germination, crop growth and maturation would have been far more variable than before the loss of groundwater. Although the increased variability in water supply and crop yield could have been partially offset by irrigation from the stream, at least until it cut too far below the terrace surfaces, farming had become a far less dependable means of sustenance than before. The conclusion is inescapable that variability in food production increased significantly during the occupation of Kiet Siel, thereby reducing the ability of the farmers to predict crop yields.

The Artificial Anasazi simulations for nearby Long House Valley clearly reflect the late-thirteenth-century environmental transformation on a geographical scale much larger than a single site. During periods of subsistence stability caused by alluvial aggradation and groundwater accretion, settlements were distributed widely across the landscape, whereas during unstable intervals of degradation and groundwater decretion, settlements were concentrated in favorable localities (Dean et al. 2000:Figure 4; Gumerman et al. 2003: Figures 1–3). The dominance of high-frequency precipitation during unstable periods is reflected in high variability in the simulated locations of fields, and the dominance of low-frequency hydrology during the stable periods is evident in low variability in the siting of fields (Dean et al. 2000:Figure 5).

FOOD STORAGE RESPONSES TO SUBSISTENCE STRESS AT KIET SIEL

Kiet Siel occupies a large rockshelter in the Kiet Siel Canyon branch of the Tsegi Canyon system. The Tsegi is drained by Laguna Creek, a perennial, spring-fed stream that empties eastward into Chinle Wash, a tributary of the San Juan River. The floor of the canyon contains deep deposits of Holocene alluvium that today are cut to bedrock by a fully integrated system of arroyos that began forming in the 1880s.

Kiet Siel, along with numerous contemporaneous sites in the area, is assigned to the Tsegi phase, which dates from A.D. 1250 to 1300 (Ambler 1985:49–57; Colton 1939:58; Dean 1969:196, 1996b, 2002; Lindsay 1969). This phase represents a substantial immigration into Tsegi Canyon that began about 1250 and culminated in the 1280s, when around 800 people resided there in more than 20 villages (Christenson 2003; Dean 1969). Tsegi phase sites in the canyon range from only a few rooms to cliff dwellings such as Kiet Siel and Betatakin,

TABLE 8.1. Attributes of Functional Room Types at Kiet Siel.

	FEATURES				
ROOM TYPE	FLOOR	FLOOR FEATURES	ROOF	WALLS	ENTRANCES
Living rooms (LR)	Leveled, plastered	Fire pit, entry box, mealing bins	Smoke-blackened	Mediocre masonry, jacal common, interior plaster, interior smoke-blackening	Doorways large, low-silled, jambs ungrooved; roof hatches common
Courtyards (C)	Leveled, plastered	Fire pit, mortars, mealing bins	Unroofed	Walls formed by surrounding rooms	Low-silled doorways
Granaries (G)	Unmodified bedrock	None	Not smoke-blackened	Superior masonry, exterior chinking, exterior plaster, jacal rare, no interior smoke-blackening	Doorways small, high-silled, jambs grooved for slab door, loops for door-closure bolt, clay coping, exterior platform with step
Storerooms (S)	Unmodified bedrock	None	Not smoke-blackened	No consistent wall attributes, no interior smoke-blackening	No consistent entrance attributes
Mealing rooms (M)	Leveled, plastered	Mealing bins, metates	Usually unroofed	Usually open on one side	No consistent entrance attributes

with more than 100 rooms. The Tsegi phase "boom" was brief, for by 1300 the canyon and its environs were abandoned.

Kiet Siel consists of 155 rooms and courtyards and four kivas ranged along three "streets" that parallel the back wall of the rockshelter (Figure 8.2). Five functionally different types of secular rooms are recognized (Dean 1969:26–34, 1970) on the basis of recurring associations of several features (Table 8.1). Living room (LR) attributes indicate that these chambers were used for domestic activities connected with the maintenance of a small residence unit, probably some sort of family group. Courtyards (C) are unroofed areas that served as loci of domestic activities similar to some of those that were carried on in the living rooms. Granaries (G) possess features designed to keep insects, rodents, and the weather out, and these rooms undoubtedly were used for storing perishable foodstuffs. Storerooms (S) lack the features typical of living rooms and granaries and probably were used for storing less perishable, perhaps nonfood, items. Mealing rooms (M) contain batteries of graded metates that were used for grinding maize and other seeds into meal. Secular rooms are grouped into larger structural units called room clusters (RC), each of which consists of at least one living room, one to several storage chambers (G + S),

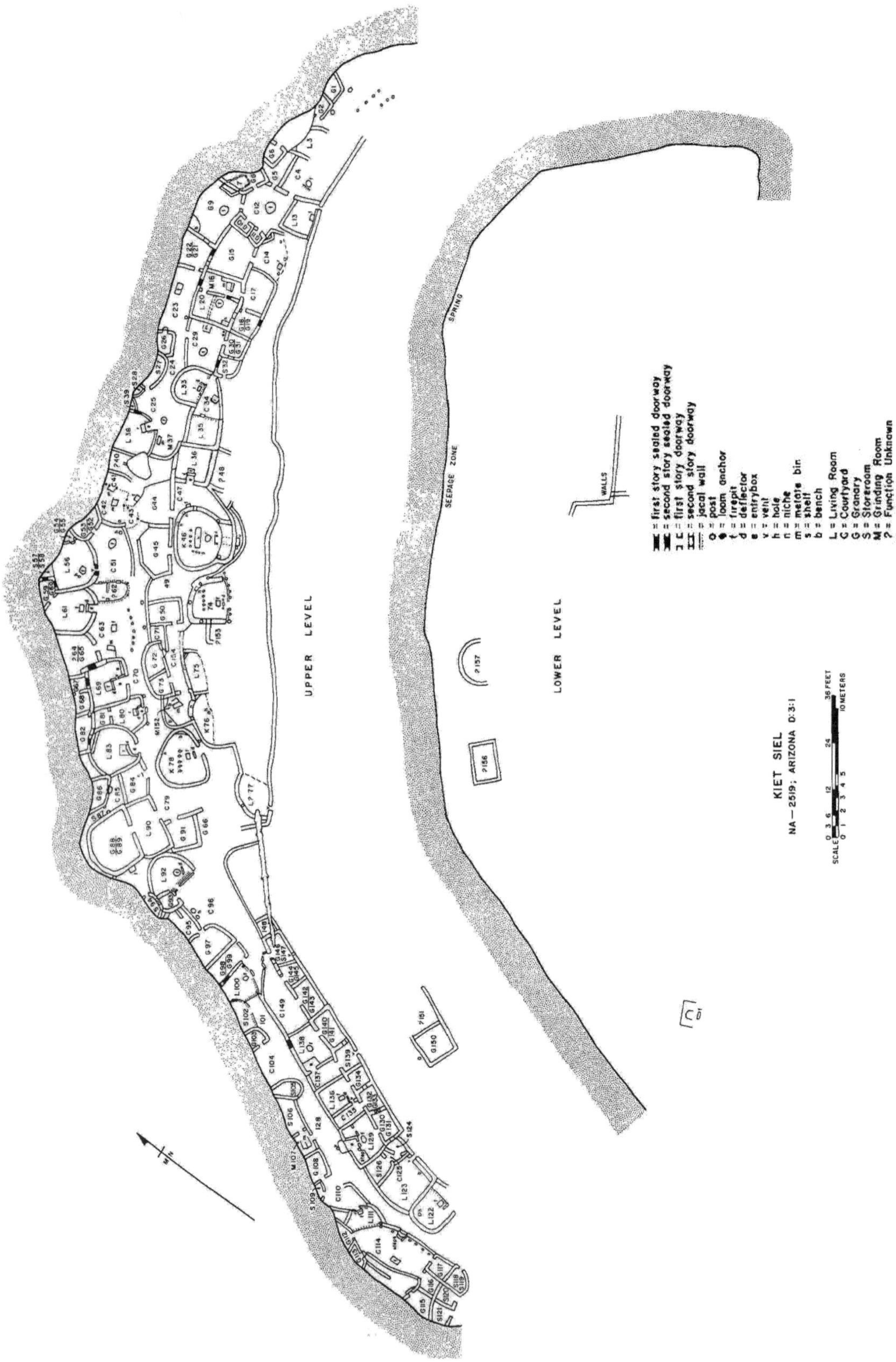

FIGURE 8.2 Floor plan of Kiet Siel, Navajo National Monument, Arizona.

occasionally a mealing room, and in all but a few cases a courtyard. The room cluster is the basic architectural unit of the site structure and was the locus of a primary social unit of food procurement, storage, and consumption. This unit of residence and economic cooperation likely was a household comprising a married couple, their unmarried children, and sometimes one or two additional adults (Dean 1969:36–39, 1970:163). Nonsecular architecture includes kivas, D-shaped kiva annexes, and a few granaries associated with some but not all of the kivas.

If augmenting production and storage is, indeed, a useful technique for offsetting the effects of increased high-frequency variability in production, the environmental changes of the late thirteenth century should have triggered adaptive responses in the storage capacity of Kiet Siel. Archaeological data from the site should reveal whether these expectations are fulfilled or contradicted. The data are organized within the physical framework of the village architectural structure and within the temporal framework provided by tree-ring dates from the site. An intensive dendrochronological-architectural study of Kiet Siel (Dean 1969:103–143) isolated five constructionally significant periods in the history of the village: 1245–1271, 1272–1275, 1276–1282, 1283–1286, 1287–abandonment. The first four of these intervals provide the chronological control for the analysis of the Kiet Siel data.

Given the adaptive situation in Tsegi Canyon in the late thirteenth century, two complementary hypotheses relating changes in storage capacity to increasing high-frequency subsistence stress can be tested against the archaeological data from Kiet Siel. Hypothesis 1 states that stress on a subsistence agriculture system promotes the development of intra- or intercommunity food redistribution to offset occasional crop failures suffered by subgroups within the redistributive system. Hypothesis 2 is that stress on a subsistence agricultural system produces an increase in stored food reserves even in the absence of large-scale redistribution of resources.

If the response predicted by Hypothesis 1 took place at Kiet Siel, we should find the innovation during the occupation of the village of storage facilities that served groups of people larger than the individual household. Such structures do exist in the form of the granaries associated with some kivas. Kiva granaries, however, seem to have been present from the founding of the village and therefore cannot be attributed to stress conditions that arose during the occupation of the site. If redistribution had become more important during the occupation of Kiet Siel, there should have been a significant increase through time in the capacity of kiva granaries. Although expansion of this type did not occur, household storage increased markedly; therefore, it is unlikely that there were any significant changes in the redistribution of food within the village. Apart from trails linking some of the sites, there is no substantive evidence for cooperation, economic or otherwise, among the Tsegi phase villages in Tsegi Canyon; therefore, intercommunity redistribution as a response to the existing stress situation can be discounted.

Table 8.2. Frequency, Area, and Volume Data for Living Rooms, Activity Spaces, and Granaries at Kiet Siel.

	PERIOD			
	1245–1271	1272–1275	1276–1282	1283–1286
Frequencies				
Number RC	12	16	17	20
Number LR	12	17	18	21
Number G	21	38	43	54
Floor area data (m2)				
Total LR area	108.5	130.3	119.8	136.9
Total AS area	141.0	213.8	264.9	343.7
Total G area	66.8	134.2	159.7	211.3
Mean LR area	9.0	7.7	6.6	6.5
Mean AS area	20.1	17.8	17.7	17.2
Mean G area	3.2	3.5	3.7	3.9
Room volume data (m3)				
Total LR volume	a	160.8	174.2	171.7
Total G volume	96.1	123.2	174.0	237.9
Mean LR volume	14.1	12.4	10.9	9.5
Mean G volume	5.1	4.4	4.8	5.3

Notes:
LR = Living room AS = Activity space (LR+C+M)
C = Courtyard RC = Room cluster (LR+C+M+G+S)
M = Mealing room
G = Granary a = Data insufficient for accurate calculation.
S = Storeroom

Hypothesis 2 predicts that increasing stress on the subsistence system will trigger corresponding increases in storage capacity. The data for Kiet Siel as a whole (Tables 8.2 and 8.3) strongly support the hypothesis. Although both granaries and living rooms increased in number through time, they did not increase at the same rate, for there are marked differences in the ratios between the living room and granary frequencies of the first period (6:10) and that of the last period (4:10). Furthermore, though the amount of floor area devoted to each class of structure increased with the growth of the village, the ratios of total living room area to total granary area and of total activity space area (LR + C + M) to total granary area drop markedly from early (16:10 and 21:10) to late (6:10 and 16:10). Similar trends are evident in the figures for living room and granary volumes and the ratios between them. In addition, the floor area and volume of the average living room declined through time while the floor area and volume of the average granary increased slightly (Table 8.2) with con-

TABLE 8.3. Frequency, Area, and Volume Ratios for Kiet Siel.

	PERIOD			
	1245–1271	1272–1275	1276–1282	1283–1286
F_{LR}:F_G	6:10	4:10	4:10	4:10
∑ area LR:∑ area G	16:10	10:10	8:10	6:10
∑ area AS:∑ area G	21:10	16:10	17:10	16:10
Mean LR area:mean G area	28:10	22:10	18:10	17:10
Mean AS area:mean G area	63:10	51:10	48:10	44:10
∑ volume LR:∑ volume G	a	13:10	10:10	7:10
Mean LR volume:mean G volume	a	28:10	22:10	18:10

Notes:
LR = Living room — AS = Activity space (LR+C+M)
C = Courtyard — RC = Room cluster (LR+C+M+G+S)
M = Mealing room
G = Granary — a = Data insufficient for accurate calculation.
S = Storeroom

comitant decreases in the ratios between mean living room area and volume and the corresponding values for granaries (Table 8.3). These data illustrate the increasing importance through time of storage capacity at the expense of living and working space, and they offer strong support for the second hypothesis.

Because the household and not the community as a whole was the primary unit of economic production and consumption, it is appropriate to look at internal changes in the room clusters, the architectural equivalents of individual households. Metrical, ratio, and percentage data on the "average" room cluster were derived by dividing the total area and volume devoted to each class of structure by the number of room clusters for which data on each structural class were available. Intraperiod, between-cluster variation in the availability of data and the omission of certain types of rooms (mealing rooms, storerooms, and unknowns) account for the fact that the total room cluster area and volume figures exceed the sums of the granary, courtyard, and living room values. The tables lack volume figures for courtyards (C) and activity spaces (LR + C + M) because these totally or partially unroofed features have no definable volume. The volume of a room cluster equals the sum of its roofed components, that is, living rooms, storerooms, and granaries.

Tables 8.4 to 8.6 confirm the hypothesis. The room cluster was not static through time but changed both in overall size and in the dimensional relationships of its components. The total floor area of the room cluster increased by 15 percent, a gain due mainly to an 89 percent expansion of granary space. The fact that average granary size (Table 8.2) was fairly stable through all periods indicates that room-cluster granary expansion was generally achieved by increasing the number of granaries per cluster (Figure 8.3), an indication that is

Table 8.4. Metric Data on the "Average" Room Cluster at Kiet Siel.

	PERIOD			
	1245–1271	1272–1275	1276–1282	1283–1286
Frequencies				
Number LR	1.0	1.0	1.0	1.0
Number C	1.0	1.0	1.2	1.2
Number G	1.8	2.4	2.7	2.7
Number S	0.6	0.7	0.8	0.8
Total number units (LR+C+M+G+S)	4.5	5.2	5.7	6.0
Floor area data (m^2)				
LR area	9.0	8.1	7.5	6.8
C area	10.1	9.2	9.6	9.6
G area	5.6	8.4	10.0	10.6
AS area (LR+C+M)	20.1	17.8	17.7	17.2
RC area (LR+C+M+G+S)	25.3	27.6	29.3	29.2
Room volume data (m^3)				
LR volume	14.1	13.4	12.4	10.1
G volume	8.7	10.3	12.4	14.0
RC volume (LR+G+S)	29.8	26.7	27.8	29.3

Notes:
LR = Living room
C = Courtyard
M = Mealing room
G = Granary
S = Storeroom
AS = Activity space (LR+C+M)
RC = Room cluster (LR+C+M+G+S)

Table 8.5. Area and Volume Ratios for the "Average" Room Cluster at Kiet Siel.

	PERIOD			
	1245–1271	1272–1275	1276–1282	1283–1286
Area LR:area G	16:10	10:10	8:10	6:10
Area AS:area G	36:10	21:10	18:10	16:10
Volume LR:volume G	16:10	13:10	10:10	7:10

Notes:
LR = Living room
G = Granary
AS = Activity space (LR+C+M)

TABLE 8.6. Percentage of Floor Area and Volume of "Average" Room Cluster Devoted to Granaries at Kiet Siel.

	PERIOD			
	1245–1271	1272–1275	1276–1282	1283–1286
% Area of rooms	35.7	46.2	51.6	54.1
% Area of room clusters	22.0	30.4	34.1	36.2
% Volume of rooms	29.3	38.5	44.8	47.8

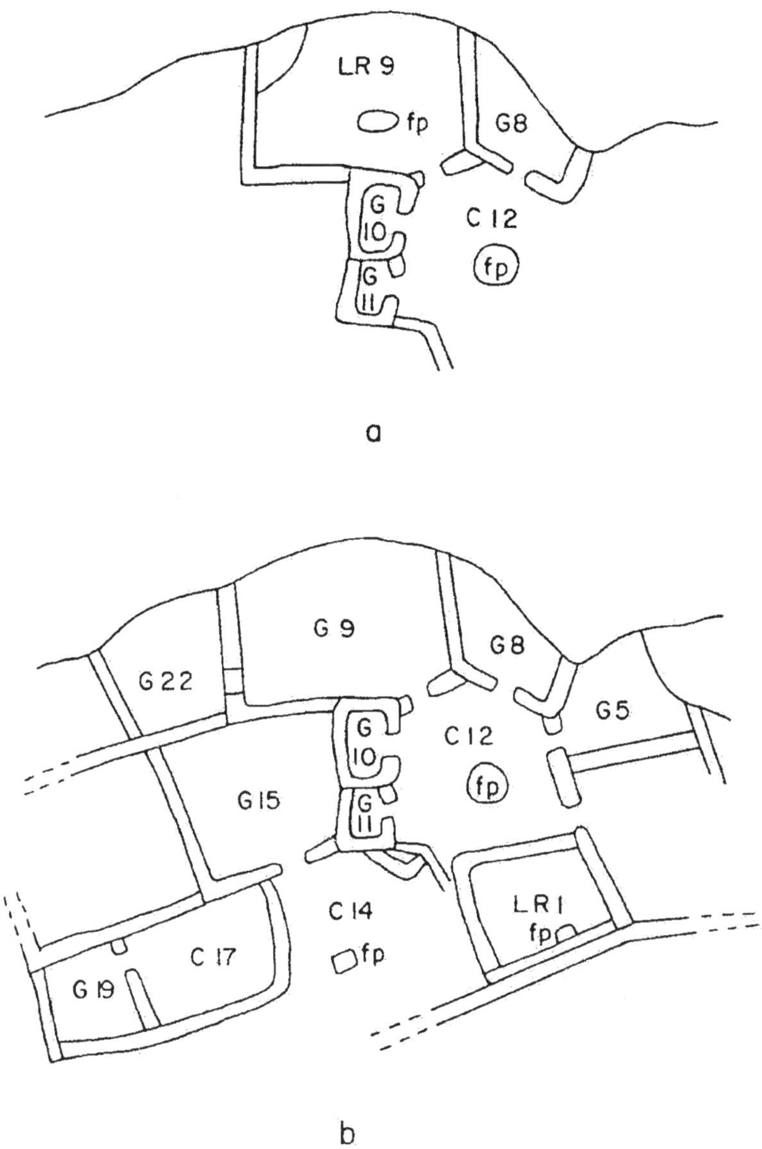

FIGURE 8.3 Room cluster 2 at Kiet Siel: (a) in A.D. 1271, when it consisted of a courtyard (C), a large living room (LR), and three fairly small granaries (G); and (b) in 1286, when it consisted of three courtyards, a small subterranean dwelling, and eight granaries, two of which (G9 and G15) were vacated living rooms converted to storage functions.

TABLE 8.7. Per Capita Area and Volume Data for Kiet Siel.

	PERIOD			
	1245–1271	1272–1275	1276–1282	1283–1286
Floor area/person (m^2)				
LR area	1.8	1.6	1.5	1.4
C area	2.0	1.8	1.9	1.9
G area	1.1	1.7	2.0	2.1
AS area	4.0	3.6	3.5	3.4
RC area	5.1	5.5	5.9	5.8
Room volume/person (m^3)				
LR volume	2.8	2.7	2.5	2.0
G volume	1.7	2.1	2.5	2.8
RC volume	6.0	5.3	5.6	5.9

Notes:
LR = Living room
C = Courtyard
M = Mealing room
G = Granary
AS = Activity space (LR+C+M)
RC = Room cluster (LR+C+M+G+S)

borne out by the changes in granary frequency shown in Table 8.4. Important changes are found in the relationships between living room volume, which decreased by 28 percent, and granary volume, which increased by 61 percent. The area and volume ratios (Table 8.5) and percentage of average room-cluster space devoted to granaries (Table 8.6) show that, from early to late, room-cluster space devoted to domestic activities was more than halved relative to that set aside for the storage of food. It appears, then, that the people of Kiet Siel, in order to increase the storage capacity of each household while remaining within the fairly constant size limits of the room cluster, were forced to reduce the size of the dwellings to accommodate the enlarged storage volume. Space limitations of the floor of the rockshelter were undoubtedly important in keeping the total size of the room cluster fairly constant even in the face of a need for more household food storage capacity.

The Kiet Siel data can also be considered in terms of space per standard unit of population. Since the "average" room cluster includes only one living room, the data presented in Tables 8.4 to 8.6 are standardized in terms of living-room units as well as room-cluster units. If each living room can be considered to represent a fairly constant number of people, these figures can be taken as representative of the architectural space devoted to a standard unit of population. Using the figure of five persons as a gross approximation of the mean number of individuals per living room (Dean 1969:76, 143), a rough estimate of the average amount of room space per person can be made. These figures (Table 8.7) show that per capita living-room area and activity space decreased by 22 percent and 15 percent, respectively, while per capita granary area increased by 91 percent during the occupation of the village. Similarly, per capita living-room volume decreased by 28 percent, and per capita granary volume increased by 65 percent during the same interval. Despite these important changes in the amount of room space devoted to a single individual, per capita room-cluster area and volume remained relatively stable throughout the history of Kiet Siel.

Additional evidence reinforces the quantitative support for the second hypothesis. Toward the end of the occupation of Kiet Siel, several living rooms that had been abandoned were converted into granaries by the inhabitants of adjacent room clusters (Figure 8.3). In addition, the granaries of several abandoned room clusters were appropriated and used by other households. Post-1275 Kiet Siel housed a population in flux. Households came and went, with those remaining in the village continually increasing food storage capacity. Furthermore, the construction in 1274 of a masonry and log sump represented an effort to conserve another critical resource, domestic water, which also would have been depleted by the lowering of the water table in the alluvium.

Other possible adaptations to subsistence stress could have occurred at Kiet Siel in addition to the expansion of food storage capacity. The evidence, however, offers little support for any of these possibilities. Decreased production can be offset by decreasing consumption of the resources involved. However, a permanent reduction in consumption is not a viable response to systemic changes in food procurement affecting people such as the inhabitants of Kiet Siel, whose nutritional intake probably was marginal much of the time. Belt tightening may have been necessary in occasional lean years, but it would not have been a satisfactory long-term solution to subsistence stress. If a situation arose in which reduced consumption became the only possible response to decreasing crop production, it would probably have caused population reduction such that total village consumption decreased while an adequate level of per capita nutritional intake was maintained. In the extreme case, depopulation would take the form of starvation or emigration sufficient to reduce the population to a supportable level. Less extreme cases would involve decreased food intake resulting in malnutrition and a corresponding increase in the mortality rate, especially among children and the aged.

There is no evidence for mass starvation or increased nutritional deficiency at Kiet Siel, but there are indications of the demise or emigration of several households late in the history of the village. This population loss was at least partially offset by the immigration of other households, so that up to 1286 there was little net decline in population (Dean 1969:Figure 52). Malnutrition and starvation may have become significant in the terminal years of Kiet Siel as the subsistence situation continued to deteriorate, but it is clear that this result, if it occurred at all, took place only after considerable adjustments had been made in storage capacity.

Another possible response to changes in the adaptive situation is to remove the affected society from the context of the stress—in other words, to emigrate. The high degree of household mobility evident at Kiet Siel and other Tsegi phase sites may be a small-scale manifestation of this type of response whereby individual families sought to better their lot by moving to more favorable localities within the Tsegi area. Ultimately, as stream entrenchment continued to its climax, internal adjustments in the procurement, storage, and consumption processes of the subsistence system proved incapable of maintaining the society, and the people of Tsegi Canyon abandoned the area in favor of a locale capable of supporting their way of life, probably the Hopi mesas–middle Little Colorado area to the south (Dean 2002). Emigration was the final response to subsistence stress in the Tsegi area and was resorted to only after other, less drastic adjustments, such as increasing storage capacity, had failed.

Based on analogy with past grain production and storage experience in the United States, one might attribute an increase in storage capacity to the production of an agricultural surplus. However, surplus production of this type, which exceeds the requirements of the consuming group, usually occurs only in the context of a market economy artificially sustained by government intervention. Prolonged surplus production within the context of a subsistence economy is unlikely, for long-term surpluses produce little return on the time and labor invested in them and contribute little to the survival of the group in question. Although occasional fortuitous surpluses can be "invested" in community projects (Burns 1983:256–258), the demand for "public works" is fairly limited in societies like those of the Tsegi phase, and the inhabitants of Kiet Siel and Tsegi Canyon soon would exhaust the possibilities of this type of investment.

Environmental amelioration that improved food production would lead not to surpluses but to a reduction in planting and a concomitant decrease in storage capacity. In any case, data from Kiet Siel fail to support the surplus production idea. It is hard to see how surpluses could have been achieved under the deteriorating environmental conditions of the late thirteenth century. Quite apart from the problems caused by arroyo cutting, no fortuitous surpluses resulting from unusually favorable weather conditions would have been likely. The very instability of the population of Kiet Siel argues against a surfeit of food overflowing the granaries of the village. If conditions were unusually favorable, we should not now see so much evidence for the movement of small household groups both within and among Tsegi phase communities. Finally, it is doubtful that the Tsegi area would have been completely abandoned by 1300 if agricultural surpluses were to be expected.

Another variable that might affect storage capacity independently of external stress on the subsistence system is change in the population of the consuming unit. An increase in the number of people in a household or a village would require a concomitant expansion of storage capacity. In this study, population variability is controlled by standardizing the data in terms of the room cluster

(Tables 8.4–8.6), which is estimated to represent an average population of five persons, and in terms of the individual (Table 8.7). Even if the "average" room cluster does in fact represent a standard number of people, it is logical to suppose that individual household size was not static through time, that it changed as children were born and as adults left, and that these changes had some effect on room-cluster storage capacity. If there was some sort of regular increase in average household size during the occupation of Kiet Siel, the figures in Tables 8.4–8.7 may not accurately represent the respective time intervals because of systematic errors in the population estimates for these periods. Nevertheless, although natural population growth may be responsible for some of the observed increase in storage, strong arguments can be raised against its being a major factor in this regard.

While it is true that household size increases as children are born and survive, it also decreases as adults leave or die. It would be stretching the limits of coincidence to assume that these changes were synchronous among all the households at Kiet Siel. On the contrary, at the same time that some households were growing, others were remaining static and others were declining. Synchroneity in domestic cycles could be expected only if all the founding households of the village had been at the same stage of development when they first took up residence in the rockshelter. The differences in size, composition, and spatial and temporal distributions of the early room clusters at Kiet Siel (Dean 1969:Figure 34) and the range of 8 to ca. 20 years in the addition of living rooms to existing room clusters (Dean 1989) refute this at best somewhat remote possibility. Furthermore, the creation of new families by marriage and the immigration of new households into the village would tend to offset any trend toward larger average room-cluster populations. The between-room-cluster variations in population prevalent at any particular point in the history of the site would have eliminated any long-term trends in average household size.

Other, more specific considerations also contradict the idea that undetected population increase could have caused the expansion of granary space at Kiet Siel. The reuse of abandoned granaries and the conversion of abandoned living rooms into granaries cannot be attributed to the natural growth of households. The decrease through time in average living room size is incompatible with the idea of significant growth in the size of the families that occupied them. One could argue that the packing effect of building in a rockshelter with a limited floor area might force a reduction in living room size, even in spite of increasing household population, in order to maintain the storage capacity necessary to support the growing household. If household population pressures were serious, however, it would have been easy to enlarge the living room at the expense of the courtyard, a feature that did not change appreciably in size through time. The decreasing ratios between activity area and volume and the area and volume devoted to food storage also militate against population increase as an important stress factor.

CONCLUSIONS

The Kiet Siel data provide a strong positive test of the proposition that storage is directly related to the degree of variability in food production. The lowering of alluvial water tables in Tsegi Canyon increased variability in food production by replacing a comparatively stable source of water (groundwater) with a highly variable (and deficient) source, precipitation. In response to these altered circumstances, storage capacity relative to habitation space was increased significantly in an effort to buffer the subsistence system against the more extreme fluctuations in crop yields induced by the greater reliance on annual precipitation. The data also indicate that this response to greater subsistence stress took place primarily on a household basis rather than at a higher level of social integration. Apparently, other means of augmenting storage as a buffer against variability in food production—such as the concentration and redistribution of food on a larger, community scale and the mobilization of intercommunity trade or cooperative relationships—were not employed at Kiet Siel and contemporaneous communities in Tsegi Canyon. In sharp contrast to the Tsegi situation, the inhabitants of nearby open pueblos in Long House, Klethla, and Laguna Creek valleys engaged higher levels of integration in the effort to adapt to the population-resource imbalances of the late thirteenth century (Dean 2002; Dean et al. 1978; Haas and Creamer 1993). Understanding these significant differences in behavior on the part of adjacent groups of people awaits further research.

The results of the Kiet Siel analysis support the idea that food storage is an important means of compensating for high-frequency fluctuations in crop production. As expected (Dean 1988a; Jorde 1977), increased annual variability in crop yield triggered a substantial buildup of storage capacity at Kiet Siel. This outcome confirms the storage system as an important locus of adaptation to certain types of subsistence stress. Given the strong link between food production and storage, adjustments in the latter should be among the first responses to fundamental changes in the former. Other, more drastic reactions probably would occur only when the initial adjustments proved insufficient to handle the altered circumstances. It seems likely, therefore, that changes in storage capacity would be sensitive indicators of significant changes in food production. Indeed, such storage changes may serve as warning signals to archaeologists of developing problems in the production aspect of prehistoric subsistence systems.

Finally, the Kiet Siel analysis illustrates how meaningful questions about human adaptive behavior can be addressed with archaeological data. Tight control of paleoenvironmental, archaeological, and chronological variability permitted the isolation and study of a specific instance of change—change in environmental factors that substantially increased high-frequency variability in crop yields—and of the behavioral reactions to this change. Similarly, this study indicates that important issues that cannot be directly resolved on large spatial scales can successfully be addressed at smaller scales of analysis down to and including that of individual sites. Finally, the Kiet Siel storage study demonstrates in a small way that archaeological data and methods are capable of elucidating behavioral processes that are vital to the change and evolution of human cultures.

ACKNOWLEDGMENTS

The fieldwork and tree-ring dating that underlie this chapter were sponsored by the Southwest Region of the National Park Service and done under the auspices of the Laboratory of Tree-Ring Research (LTRR) at the University of Arizona. I am indebted to Erik K. Reed, R. Gordon Vivian, and Bryant Bannister for their support of this phase of the research. Personnel of Navajo National Monument provided logistic and moral support for the work, particularly Superintendent Arthur H. White, archaeologist Keith M. Anderson, administrative assistant Robert J. Holden, rangers Robert Laughter, Buddy D. Martin, and Larry Powers, and crew members Ake Bigman, Bob Black, Keith Boone, and Terry Gray Mountain. Eddie K. Austin supplied the horses and packing expertise to transport equipment to and from Kiet Siel along the rugged trails of Tsegi Canyon. I am, as always, especially indebted to Alexander J. Lindsay, Jr., for sharing with me his unexcelled knowledge of Kayenta Branch archaeology. John W. Hannah and Gary Funkhouser produced the figures, and Carrie M. Dean made the tables used herein. The LTRR's Southwest Paleoclimate Project, which created the Tsegi Canyon dendroclimatic reconstructions, has enjoyed the support over the last thirty years of the National Park Service, the Southwestern Parks and Monuments Association, the Department of Defense Advanced Research Projects Agency, and the National Science Foundation.

9

The Effects of Environmental Fluctuations on Ancient Livelihoods:

Implications of Paleoeconomic Evidence from the Upper Basin, Northern Arizona

Alan P. Sullivan III and Anthony H. Ruter

INVESTIGATION OF PREHISTORIC LIVELIHOOD STRATEGIES AND OF THE environmental factors that influenced how self-sustaining societies provisioned themselves has deep roots in American anthropology (Bohannon 1965; Feldman 1975; Kohl 1981; Moran 1990; Watson and Watson 1969). Archaeological research inspired by this tradition has been pursued no more vigorously than in the American Southwest (e.g., Gumerman 1988; Schwartz 1957). Despite its near-sacred place in western Americana (Pyne 1998), the Grand Canyon has not been immune to scientific controversies that entail considerations of aboriginal livelihood there (Schwartz 1966). One such controversy, with roots that extend to J. W. Powell (Euler 1969), involves the extent to which Grand Canyon's prehistoric societies were agriculturally dependent (Effland et al. 1981; Sullivan 1987).

We contribute to this ongoing dialogue by exploring the implications of Upper Basin paleoeconomic data for the Anasazi Behavioral Adaptation Model, or ABAM (Dean 1988b). Located just south of Grand Canyon, a major data point for the ABAM and its various iterations (Dean 1996a; Dean et al. 1985, 1994; Euler 1988; Euler et al. 1979), the Upper Basin is strategically situated to test many ABAM predictions (Plog et al. 1988; also Gumerman and Dean 1989). Specifically, we focus on paleoeconomic data recovered from a variety of contexts and evaluate their consequences for explaining how variation in food-provisioning systems may have been affected by environmental fluctuations. We accomplish these objectives with the following strategy. First, we present a brief history of the modes of inquiry and standards of evidence that have been used to support inferences about ancient Grand Canyon livelihood. Second, we analyze modern weather data and tree-ring-based paleoclimatic data to gain an appreciation for the planning and scheduling problems posed by Grand Canyon's inconstant temperature and precipitation patterns. Third, we discuss the results of the analyses of macrobotanical and pollen data recovered from a variety of resource production and consumption contexts in the Upper Basin that bear on the problem of determining subsistence mix. The bulk of the evidence indicates that economic activities focused on the acquisition, processing, and consumption of wild, not domesticated, plants. Finally, we evaluate some predictions and basal assumptions of the ABAM in light of Upper Basin paleoeconomic data and conclude that because of an economic focus on wild plants, Upper Basin prehistoric livelihood strategies were largely unaffected by environmental fluctuations that occurred between A.D. 1050 and 1200.

GRAND CANYON ARCHAEOLOGY AND THE INVESTIGATION OF PREHISTORIC PUEBLOAN LIVELIHOOD

The history of Grand Canyon archaeology illustrates the tenacious hold of Puebloan ethnography on conceptions of prehistoric subsistence in the Grand Canyon region (Schwartz 1966, 1990). Also embedded in Grand Canyon's archaeological history is a record of the methodological struggle to isolate relevant subsistence evidence from various types of archaeological data (Jones 1986).

Ever since Powell's daring Colorado River expeditions in 1869 and 1871–72 (Dolnick 2001; Stegner 1954), the numerous masonry ruins found within the Inner Gorge, on the Walhalla Plateau, and in the Upper Basin have been interpreted, until recently, as evidence of a once-productive agricultural economy. Such inferences, premised largely on analogical arguments with the subsistence practices of Pueblo peoples of the late nineteenth to mid-twentieth century (e.g., Bradfield 1971; Hack 1942; Morgan 1881; Powell 1875), reflect a protocol that is based on little or no direct evidence of food remains themselves (e.g., Hough 1930:67)—a practice that continues to this day (e.g., Snow 1990:289; Van West 1996:216). Moreover, the Puebloan agricultural model, as an institutionalized interpretive paradigm of prehistoric Grand Canyon life (Coder 2000), has been used to sustain the conclusions of extensive (Hall 1942; Judd 1926; Schwartz 1963, 1965; Stewart and Donnelly 1943; Taylor 1958) and intensive (Effland et al. 1981; Rice et al. 1980) surveys of presumed ancient farming landscapes.

The inauguration of excavation projects did little to dislodge the entrenched Puebloan agricultural model, even though direct evidence of agricultural production is highly variable, if not scant. For instance, for the Inner Canyon, cultigen pollen was found in 37 of 75 samples (49.3 percent) from 19 sites at Unkar Delta, although in unknown quantities (Schwartz et al. 1980:40–44; a large volume of corn pollen was recovered from "ritual" contexts, however). In contrast, Hutira (1986) reports a single pollen grain each of squash and corn from two sites along the Colorado River (more than 6,300 pollen grains from 38 samples from five sites were analyzed). Exactly the same results, coincidentally, have been reported by Schoenwetter and DaCosta (1976), who analyzed one sample from a check dam at the Marble Canyon Archaeological Community (Euler 1987). Similarly, of 32 samples examined from 13 sites on the Walhalla Plateau (Schwartz et al. 1981:35–36), 13 samples (40.6 percent) from nine sites yielded a grand total of 36 cultigen pollen grains (from several thousand that were examined).

Macrobotanical remains of domesticated plants (complete or fragmentary specimens) occur sporadically and unevenly. Three rooms at three sites on Unkar Delta and four rooms at one site (GC 212/212A) on Walhalla Plateau yielded a total of 6 and 30 maize specimens, respectively (Cutler and Blake 1980:211). From five sites along the Colorado River, Hutira (1986) noted the occurrence of about three dozen maize remains recovered from 11 of 44 samples. Even if clusters of corncobs are included from Inner Gorge "granaries" (e.g., Nankoweap Creek [*n* = 6; Cutler 1963:301], UN-46 [*n* = 19; Cutler and Blake 1980:211]), and "storage cists below the rim of Walhalla Plateau" (Judd 1926:149), domesticated plant remains are overwhelmed by those of wild plants. In fact, one "granary" located near Bright Angel Pueblo in the Inner Canyon produced only wild-plant remains (Schwartz et al. 1979:90).

From these data, it seems a stretch to argue that maize agriculture was a consistent, dominant economic pursuit during the principal period of Grand Canyon's prehistoric Pueblo (Anasazi) occupation, A.D. 1050–1150/1190 (Lyneis 1996:23; Schwartz 1965:295). Significantly, however, the recovery of direct evidence, in either macrobotanical or pollen form, initiated a mode of inquiry that has led to considerations of the importance of "subsistence mix" (e.g., Powell 2002:90–93). For our purposes, this term refers to the proportional contributions that wild plants and domesticated plants make to the annual food supply (Gasser 1982:21). It has been hypothesized, for instance, that variation among paleoeconomic assemblages may be attributable to variation in the degree to which agricultural populations were buffering low or unpredictable agricultural yields caused by high-frequency environmental fluctuations (Dean 1994). As such, the occurrence of wild resources, often in large quantities, is thought to be related to their role as backup or famine foods (Minnis 1996) for those times when agricultural systems failed or underproduced (Dean 1988b:35). We illustrate below that the paleoeconomic record of the Upper Basin, which features high frequencies of wild-plant remains recovered from a variety of archaeological phenomena, offers a different perspective on how some prehistoric Puebloan groups contended with high-frequency environmental fluctuations.

UPPER BASIN ENVIRONMENT AND CLIMATE DURING THE TWENTIETH, ELEVENTH, AND TWELFTH CENTURIES

Geology, Vegetation, and Hydrology

The Upper Basin is a flexed, faulted, and tilted block of the northeastern Coconino Plateau that is approximately 140 square miles in area (Babenroth and Strahler 1945:135; Barnes 1987:381; Huntoon 1990:265). Varying in elevation from 7,400 feet asl at Desert View along the South Rim to 6,100 feet asl at Lee Canyon (Figure 9.1), the Upper Basin is covered by a dense pinyon-juniper woodland that is punctuated by patches of sagebrush and grass (Rand 1965;

TABLE 9.1. Physical and archaeological attributes of Black Mesa and the Upper Basin.

ATTRIBUTE	BLACK MESA	UPPER BASIN
Near-surface bedrock	Sandstone	Limestone
Seeps, springs	Present	Absent
Holocene alluvium		
Valley bottoms	Common	Rare
Alluvial fans	Common	Rare
Buried features		
Archaeological sites	Present	Absent
Trees	Present	Absent
Pre–Basketmaker III Occupation	Common	Absent
Basketmaker III–Pueblo I Occupation	Common	Uncommon

Strahler 1944:376–377). In lower elevations, the woodland vegetation thins to scattered juniper trees and ultimately to sagebrush and grasslands (Brewer et al. 1991; Euler et al. 1979:1095). Paleoecological studies (Cole 1990) indicate that this vegetation pattern developed well before the earliest traces (ca. 2000 B.C.) of humans in the Grand Canyon area (Emslie et al. 1987, 1995; Euler 1983; McNutt and Euler 1966) and, with the possible exception of the effects of widespread burning (Sullivan 1996), has remained stable until historic times (Bonnicksen 2000:236–240).

The area is notorious for having no surface water whatsoever (Metzger 1961). Because the Upper Basin is capped by porous, heavily fractured limestone (Hopkins 1990), whatever precipitation falls is absorbed rapidly. Hence, in contrast to other areas of the southern Colorado Plateaus, such as Black Mesa, no seeps, springs, or streams occur in the Upper Basin (Table 9.1). We surmise that two solutions were developed to deal with the lack of surface water. The first was to harvest runoff from slick-rock exposures (Norr 1997), a technique that worked well during those portions of the year when precipitation, in the form of rainfall or snow, actually materialized. The second solution was to abandon the Upper Basin during the spring drought, when hardly any rain falls (see below), and move to the Inner Canyon or to the wetter North Rim (for details of the "cross-canyon" settlement model, see Sullivan et al. 2000; cf. Geib 1996).

Twentieth-Century Temperature

Sellers and Hill (1974:240) note that at least 148 days elapse, on average, between frosts in the region—a value that fits squarely within Hendricks's (1985:12) range of 130–170 frost-free days. Weather data from 1948–2000, involving four stations (average elevation is 6,900 ft asl) in the vicinity of Grand Canyon National Park,[1] reveal the frost-free window (Figure 9.2).[2] By modern Puebloan agricultural standards, this period is more than sufficient to ensure

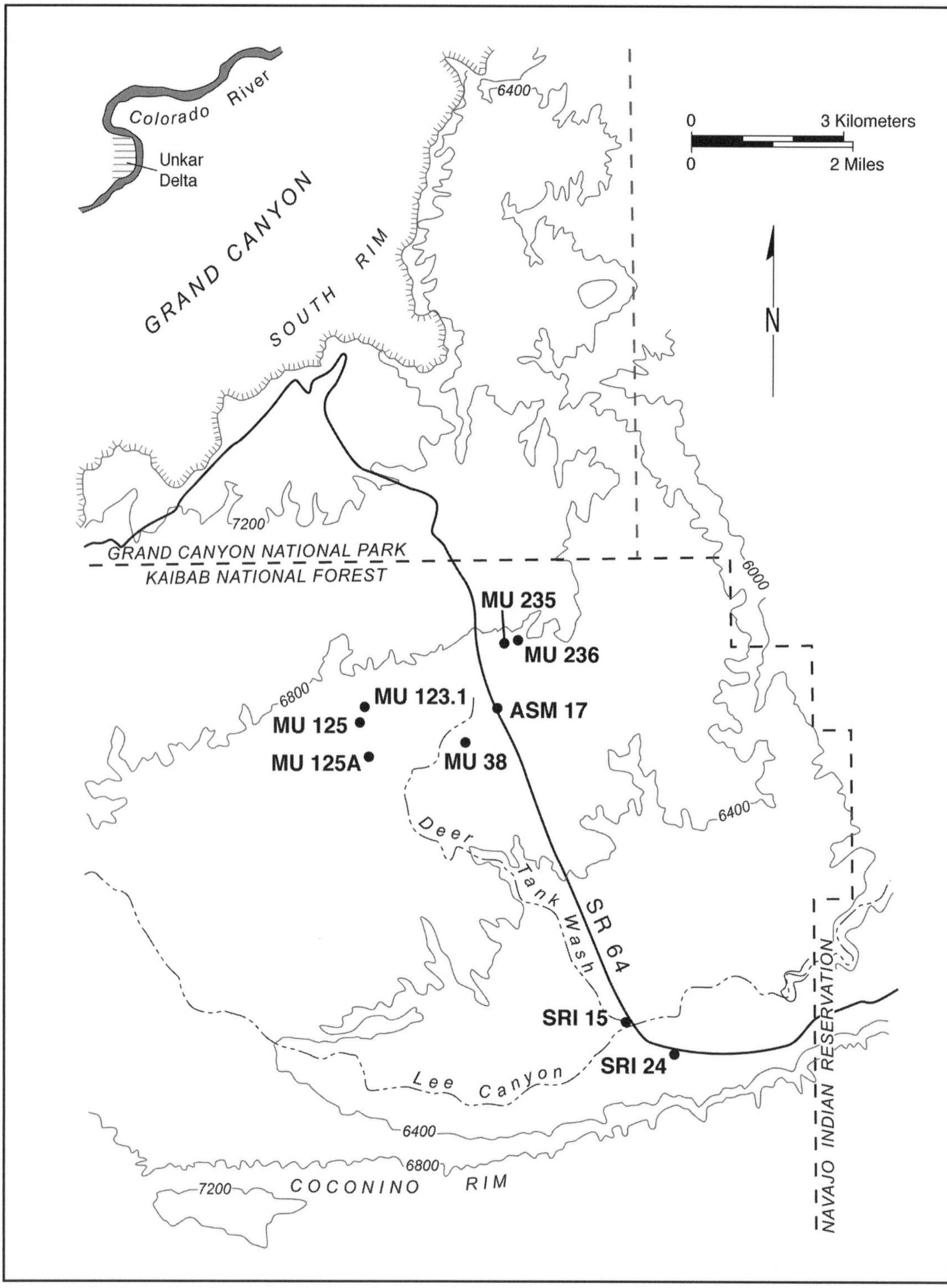

FIGURE 9.1. The Upper Basin: locations of archaeological sites. Contour interval is 400 ft.

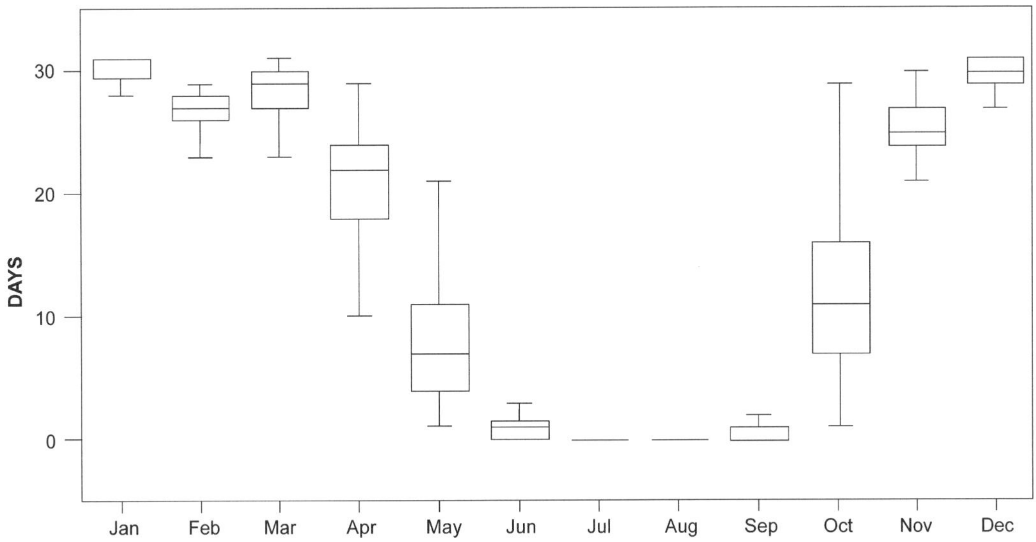

FIGURE 9.2. Means and standard deviations of the number of days below freezing per month for the period 1948–2000 based on Grand Canyon weather-station data.

against crop loss caused by freezing temperatures (Muenchrath and Salvador 1995).

Twentieth-Century Precipitation

An 80-year run (1904–1983) of Grand Canyon weather-station data indicates that average annual precipitation is 16.06 inches (standard deviation = 5.11 in). In view of its lower elevation, the Upper Basin in all likelihood receives slightly less (e.g., 0.5–1.0 in less) average precipitation annually. Also, the 80-year precipitation record clearly shows the spring drought (Figure 9.3), a time when little or no precipitation falls across the southern Colorado Plateaus (Dean 1988a:123).

Eleventh- and Twelfth-Century Precipitation

Dendrochronological data from the Laboratory of Tree-Ring Research's Coconino Plateau Station provide a record of high-frequency paleoenvironmental variation (Dean 1988a:135–137).[3] The Coconino Plateau tree-ring data represent "average" conditions across an amorphous area north of the San Francisco Peaks, south and east of the Colorado River, and west and south of the Little Colorado River (J. S. Dean, personal communication 2001). Here we focus our analyses on the period between A.D. 1049 and 1190/1200, which represents the tree-ring-dated span of prehistoric Puebloan occupation of the Upper Basin (Haury 1931; Robinson and Cameron 1991:1, 6; Sullivan 1987). According to the ABAM, with the exception of the latter half of the twelfth century (Dean et al. 1985:543), this late Pueblo II–Pueblo III time period is characterized by increasing effective moisture, aggrading floodplains, and high water tables (Euler et al. 1979:1099).

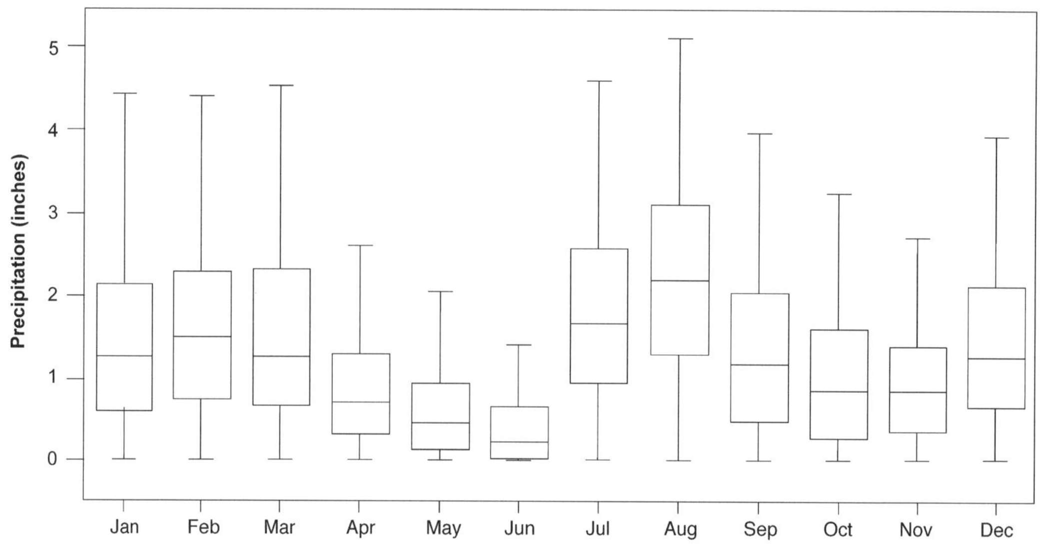

FIGURE 9.3. Means and standard deviations (in tenths of inches [.10 in]) of monthly precipitation for the period 1904–1983 based on Grand Canyon weather-station data.

Analysis of annual precipitation during this period shows that, on average, 14.44 inches were available and no year received less than 10.0 inches (Figure 9.4). Focusing only on below-average years, we find that one or two years of below-average precipitation were followed by a year or two, if not more, of above-average precipitation (Figure 9.5). Two exceptions to this predictable pattern are a three-year period (A.D. 1093–1095) and a five-year period (A.D. 1111–1115)—both of which were followed by multiyear periods of above-average precipitation. It should be noted that, based on relatively imprecise ceramic-group dating (Sullivan et al. 1995), two excavated single-room masonry structures (MU 38 [Becher 1992] and SRI 24 [Whittlesey 1992]) in the Upper Basin were built and used during these two short-term "drier" periods.

In marked contrast, however, are three periods of protracted below-average precipitation: A.D. 1064–1075 (11 of 12 years were below average), A.D. 1124–1136 (11 of 13 years were below average), and A.D. 1185–1193 (7 of 9 years were below average). It is important to note that with the exception of the second dry period (A.D. 1124–1136), these periods are accompanied by tree-ring-dated building activity in the Upper Basin: site MU 125 (Sullivan 1997) was built during the earlier period and the Tusayan Ruin (Haury 1931) during the latter. Evidently, these periods of below-average precipitation were insufficient to discourage construction of some of the larger settlements recorded in the Upper Basin (5 and 18 rooms, respectively; Dean et al. 1994:62). In addition, with the exception of the A.D. 1185–1193 dry period, the post–A.D. 1150 period appears to have been relatively normal with respect to interannual oscillations of precipitation (see Dean et al. 1985:543). And the region appears to have been abandoned, that is, perennial settlements no longer were built, at a time (ca. A.D. 1200) when climatic conditions were average (see Dean 1996a:46).

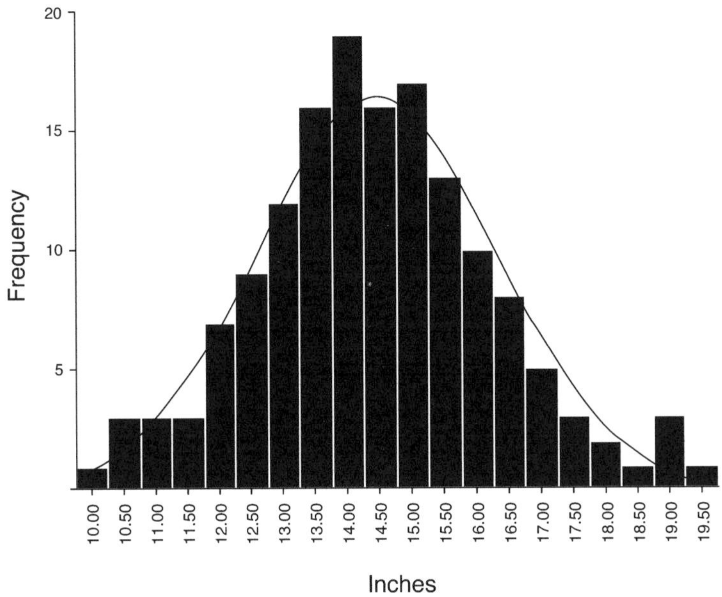

FIGURE 9.4. Histogram of precipitation values for the period A.D. 1049–1200 reconstructed from Coconino Plateau tree-ring data (mean = 14.44 in, standard deviation = 1.84 in).

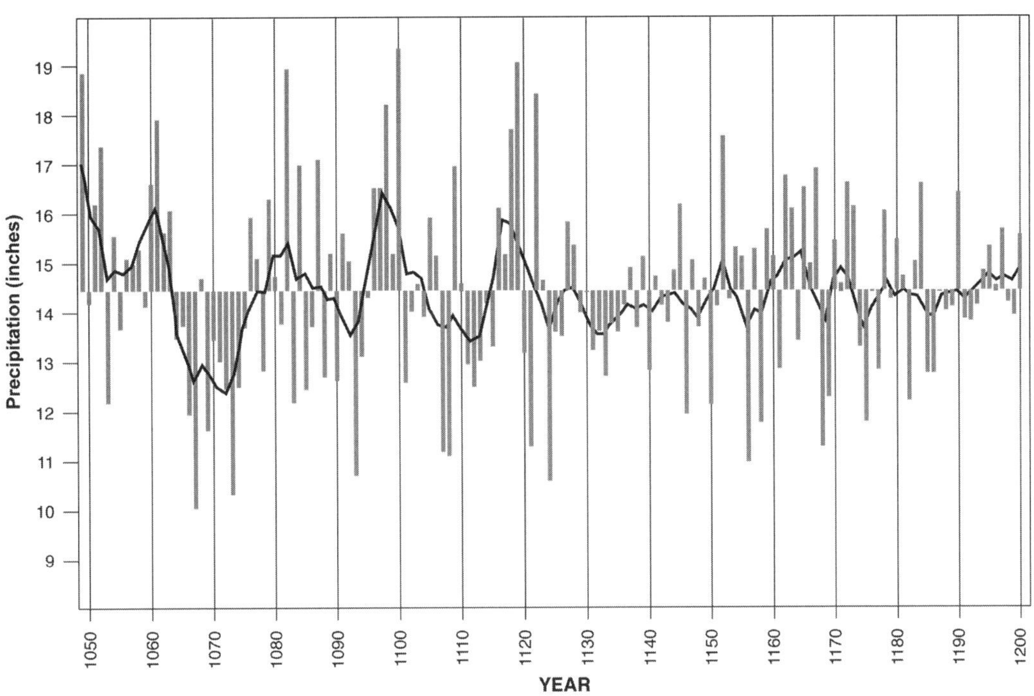

FIGURE 9.5. Histogram of annual precipitation for the period A.D. 1049-1200 reconstructed from Coconino Plateau tree-ring data. Dark line is a three-year running average.

Discussion

Assuming that the annual distribution of paleo-precipitation did not differ dramatically from the modern pattern, agriculturalists would still have had to overcome short-term, seasonal deficits represented by the spring drought, which can be devastating for maize production (Rose 1989:49–52). Crops planted in the historically dry months of April–June require sufficient soil moisture to germinate as well as adequate monsoonal moisture to thrive and bear fruit (i.e., cobs with kernels). Insufficient winter precipitation or increased evaporation during the spring drought would curtail germination (Cutler 1963:301). As the following analyses of direct subsistence evidence imply, however, high-frequency environmental fluctuations may have had less profound effects on those livelihood strategies that were not dependent on maize production (Gaines and Gaines 2000:115, 120).

DIRECT MEASURES OF SUBSISTENCE VARIATION

Macrobotanical assemblages differ enormously with respect to yield (i.e., the number of plant remains) and taxonomic diversity, which makes it difficult to compare them unproblematically (Miksicek 1987). In the investigation of the extent to which taxa vary with respect to one another (Popper 1988:60), a crucial matter in determining whether paleoeconomies were based on wild or domesticated plants (King 1994:185), two measures—relative abundance and ubiquity—dominate the literature (Wagner 1988:29–30). In addition to gauging floral importance (Miksicek 1983; Pearsall 1988:110), relative abundance, which is the percentage of the remains of a taxon in an assemblage, standardizes for different sample sizes (Miller 1988:74). Ubiquity, which is the percentage of samples that disclose the remains of a specific taxon, accounts for variation in taxa spatial distributions (Minnis 1980:380) and is thought to measure taxon use (Minnis 1985b:104; see Popper 1988:64) as well. Fundamentally, however, both measures depend on which plant remains are counted: charred seeds only (Minnis 1981) or the sum total of charred and uncharred remains (Toll 1988:39). Though seemingly a trivial methodological point, variation between such measures affects impressions of whether patterning among higher-level units of analysis (i.e., categories of archaeological phenomena) is attributable merely to the method of quantification or to systemic economic differences (Adams 1980:9; Rocek 1995).

SEARCHING FOR EVIDENCE OF UPPER BASIN LIVELIHOOD STRATEGIES

Excavations at eight sites in the Upper Basin (Figure 9.1), lying at elevations between 6,780 and 6,140 feet asl, have produced a suite of well-dated contexts that were sampled extensively for archaeobotanical remains (Table 9.2). Two of the sites (SRI 15 and SRI 24) are located in a sagebrush-grassland setting

Table 9.2. Archaeological Characteristics of Eight Excavated Sites in the Upper Basin.

SITE	TYPE	STRUCTURES	ROOMS	DATES (A.D.)	DATING METHOD	OCCUPATION MODE	ABANDONMENT MODE	REFERENCES
ASM 17	Masonry and timber ruin	4	5	1049/1054–1064	Tree-ring	Continuous	Catastrophic	Sullivan 1987
MU 38	Masonry ruin	1	1	1061–1080	Ceramic crossdating	Discontinuous	Gradual/Planned	Becher 1992
MU 123.1	Fire-cracked-rockpile			1050–1291	Radiocarbon	Episodic	Episodic	Cook 1995
				1275–1408	Radiocarbon			
				1447–1650	Radiocarbon			
MU 125	Masonry ruin	2-3	6	1070–1080	Tree-ring	Continuous	Semicatastrophic	Sullivan et al. 1995
MU 235	Fire-cracked-rockpile			417–559	Radiocarbon	Episodic	Episodic	Sullivan 1992
				687–963	Radiocarbon			
MU 236	Fire-cracked-rockpile			1288–1414	Radiocarbon	Episodic	Episodic	Sullivan et al. 2001
				1299–1409	Radiocarbon			
SRI 15	Masonry ruin	2	3	750; 1150	Mean ceramic dating	Discontinuous	Unknown (gradual?)	Christenson 1992
				731–1152	Radiocarbon			Whittlesey 1992
				1090–1130	Ceramic crossdating			Sullivan et al. 1995
SRI 24	Masonry ruin	1	1	1105–1115	Mean ceramic dating	Discontinuous	Unknown (gradual?)	Christenson 1992

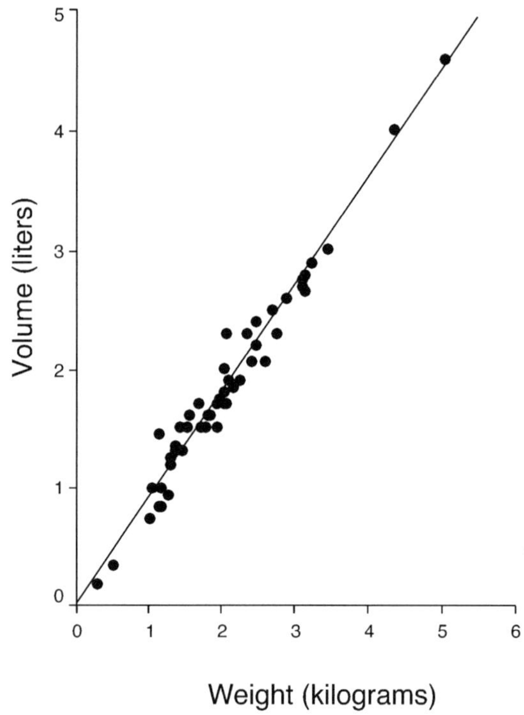

FIGURE 9.6. Scatterplot of flotation-sample weight (kg) and volume (l). A regression line has been superimposed on the distribution of the cases ($N = 53$).

(Whittlesey 1992), and the other six are in dense pinyon-juniper woodland (Sullivan 1992, 1997; Sullivan, ed. 1986).

Of 82 macrobotanical samples (Table 9.3), exactly 50 percent have information regarding weight, which is a crucial variable for calibrating differences in yields and for estimating aspects of preservation (Ford 1988). Weight rather than volume is used because it is measured rapidly with electronic balances and, perhaps more important, reduces the possibility of specimen loss and fragmentation as soil samples are decanted into and subsequently emptied from volumetric receptacles. Moreover, because weight and volume are highly correlated ($r^2 = .966$, $p < .001$; Figure 9.6), values of either variable can be estimated reliably (Miller 1988:73).

The following procedures were routinely employed by paleobotanists searching for plant remains in Upper Basin archaeobotanical samples (Brandt 1991; Cummings 1986; Cummings and Puseman 1995, 1997; Huckell 1992). First, the "vortex" water-screening method was used to obtain the organic light-fraction from soil samples, which was then air-dried. Second, the light-fraction was passed through a series of progressively smaller screens (range of openings = 4.0–0.25 mm; see Toll 1988). Finally, the contents of each screen were examined and remains were identified and enumerated by taxon.

To assess depositional and preservational variability (Miller 1988:83), we calculated two density or seed-concentration indices (i.e., the number of macrofloral remains divided by the number of samples or by sample mass). The average yield per sample is 36.5, 13.5, and 66.1 specimens for edible remains, inedible remains, and total plant remains recovered, respectively. A similar pattern is obtained when the average yield per kilogram is considered: 15.4, 0.8, and 18.0 specimens for edible, inedible, and total remains, respectively. Significantly, both sets of values fall toward the upper end of the range for semiarid environments (Miksicek 1987:217). Thus, preservation appears to have been at least moderately good but not extraordinary (Cummings and Puseman 1997:5), which is not uncommon for areas like the Upper Basin that are characterized by pronounced seasonal fluctuations in temperature and precipitation.

INFERRING UPPER BASIN LIVELIHOOD STRATEGIES

It is useful to conceive of food-supply systems as consisting of areas, features, and technologies designed to (1) produce and initially process edible resources and (2) store, prepare, and consume them (Ellen 1982; Fish and Donaldson 1991). The logic of inference entailed by this conceptualization posits that patterned similarities and differences among plant taxa from (1) resource-production and (2) post-resource-production contexts will disclose system-wide modalities in livelihood strategies (Sullivan 1996).

To illustrate the interpretive advantages of this approach and to dampen the effects of sample attenuation (i.e., where many categories have few or no specimens each), we have configured the sampling contexts in Table 9.3 into three groups, as follows. Group A consists of three sites, MU 123.1, MU 235, and MU 236, that produced samples related to resource-acquisition activities exclusively. These fire-cracked rockpiles and their associated artifact scatters disclosed no evidence of architecture or other traces of habitation (Cook 1995; Sullivan 1995; Sullivan et al. 2001). Group B consists of three sites, SRI 15, SRI 24, and MU 38, that produced samples from post-resource-production contexts. These three masonry ruins are the remains of nonperennial settlements with two or fewer rooms and no evidence of posthole patterns (Becher 1992; Whittlesey 1992). Group C consists of two sites, ASM 17 and MU 125, that produced samples from post-resource-production contexts. These perennial settlements had four or more rooms and definitive posthole patterns (Sullivan 1997; Sullivan, ed. 1986; Sullivan et al. 1995).

Relative abundance and ubiquity values of various plant taxa for these three groups were calculated in two ways to cover the range of depositional and preservational conditions that might affect the "descent of evidence" regarding wild-plant and domesticated-plant use patterns (Dennell 1979; Hastorf 1988:120–121). First, the least conservative and most inclusive calculations involve relative abundance and ubiquity of all plant remains, which includes charred and uncharred specimens representing edible and inedible plant parts. Second, the most conservative and least inclusive calculations are for relative abundance and ubiquity of charred, edible plant parts only (see Nelson and Diehl 1999:144–145).

Table 9.3. Archaeobotanical Data from Eight Excavated Sites in the Upper Basin.

SITE NO.	STRUCTURE/ FEATURE NO.	ROOM/ FEATURE NO.	TYPE	EVIDENCE OF BURNING	SAMPLE CONTEXT	NO. OF SAMPLES	WEIGHT (KG)	VOLUME (L)	PLANT REMAINS[a]		
									EDIBLE	NON-EDIBLE	TOTAL
ASM 17	1		Room	Yes	Pit ($n = 4$) contents	10		>40	219	2	318
	1		Room	Yes	Hearth ($n = 2$) contents	2		8	327	2	429
	2		Room	Yes	Hearth contents	2		4	3	0	17
	3		Room	Yes	Hearth contents	1		4	0	0	7
	3		Room	Yes	Vessel ($n = 3$) contents	3		12	106	0	216
	3		Room	Yes	Floor sediments	2		8	6	0	6
	4		Room	Yes	Hearth contents	1		4	0	0	0
	4		Room	Yes	Floor sediments	1		4	13	0	13
	4		Room	Yes	Vessel ($n = 2$) contents	2		8	206	0	206
MU 38	1		Room	Yes	Floor sediments	2	1.12		3	1	7
		1	Exterior stone-lined feature	Yes	Contents	2	8.97		1	8	59
MU 123.1		1	Pit	Yes	Contents	1	3.57		39	11	53
		2	Ash concentration	Yes	Contents	2	7.56		2	0	5
		2	Ashy area near pit	Yes	Contents	2	7.14		3	0	3
MU 125		2	Room	Yes	Metate debris	2	4.20		44	2	47
		2	Room	Yes	Hearth contents	4	9.33		67	3	72
		2	Room	Yes	Ash-filled pit	1	2.85		9	0	9
		2	Room	Yes	Posthole ($n = 5$) contents	5	13.80		212	21	246
		3	Room	No	Metate debris	2	1.59		4	0	72

Table 9.3. *(continued)*

SITE NO.	STRUCTURE/ FEATURE NO.	ROOM/ FEATURE NO.	TYPE	EVIDENCE OF BURNING	SAMPLE CONTEXT	NO. OF SAMPLES	WEIGHT (KG)	VOLUME (L)	PLANT REMAINS[a] EDIBLE	NON-EDIBLE	TOTAL
		3	Room	No	Floor sediments	1	0.78		6	1	13
		3	Room	No	Hearth contents	1	2.21		4	0	4
		5	Room	No	Shallow burned pit	1	2.87		15	2	17
		6	Room	No	Deep unburned pit	1	2.45		22	7	29
		96-3	Exterior pit	Yes	Contents	2	5.22		20	0	21
		96-14	Fire-cracked-rock	Yes	Ash	1	2.70		132	0	150
MU 235			Artifact arrays	Yes	Occupation-surface sediments	3	3.61		5	0	11
		Test Trench 1	Ash concentration	Yes	Contents	1	2.01		15	12	27
		5 & 6	Fire-cracked-rock	Yes	Ash	2	3.75		9	8	21
MU 236		3, 4, & 6	Pits	Yes	Contents	3	8.62		2	0	12
		1	Fire-cracked-rock	Yes	Ash	1	3.34		12	1	16
SRI 15	1	A	Room	No	Floor sediments	2			8	3	77
	1	B/C	Room	Yes	Floor sediments	3			26	11	84
	2		Feature	No	Floor sediments	1			0	5	9
SRI 24	3		Room	Yes	Floor sediments	9			62	126	319
	3		Room	Yes	Hearth contents	1			409	833	1,700
	3			n/a	Exterior features	2			50	50	193

[a] Edible remains include the following plant parts: nuts, seeds, kernels, achenes, tissue, caryopses, and embryos. Non-edible remains include the following plant parts: cone scales, seed coats, cupules, nutshells, glumes, rachis fragments, culms, fibers, stems, and spine bases. Total remains include needles (e.g., evidence of needle-based decoction at ASM 17) but exclude wood, bark, twigs, leaf, rootlets, and charcoal.

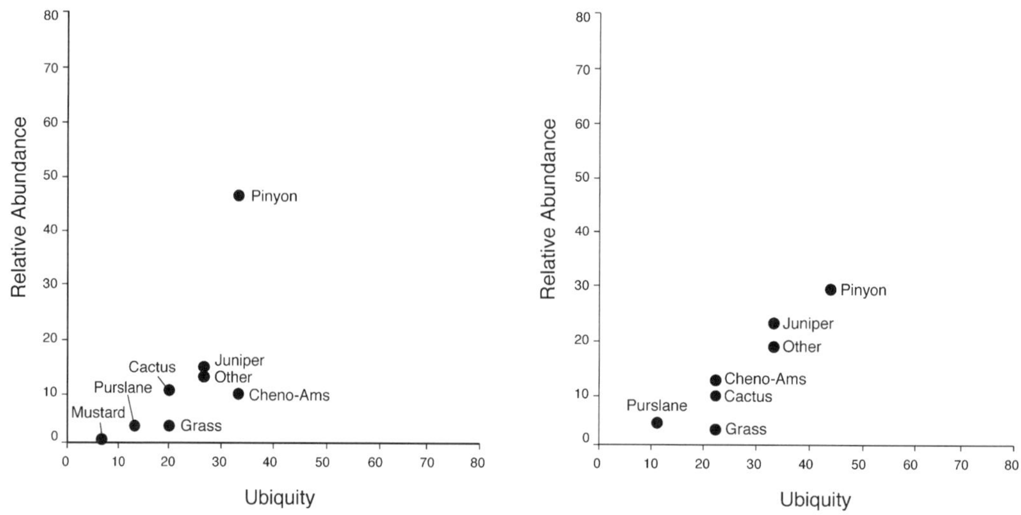

FIGURE 9.7. Ubiquity and relative-abundance scatterplots of plant taxa in samples recovered from three resource-acquisition and processing sites (Group A: MU 123.1, MU 235, and MU 236) in the Upper Basin. Left: all remains (charred and uncharred edible and non-edible plant parts); right: charred, edible remains only. Other includes unidentifiable ($n = 6$), processed edible tissue ($n = 2$), Fabaceae ($n = 1$), Physalis ($n = 1$), Solanaceae ($n = 1$), and Portulaca ($n = 1$).

Patterns of Plant Use: Macrobotanical Evidence

For Group A sites (Figure 9.7), with the exception of "other" plant remains (which includes nine unidentified specimens and single occurrences of six rare taxa), pinyon is both abundant and widespread compared with all other taxa when either all remains or edible, charred remains only are considered. Juniper, cheno-ams, cactus, and grass are less abundant and widespread, although absolute specimen quantities are low ($n = 119$), especially when edible, charred remains are taken into account ($n = 65$).

When relative abundance and ubiquity values for all remains (n = 1,595) recovered from the Group B sites are examined (Figure 9.8), pinyon again is abundant and widespread; cheno-ams and maize are fairly ubiquitous but, in comparison with pinyon, not very abundant. Note, however, that when relative abundance and ubiquity values for edible, charred remains only are examined ($n = 558$), pinyon and maize virtually disappear; in this rendering, cheno-ams and, to a lesser extent, juniper and grass are comparatively abundant and widespread. Such volatility illustrates the effects that units of analysis have on judgments regarding the degree to which different plant taxa contributed to the food supply.

Yet another picture emerges for the plant remains recovered from the Group C sites (Figure 9.9). Either pinyon or cheno-ams are abundant or widespread whether all remains (n = 2,398) or just edible, charred remains (n = 1,259) are considered. All other taxa, including maize, are not very abundant. When this pattern first crystallized (Sullivan 1987), it was thought that perhaps the rarity of maize could be attributed to its role as a dietary staple. That is, maize had been so thoroughly consumed (i.e., drawn down) that by-products of its usage rarely found their way into sampled contexts, thereby elevating the ubiquity and relative abundance values of wild-plant taxa. However, now that excavation data are available from a broad range of contexts, including deposits that are consequences of catastrophic and noncatastrophic abandonment

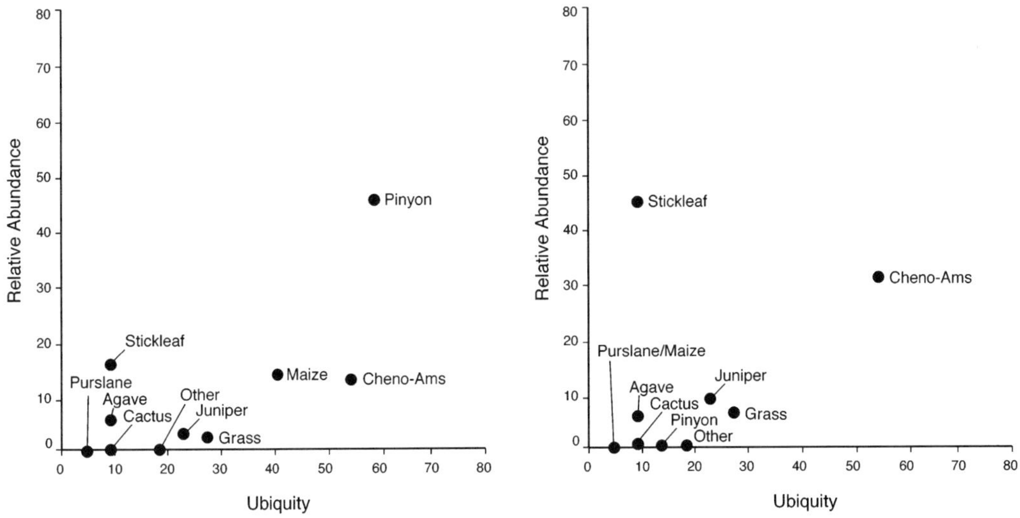

FIGURE 9.8. Ubiquity and relative-abundance scatterplots of plant taxa in samples recovered from three nonperennial sites (Group B: MU 38, SRI 15, and SRI 24) in the Upper Basin. Left: all remains (charred and uncharred edible and non-edible plant parts); right: charred, edible remains only. Other includes plantain ($n = 1$), Compositae ($n = 1$), Corispermum ($n = 1$), and Sphaeralcea ($n = 1$).

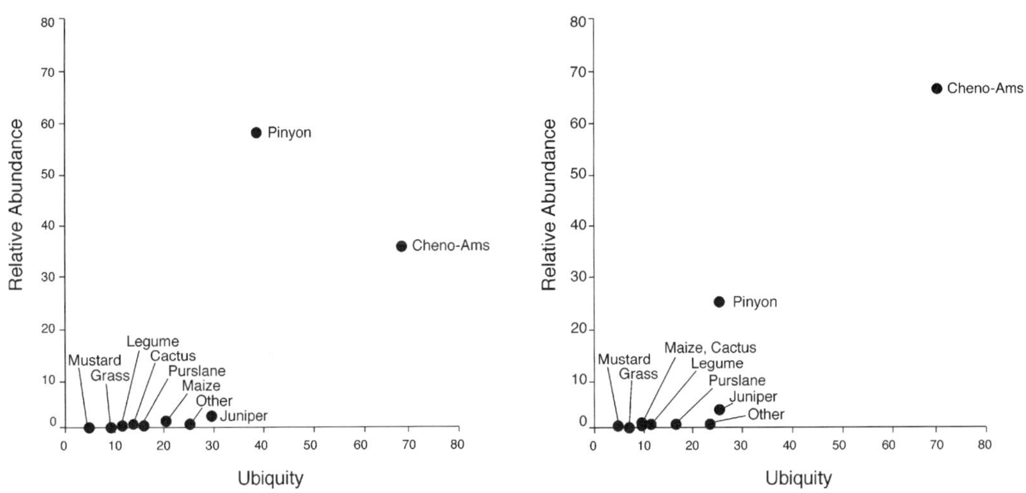

FIGURE 9.9. Ubiquity and relative-abundance scatterplots of plant taxa in samples recovered from two perennial sites (Group C: ASM 17 and MU 125) in the Upper Basin. Left: all remains (charred and uncharred edible and non-edible plant parts); right: charred, edible remains only. Other includes Trifolium ($n = 1$), Vitis ($n = 1$), unknown ($n = 3$), processed edible tissue ($n = 5$), monocot ($n = 1$), Fabaceae ($n = 1$), and unidentified ($n = 6$).

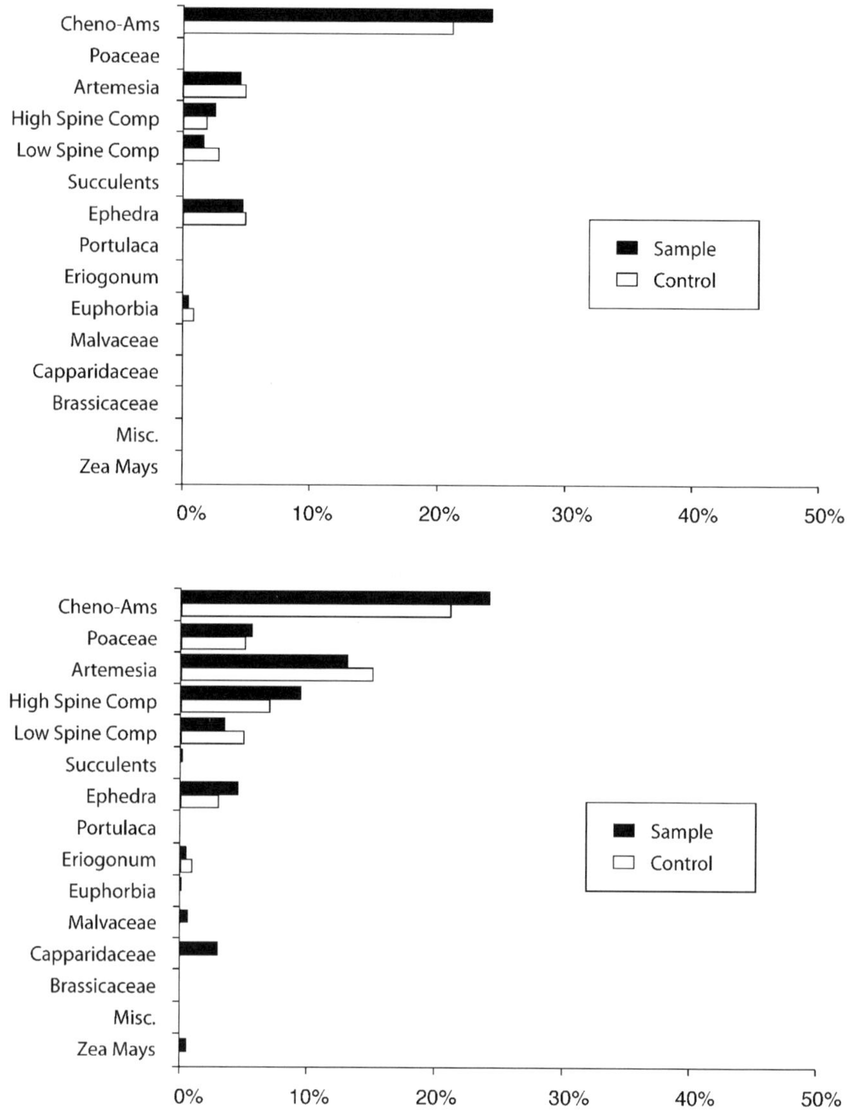

FIGURE 9.10. Taxa percentages of non-arboreal pollen recovered from two rock-alignment complexes in the Upper Basin. Top: SRI 24; bottom: MU 125A. Note the absence of maize pollen at SRI 24.

modes, as well as continuous and discontinuous occupation modes of varying lengths (Table 9.3), the "drawdown" hypothesis is untenable.

Discussion

No matter how relative abundance and ubiquity values are calculated (see Rocek 1995:229), wild-plant taxa dominate Upper Basin macrobotanical assemblages (Cummings 1996). Based on these data, pinyon nuts and seeds from chenopods and amaranth, which have millennia-long use histories in the American Southwest and the Great Basin (Bohrer 1975; Doebley 1984; Dunmire and Tierney 1997; Lanner 1981; Madsen 1986; Minnis 1991; Thomas 1983), represent key resources that prehistoric societies depended on long after maize became available (Wills and Huckell 1994; see Reed 2000). Moreover, sub-average returns of wild resources could have been supplemented by maize

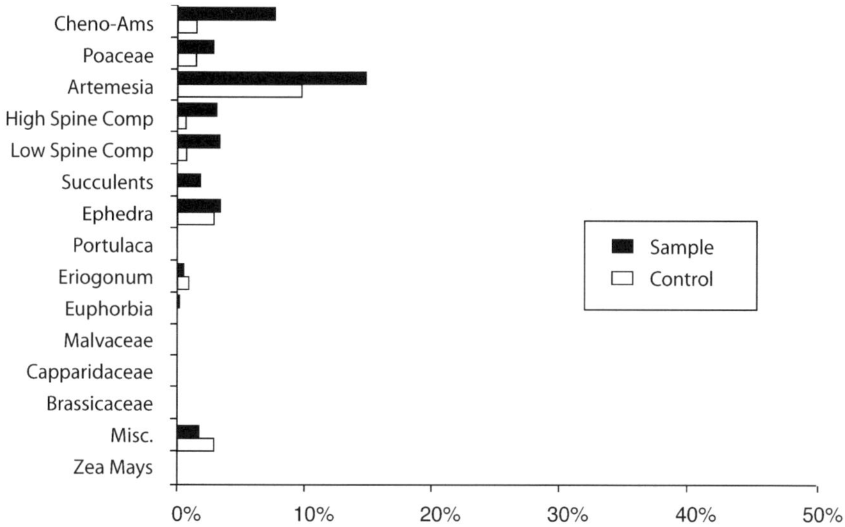

FIGURE 9.11. Taxa percentages of non-arboreal pollen recovered from two manos and two slab metates in the northern array of artifacts at MU 235 (a resource-acquisition and processing site) in the Upper Basin. With the exception of Eriogonum, note the elevated percentages of all other pollen types.

production (Minnis 1985a:316; Whittlesey 1993:267) rather than vice versa, which is commonly presumed (e.g., Cummings 1995:349; Schwartz 1963:295).

Patterns of Plant Use: Palynological Evidence

The foregoing analyses and conclusions are supported by pollen data from the Upper Basin and beyond. For example, analysis of nine samples from "agricultural" terraces near SRI 24 failed to disclose any maize pollen (Bozarth 1992:144). Note, however, the elevated percentages of cheno-am and high-spine composite pollen relative to the control samples (Figure 9.10)—a situation that differs insubstantially from that documented at the Sunset Crater "ash field" near Flagstaff, Arizona (Berlin et al. 1990). Back in the Upper Basin at MU 125A (Sullivan 2000), maize pollen was found in four of eight samples from another set of rock-alignment terraces but accounted for less than 0.5 percent of all plant pollen (Figure 9.10). Again, the percentages of high-spine composite and grass (Poaceae) pollen are elevated with respect to controls. Even though percentages of some taxa (e.g., cheno-ams and low-spine composites) are higher in the control sample, the presence of aggregates argues strongly for their propagation on the terraces behind the alignments (Bohrer 1981; Brown 1996:67). Investigations at the Merriam Crater "ash field" east of Flagstaff produced strikingly similar results (Berlin et al. 1977). It is difficult to avoid the conclusion, based on the results of four independent studies, that landscape features often interpreted as locales for the production of maize (e.g., Stewart and Donnelly 1943; Woodbury 1961) may have been used principally for the production of wild plants. Hence the presence of cheno-am, grass, and high-spine composite pollen in such contexts appears to be indicative of food production rather than the invasion of terraces and fields by non-economic agricultural "weeds" (e.g., Fish 1994:47, 49).

The pollen record for this terrace/field pattern is analogous to that determined from sediment samples collected beneath two manos and around the

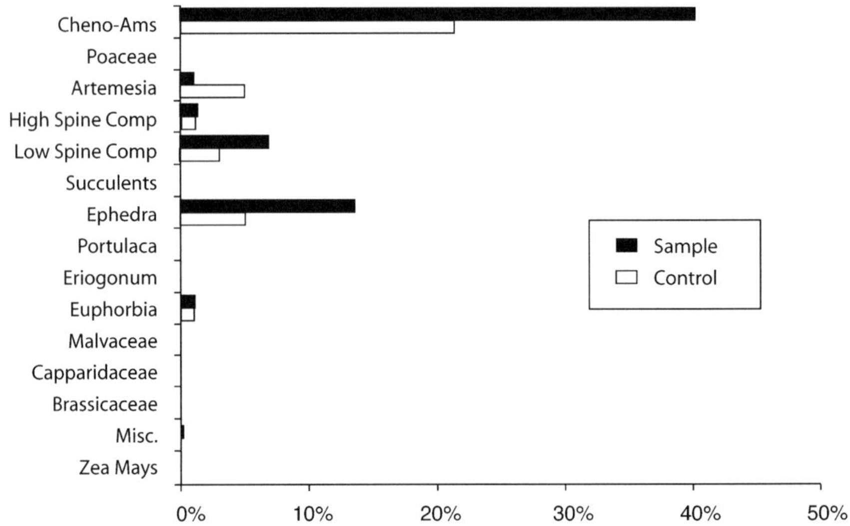

FIGURE 9.12. Taxa percentages of non-arboreal pollen recovered from a nonperennial site (SRI 24) in the Upper Basin. Note the elevated percentages of cheno-ams, low-spine composite, and Ephedra pollen types.

perimeters of two slab metates that reposed directly on a prehistoric surface at MU 235 (Figure 9.11), one of the Upper Basin resource-acquisition (Group A) sites (Sullivan 1992). Moreover, the comparatively high frequencies of cheno-am, grass, artemesia, and high-spine composite pollen, as well as aggregates of these taxa, support the inference that processing of these plants occurred in close proximity to the ground stone artifacts (Cummings and Puseman 1995; see Adams and Bohrer 1998:134).

Analysis of pollen samples recovered from various features at SRI sites 15 and 24 (Group B) disclosed no maize pollen (four bilobate phytoliths were noted [Bozarth 1992:143], but disagreement exists as to whether such forms pertain to *Zea mays* exclusively [Doolittle and Frederick 1991]). Yet cheno-am and ephedra pollen grains were particularly abundant, especially at SRI 24 (Figure 9.12). Clearly, even if these two nonperennial sites represent the remains of conventional agricultural "field houses," maize processing and storage, and presumably consumption, were far overshadowed by subsistence pursuits that focused on wild resources (Whittlesey 1992). Similarly, no maize pollen was recovered from ASM 17 (Group C) (Davis 1986), and only a few maize pollen grains (< 0.5 percent) were recovered from MU 125 (Group C) (Cummings and Puseman 1995, 1997).

Discussion

Remains of maize in any form have been difficult to find in the paleoeconomic records of Upper Basin sites (see Gumerman 1984). That wild-plant macrobotanical remains and pollen have been recovered in abundance from episodically used arrays of resource-acquisition tools and features unassociated with settlements (Group A), from nonperennial settlements (Group B), and from perennial settlements (Group C) argues rather convincingly that inferred patterns of plant use cannot be attributed to bias in recovery or analysis (see Hard et al. 1996). Therefore, assuming that all these remains, which were recovered

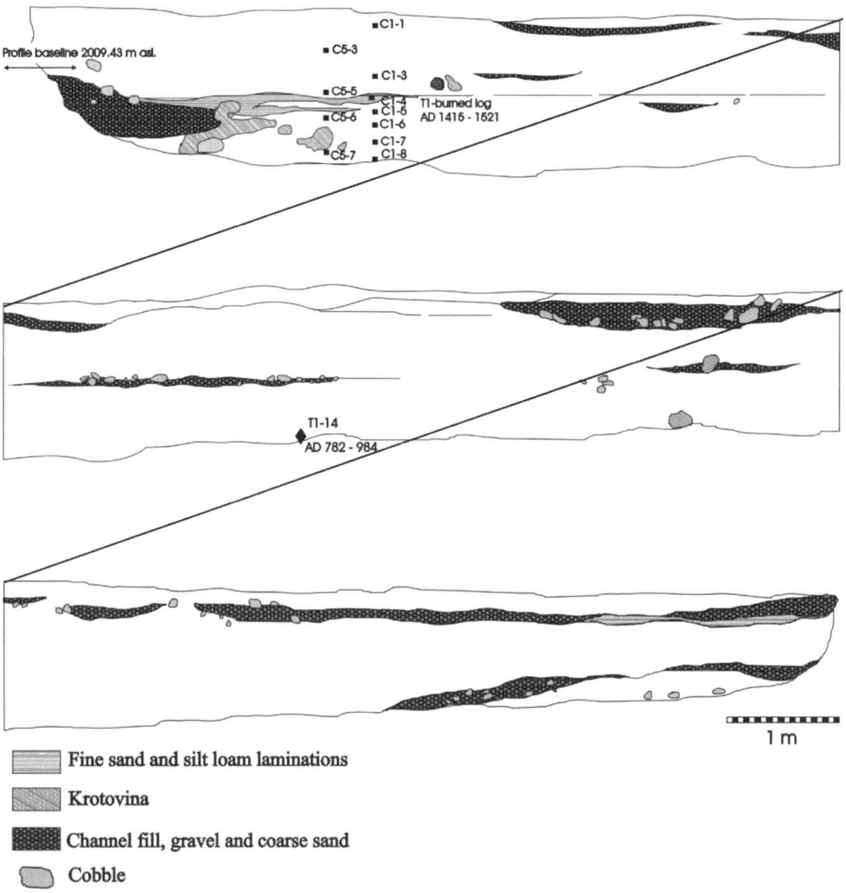

FIGURE 9.13. Staggered profile of Trench 1 at Simkins Flat showing braided channels, variation in channel fills, krotovina, and pollen-column (C1-3) and charcoal (T1) sampling locations; specimens "T1-14" and "T1-burned log" yielded 2-sigma radiocarbon determinations of A.D. 782–984 (AMS) and A.D. 1415–1521, respectively.

from diverse contexts and analyzed by several different specialists, have not conspired to mislead us, it is difficult to see how subsistence economies in this part of the Southwest were committed to a maize-dependent lifestyle.

THE UPPER BASIN AND THE ABAM

Several predictions of the ABAM (Plog et al. 1988) can be assessed with Upper Basin paleoeconomic and alluvial geological data, which we now describe briefly. Of four backhoe trenches placed in Simkins Flat, an alluviated valley in the center of the Upper Basin, the stratigraphy revealed in Trench 1 proved to be the most informative regarding long-term depositional processes (Figure 9.13). The exposed deposits contained braided-channel material ranging from cobbles, in poorly sorted gravel and coarse sand, to fine sand and silt alluvium deposited laterally to these channels (see Huckleberry and Billman 1998:599, 603). A microcharcoal sample ("T1-14" in Figure 9.13) collected from the base of Trench 1 yielded an AMS age of 1,141 ± 36 years B.P. (2-sigma calibrated data range is A.D. 782–984). A conventional radiocarbon determination obtained from the outer rings of a buried, partially charred pinyon ("T1-burned log" in Figure 9.13), located 60 cm higher in the deposits than the microcharcoal

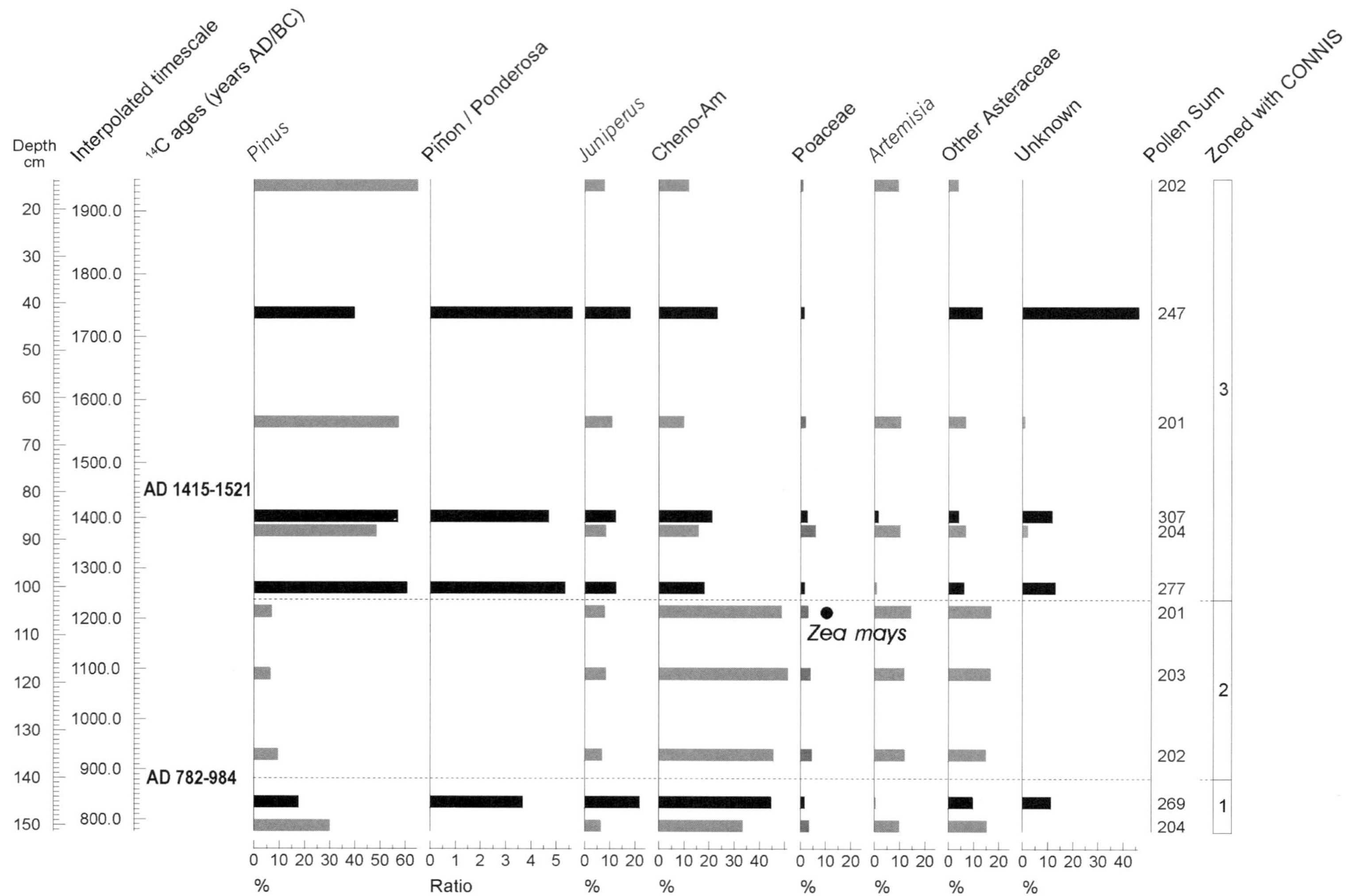

FIGURE 9.14. Taxa variation for 11 pollen samples recovered from various depths at Simkins Flat. Note that cheno-ams were most abundant when the Upper Basin was occupied perennially. These and other noncultigen annuals were part of an anthropogenic ecosystem that thrived in flood-borne alluvium.

sample, disclosed an age of 425 ± 40 years B.P. (2-sigma calibrated data range is A.D. 1415–1521). These two dates, which are barely adequate to support a local alluvial chronology, suggest nonetheless that the Red Hawk Wash catchment (11.13 mi^2) deposited at least 1.4 m of alluvium across Simkins Flat during the last 1,000 to 1,200 years (see Karlstrom 1988:51–55).

Such embryonic evidence indicates that when the Upper Basin was occupied, Fluvial States 2 and 3, which were periods when water tables were rising or were high (Plog et al. 1988:242), would have characterized the local paleohydrology. We find no support, however, for the ABAM prediction that during Fluvial State 2 agriculture was more important than hunting and gathering (Plog et al. 1988:247) and that reliance on annual plants decreased. As enumerated above, the Upper Basin paleoeconomic data indicate just the opposite. Furthermore, analysis of eight pollen samples from Trench 1 (C1-4 to C1-8, C5-5 to C5-7; Figure 9.13), which are bracketed by the two radiocarbon dates mentioned above, disclosed only one grain of maize from 1,867 that were identified (Cummings 2001; Smith 2003). Cheno-ams were most abundant when the Upper Basin was occupied perennially (Figure 9.14). An inescapable interpretation is that these and other noncultigen annuals were encouraged in anthropogenic vegetation communities that had been deliberately established on Simkins Flat. Hence the founding of four nearby one-room structures arguably was unrelated to ancient people's considerations of the suitability of Simkins Flat for maize farming (as posited by Lipe and Matson 1975:68).

For Fluvial State 3, we have no evidence that farming is concentrated along stream channels or in deeper alluvium (Plog et al. 1988:247). Studies of farming technology in the Upper Basin (Homburg 1992; Sullivan 2000) indicate that features such as agricultural terraces are rare, especially in comparison with the North Rim (Hall 1942; Schwartz et al. 1981) and the Inner Gorge (Schwartz 1963:295; Schwartz et al. 1980). In fact, what few terraces have been found were placed upslope from alluvial valley bottoms.

It is difficult to extend to the Upper Basin the Black Mesa–based observation that during Fluvial States 2 and 3, "people were more selective in their diet, relying mostly on cultivated crops and large mammals" (Euler et al. 1979:1096; also see Gumerman and Dean 1989:111). Actually, the Upper Basin data could support this inference but only if wild plants are considered "crops" (e.g., Bye 1981). With respect to land use and mobility in the Upper Basin during Fluvial States 2 and 3, our intensive survey (Sullivan et al. 2000) has revealed that settlements were placed across the landscape without regard to "increased availability of valley floor farmland" (Plog et al. 1988:254), which was never extensive during any period, and were abandoned for reasons unrelated to variation in effective moisture or floodplain aggradation (Huckleberry and Billman 1998).

The analysis of the extensive paleoeconomic and limited geomorphological data from the Upper Basin indicates that high-frequency paleoenvironmental fluctuations between A.D. 1050 and 1200 may have had little effect on the decision making of the area's prehistoric occupants, especially with regard to livelihood options. Because subsistence focused on the production of wild resources in bulk quantities, the food supply was largely immune from factors such as reduced effective moisture and floodplain erosion, factors that had major impacts on agricultural systems elsewhere on the southern Colorado Plateaus (Dean et al. 1985). Determination of subsistence mix is a critical consideration in ascertaining how dependent prehistoric Pueblo societies were on agricultural production (Plog et al. 1988:263). Perhaps, as Kohler et al. (2000:173) and others (e.g., Dobyns 1972; Sullivan 1996) have argued, the extent to which these groups were agriculturalists has been overestimated, a consequence, largely unintentional, of applying the Puebloan agricultural model too uncritically (Schoenwetter and Dittert 1968:43).

Another factor that may have contributed to the ABAM's meager predictive success in the Upper Basin is the difficulty involved in fully appreciating the effects of anthropogenic environments on estimates of carrying capacity (Cordell 1984:315). Because people create and manage ecosystems, they can dramatically affect the density, distribution, and accessibility of edible wild plants across the landscape (Sullivan et al. 1999). Theoretically, the production of wild plants could be maintained or even intensified in the face of high-frequency environmental fluctuations that otherwise would prove catastrophic to agricultural systems (Sullivan 1996). The Upper Basin data suggest that we broaden our economic (Wills 1992), adaptationist (Dean et al. 2000), and selectionist (Leonard and Reed 1993) models to include the possibility that local variation in wild-resource production promoted the changes in regional settlement systems and population levels that we seek to understand (Gumerman and Gell-Mann 1994; Powell 2002).

RETHINKING PREHISTORIC SOUTHWESTERN ECONOMY AND SOCIETY

With little doubt, selective obsession with Western Pueblo (principally Hopi) agrarian ecology has constrained alternative interpretations of paleoeconomic assemblage variation. Edible wild resources, in particular, often have been explained away as epiphenomena (e.g., contaminants; Hutira 1986:300–301) despite their long, undisputed use histories in the Southwest and elsewhere (Dobyns 1972). Conversely, rare occurrences of cultigens are elevated to the status of staples even though they are swamped numerically by wild plants (Rocek 1995).

It is sobering, therefore, to consider R. M. Bradfield's (1995:234) retrospection of several decades of ecosystemic research among the Hopi: "Gradually, I came to regard the Hopi, both in point of historical development and in the image which they hold of themselves in relation to their environment, primarily as a hunting-and-gathering people who incidentally practice some agriculture" (see also Eggan 1989:30; Winter and Hogan 1986). Lamentably, we can only speculate on the direction that studies of ancient southwestern subsistence would have taken had Bradfield's views, rather than Powell's or Morgan's, been adopted (see Gasser 1982:45–52; also Kehoe 1981). Conceivably, the Upper Basin economic pattern could have been practiced widely throughout the woodlands of the American Southwest (Tainter 1984) far more commonly than current models imply (Powell and Gumerman 1987). At the least, the Upper Basin case provides an opportunity to challenge the orthodoxy that has constrained our thinking about the range of livelihood options available to ancient southwestern societies on the Colorado Plateaus.

NOTES

1. Meteorological data for Grand Canyon (downloaded from NOAA's National Climatic Data Center website [http://cdo.ncdc.noaa.gov]) are an aggregate of observations from four stations that cover different periods of time between 1904 and 1983. Data for 1904–1943 come from a station, presumably in Grand Canyon National Park ("Grand Canyon Headquarters"), located at 6,891.3 ft asl. A station called "Grand Canyon Airways," presumably located outside the park, was at 6,973.3 ft asl and covers the period 1944–1957. Another station, presumably in Grand Canyon National Park, situated at 6,953.3 feet asl, covers the period 1957–1976. For the period 1977–1983, yet another station in the park ("Grand Canyon National Park 2") was located at 6,783.4 ft asl. Inexplicably, all stations have the same position (36'03" N, 112'08" W), with the exception of Grand Canyon National Park 2 (36'03" N, 112'09" W).
2. For the period 1948–2000, there are missing frost-free data for the following months: January ($n = 2$), February ($n = 2$), March ($n = 1$), April ($n = 1$), May ($n = 3$), June ($n = 1$), July ($n = 2$), August ($n = 3$), September ($n = 4$), October ($n = 3$), November ($n = 5$), and December ($n = 3$).
3. Dendrochronological data were generously provided by Jeffrey S. Dean, Laboratory of Tree-Ring Research, University of Arizona.

ACKNOWLEDGMENTS

A.P.S. thanks the late Robert C. Euler and Jack Rossen for commenting on an earlier draft of this chapter and James Schoenwetter for discussing the intricacies of interpreting archaeopalynological data. Patrick M. Uphus downloaded the NOAA weather data and produced Figures 9.2–9.5, for which we are most grateful. The Upper Basin Archaeological Research Project has been supported by grants from the University Research Council (University of Cincinnati), C. P. Taft Memorial Fund (University of Cincinnati), McMicken College of Arts and Sciences (University of Cincinnati), Center for Field Research (Earthwatch), and USDA Forest Service (Kaibab National Forest). We thank John A. Hanson, Kaibab National Forest, for his long-term intellectual and administrative support of the Upper Basin Archaeological Research Project.

10

Environmental Variation and Prehistoric Culture in the Mimbres Area, Southwestern New Mexico

Darrell Creel

ARCHAEOLOGISTS HAVE LONG RECOGNIZED THAT ENVIRONMENTAL variability must have imposed constraints on the ancient inhabitants of the Mimbres River valley in southwestern New Mexico (Figure 10.1), but comparatively little research has been directed toward the nature of this relationship. The work by Minnis (1985b) and the Mimbres Foundation in the 1970s stands as the most important and has been pivotal in reconstructions of prehistoric cultural change in the area. Although much new information on cultural manifestations has since been acquired, recovery of useful paleoenvironmental data unfortunately has not been commensurate (see, however, Woosley and McIntyre 1996).

For the last several decades, Mimbres researchers have had access to local long-term tree-ring records documenting variability in precipitation (a high-frequency environmental process), but we are still without records of low-frequency processes that may also have affected prehistoric lifeways. Low-frequency processes include stream-channel entrenchment and alluvial deposition, but such processes may not be perceived by the human population in a given area. As Dean (1988a:30) notes, the impact of changes in low-frequency processes relative to the impact of changes in high-frequency processes is likely to vary according to a number of factors, perhaps the most important of which is how close a particular subsistence system is to carrying capacity. That is, the closer a system is to carrying capacity, the greater the potential impact of fluctuation in high-frequency processes. The possibility that the Mimbres population exceeded long-term carrying capacity was addressed by Minnis (1985b), but he did so using only data on high-frequency processes. In this chapter, I reexamine what has been referred to as the "collapse" of the Mimbres system (LeBlanc 1983:159) about A.D. 1130, when carrying capacity may have been exceeded. First, it is useful to summarize the preceding centuries of cultural change in the Mimbres valley.

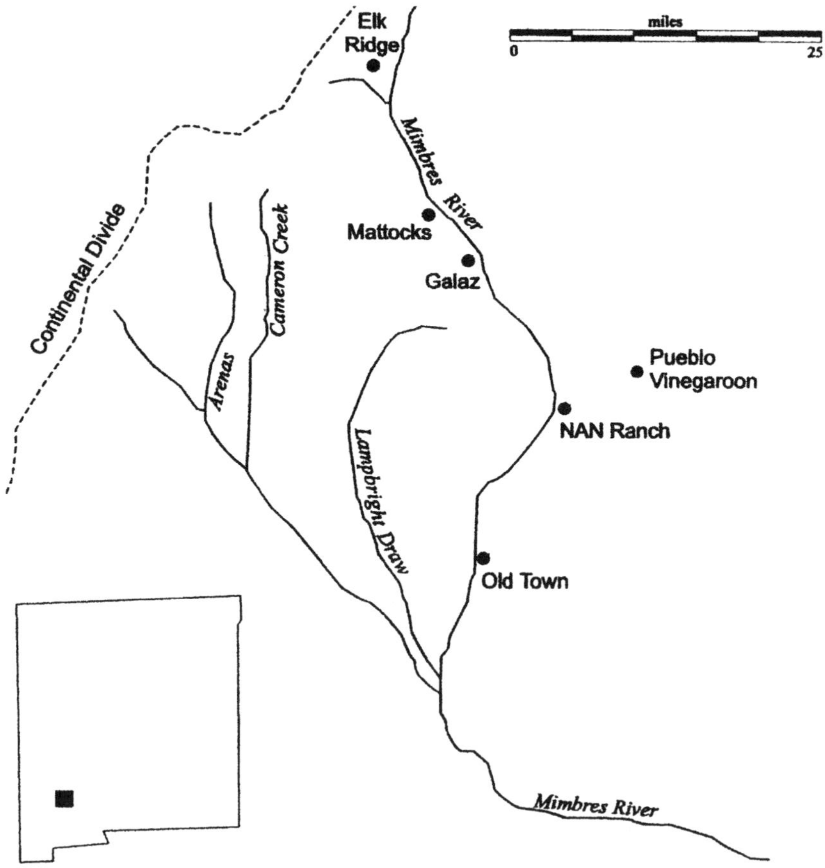

FIGURE 10.1.
The Mimbres drainage: locations of sites and streams.

CULTURAL SEQUENCE

The earliest period of Mimbres valley occupation considered here is the Late Pithouse period, ca. A.D. 550–1000. During this long period, population apparently increased, as did the degree of dependence on agriculture (Blake et al. 1986). Along the Mimbres River, settlement location was restricted to lower landforms, although there seems to have been more variation in landform selection along at least some of the major tributaries. The shape of pit structures, both domestic and communal/ceremonial, changed from round to square or rectangular; and late in the period, some pithouses were remodeled by closing off the lateral entryways and opening a ceiling hatchway accessed via ladder (Anyon et al. 1981; LeBlanc 1983; Shafer 1995). In addition, excavations at the NAN Ranch Ruin found several "sunken floor" rooms with ceiling entries built late in this period (Shafer 1995). Shafer (1995) summarizes the changes in hearth type, mortuary customs, and ceramics that occurred during the later part of the Late Pithouse period (see also Anyon and Leblanc 1984; Cosgrove and Cosgrove 1932; Haury 1936; LeBlanc 1983; Lucas 1996). Among

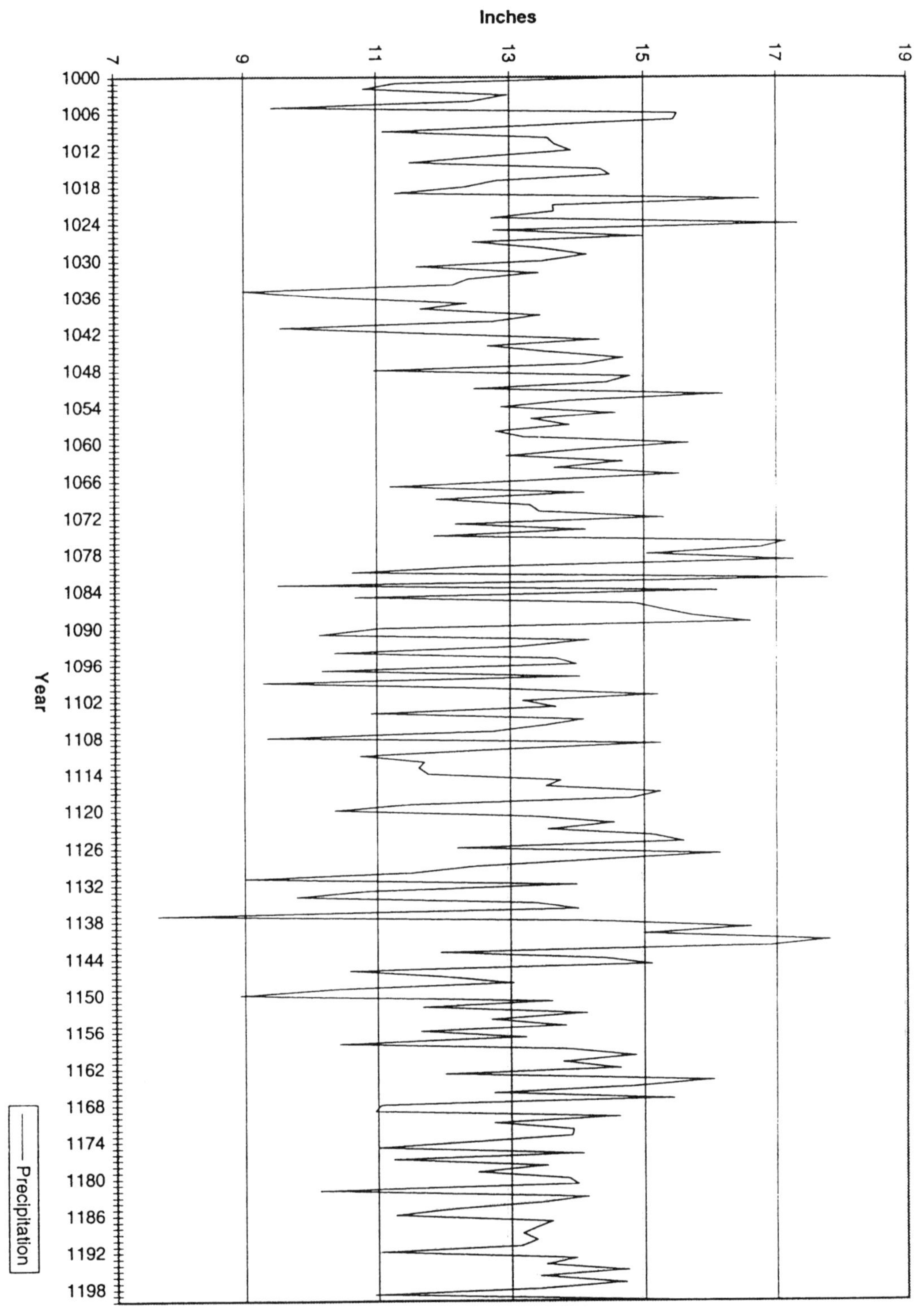

FIGURE 10.2. Upper Mimbres valley precipitation (prior August–current July) from tree rings, A.D. 1000–1250.

the more notable changes is a shift to interment of the dead beneath the floors of occupied houses. The practice of "killing" ceramic vessels placed with the dead, first by smashing them, later by punching a hole in them, also began in this period.

Using data from the Mimbres Foundations survey and excavation, Minnis (1985b) estimates the Mimbres valley Late Pithouse population maximum at about 1,200, a level he feels was too low to have exceeded carrying capacity (see Blake et al. 1986 and Ellis 1998:Chapter 8 for alternate population assessments). His estimate may be judged too high, but this does not detract from the implication that fluctuations in precipitation, water table levels, and surface streamflow would probably have been of less consequence for a smaller than for a larger population. As Minnis points out, there were several notable dry intervals during the Late Pithouse period, including the A.D. 820s, 840s, 880s, and 920s. Notable periods of high temporal variability were the A.D. 760s and 970s. It is difficult to know how the Mimbreños responded to these stressful times, but the responses presumably changed as the population increased.

There is clear cultural continuity from the Late Pithouse period to the Classic Mimbres phase (ca. A.D. 1000–1130), but one of the most important changes was a shift from pithouses to surface pueblos (Anyon and LeBlanc 1984; LeBlanc 1983; Shafer 1995). Following Gilman (1987), Lekson (1992) argues that full sedentism did not occur until the Classic period, when the population became dependent on agriculture. Not all researchers subscribe to this interpretation (e.g., Diehl 1996, 1997). Much of the population lived in large multi-room-block villages along the Mimbres River and its major tributaries, but there also seems to have been major population growth along lesser tributaries, where small plots of good farmland as well as lower-quality agricultural land were brought into production (Anyon and Leblanc 1984; Blake et al. 1986; Cosgrove and Cosgrove 1932; LeBlanc 1983; Shafer 1995).

It has been argued that the Mimbres and other streams were used for irrigation, but physical remains of irrigation ditches confidently dated to the Classic have so far been found only along the Rio Arenas near Silver City (Herrington 1979, 1982). Ditches and other water-control structures probably dating to the Classic are also known along Cameron Creek and Lampbright Draw (Ellis 1995; Herrington 1982).

Archaeologists have markedly divergent views of events occurring after the A.D. 1130s, when drought conditions apparently lowered the region's carrying capacity, but all agree that major changes in settlement pattern, building construction methods, and ceramics occurred between about A.D. 1130 and perhaps 1200 (Hegmon et al. 1999). Some villages were abandoned, particularly those in higher elevations, and it is possible that the entire area was abandoned for a period of time. Recently, Hegmon et al. (1999) have proposed revised phase systematics for this period; notably, they discuss the case for a brief "Terminal Classic," following my argument (Creel 1999a).

Sometime between the mid- to late A.D. 1100s and early 1200s, new villages assigned to the Black Mountain phase (Anyon and LeBlanc 1984; LeBlanc 1983, 1989) were established along the river and its tributaries. Unlike in the preceding Classic, however, these were confined to the middle and lower reaches, occurring well out into the desert basins (Lekson 1992; Ravesloot 1979). In contrast to the masonry construction of the Classic, Black Mountain phase pueblos were built mainly of puddled adobe. The famous Mimbres painted pottery was no longer made, and except for the utility ware, the Black Mountain phase ceramics were generally different from those of the Classic.

At one extreme is the belief that the Classic system collapsed and the Mimbres drainage was abandoned; at the other extreme is the suggestion that the region saw only partial abandonment and that there was ethnic continuity between the Classic and the following Black Mountain phases (Creel 1999a; Hegmon et al. 1999; Nelson 1999). It is relevant to note here that there has been relatively little investigation of Black Mountain phase sites and that we do not have adequate data at present to resolve the matter. Nevertheless, we can clarify some issues that have, in my opinion, clouded our interpretations and research questions. One of these issues is how we identify Black Mountain phase remains and the implications this identification has for how we perceive the changes that may have affected the late Classic population. In addition, I assess how the precipitation regime and streamflow characteristics may have affected the Mimbres population during the A.D. 1130s drought.

TWELFTH-CENTURY CERAMIC CHANGES AND PHASE ASSIGNMENTS

It has long been recognized that pottery characteristic of the Black Mountain phase occurs in many Classic Mimbres sites, but such occurrences have generally been attributed to brief reoccupations by non-Mimbreños (LeBlanc 1983:164). I have argued elsewhere that in some cases the context of these ceramics may in fact be Classic (Creel 1997, 1999a). For example, Room 97A at the Galaz site had subfloor burials, one with an associated Mimbres Black-on-white Style III vessel and one with an associated El Paso Polychrome vessel (Anyon and LeBlanc 1984:Appendices I–III). Since there is no reported evidence of temporal/stratigraphic differences between these burials, I tentatively conclude that they were contemporaneous and related to the Classic occupation. Other rooms at Galaz have similar sets of burials. In their analysis, Anyon and LeBlanc (1984:316) acknowledge the plausibility of this interpretation. To be fair, though, I should repeat their cautionary assessment that documentation of the early Galaz excavations is not definitive in matters of context. On the other hand, the intrasite occurrence of rooms with such sets of burials is more consistent with my interpretation than it is with one that posits reoccupation of these rooms during the Black Mountain phase.

The question then becomes, are the burials with Black Mountain phase pottery in these rooms Classic or Black Mountain? The answer has significant implications. If I am correct that the interments with ceramics otherwise characteristic of the Black Mountain phase were actually Classic Mimbres, then we must acknowledge that pottery types such as El Paso Polychrome, Chupadero Black-on-white, and Playas Red were being made and/or acquired by Mimbreños during the Classic period. Based on the few sites with late Classic tree-ring-dated rooms (Mattocks, NAN Ranch, Pueblo Vinagaroon, Elk Ridge), these pottery types did not appear in the area until at least ca. A.D. 1120, perhaps as late as A.D. 1150. Therefore, the best we can say is that the introduction of these types occurred just prior to, during, or perhaps after the drought of the A.D. 1130s. Strictly speaking, then, the new ceramics would not reflect a new population, especially since the Mimbres valley was only one area in the southern Southwest where this new set of pottery types was replacing older suites. The period of time from the introduction of these ceramic types until the construction of Black Mountain phase pueblos has been referred to as the Terminal Classic (Creel 1999a; Hegmon et al. 1999).

DROUGHT AND THE END OF THE MIMBRES CLASSIC

Minnis (1985b:116) argues that the Classic Mimbres population may have equaled or exceeded the system's carrying capacity (see Ellis 1998 for discussion). If so, how may the drought of the A.D. 1130s have affected the Classic Mimbreños? As shown in Figure 10.2, the 1120s in the Mimbres valley were, on the whole, good years, the 1130s dry overall, and the 1140s well above average (data from Laboratory of Tree-Ring Research's "Dendroclimatic Characterization of Southwestern Paleoclimate During the Past 2000 Years" project [NSF Grants BNS-8818628 and DBS-9205968]). A key feature of the 1130s record was a multiyear dry period. From 1129 to 1131, there were three consecutive years of below-average precipitation, but only 1131 was more than one standard deviation below average. The other dry years in the 1130s were similar except for 1137, which was the driest year in several centuries at nearly 3 standard deviations below average. Each of the one-to-three-year dry periods in the 1130s was followed by one or more years of above-average precipitation. Given what we know, the so-called drought of the A.D. 1130s was fairly severe but perhaps only slightly worse than the drought of the A.D. 1030s–early 1040s. However, its effect on what may have been a larger population was likely more pronounced.

Given the positive relationship between precipitation variability and modern streamflow fluctuation in the Mimbres River, as shown in Figure 10.3 (streamflow recalculated to reflect the same 12 months as reconstructed precipitation), it is likely that agricultural production, even from irrigated farmland, was significantly reduced during the dry years of the A.D. 1130s. However, it is difficult to relate modern streamflow to prehistoric conditions. First, the diversion of water for irrigation differentially affects measured streamflow at the main gaging stations (Hanson et al. 1994:16). Second, the presence of

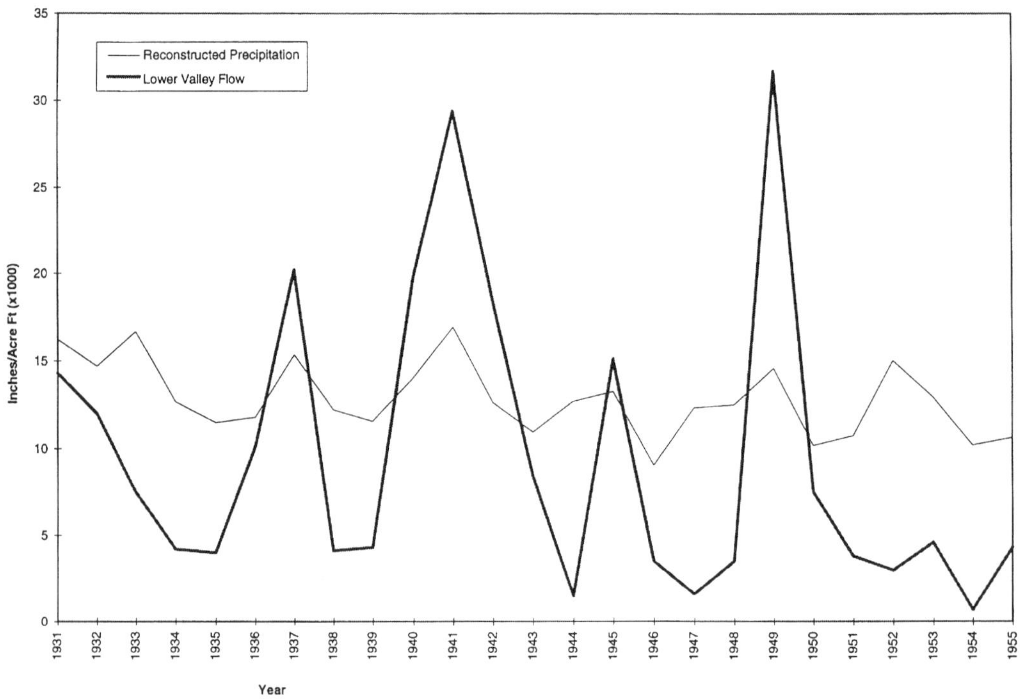

FIGURE 10.3. Upper Mimbres valley precipitation from tree rings and streamflow at lower valley gaging station, 1931–1955 (both prior August–current July). The high streamflow volume for 1937, 1941, and 1949 was primarily from heavy winter (January–April) precipitation.

volcanic formations at or near the surface creates complex hydrologic conditions in the Mimbres valley. Some of these volcanic formations (also known as bedrock masses) act as thresholds that force groundwater up to or near surface and thus significantly affect both water table and surface flow levels.

The locations of modern streamflow gaging stations do not permit adequate documentation of the small-scale variation in surface flow inherent in this hydrologic setting, variation that must have been important to prehistoric agriculturalists. For example, no streamflow was recorded at the Faywood gaging station in the lower Mimbres valley on January 14, 1937, nor had there been any for at least four months, and precious little was recorded in the five months before that (U.S. Geological Survey 1960:617–619). But aerial photography clearly reveals that there was surface flow three miles downstream at Old Town, where the water table is shallow because of a nearby threshold.

The modern records reveal monthly and yearly streamflow variability that would have been important to prehistoric farmers (U.S. Geological Survey 1960:617–619, 1964:555–557). Summer runoff would have been critical, and in some years of low overall streamflow, most of the runoff came in July–September, when it was most needed. There have been years of low streamflow when the critical July–September runoff was also low. Conversely, high runoff values in winter months, fairly common in wetter intervals, would probably have been of relatively little consequence other than to recharge the water table and perhaps wash out diversion dams. Similarly, low winter runoff would have had comparatively little direct impact the following growing season except as it may have led to somewhat lowered water table levels and surface flow. Finally, it is likely that more precipitation infiltrated the soil and recharged groundwater supplies before the degradation of vegetation by domestic livestock after its introduction in the late 1800s.

As a result, it is difficult to retrodict Mimbres River streamflow variability as it may have affected prehistoric farmers. What may have been dry years based on the tree-ring record may not have been so bad in much of the Mimbres valley, especially if the preceding year was wetter and the water table levels were high. It seems probable, though, that successive dry years had a perceptible negative impact. Indeed, the dry six-year period of A.D. 1129–1134 (Figure 10.2) must have been difficult for the Mimbres population since surface flow that could be diverted for irrigation probably diminished or ceased along some segments of the valley. The slightly above-normal precipitation for A.D. 1132 probably did little to recharge the water tables, which were presumably much reduced after three dry years. Therefore, the following two dry years, 1133–1134, may well have been devastating for much of the population. As drought conditions progressed and surface flow ceased, the amount of irrigated land in the Mimbres valley presumably decreased as diversion became impossible in those sections of the valley most susceptible to a lower water table.

What options did the Mimbreños have in those dry years? Minnis's (1985b:150) data indicate that cultivated and native food supplies together may have been inadequate to sustain the population during the dry years of the A.D. 1130s (but see Ellis 1998). The only response for which archaeologists have any solid data is emigration or abandonment, the traditional explanation. The Classic Mimbres pueblos above an elevation of about 6,000 feet were apparently abandoned, but it is not clear that those at lower elevations were abandoned as well. Indeed, there appears to have been major construction at Old Town at about this time (Creel 1999b). If so, it implies that the Black Mountain phase began somewhat later than the often cited date of A.D. 1130 and that the changes leading to the Black Mountain phase were not directly related to the drought.

There are substantial Black Mountain phase pueblos on or near most of the Mimbres pueblos in the middle and lower portions of the valley. In no case (except possibly Galaz) is there substantive data on the temporal difference between the Classic and Black Mountain phase room blocks, but it is possible that some of the latter were built as, or shortly after, the nearby Terminal Classic room blocks were abandoned. The question for which I have no answer is, were the new room blocks part of the same communities as the Classic room blocks? Perhaps it would be worth reconsidering our definition of community and the conceptual problems that can result from arbitrary site designations.

Beyond that, it is instructive to consider the possible abandonment of the Mimbres valley in the A.D. 1130s. Times may well have been bad during some years, and movement elsewhere seems a practical response if there is somewhere else to go. We should recognize that conditions in the Mimbres valley, as

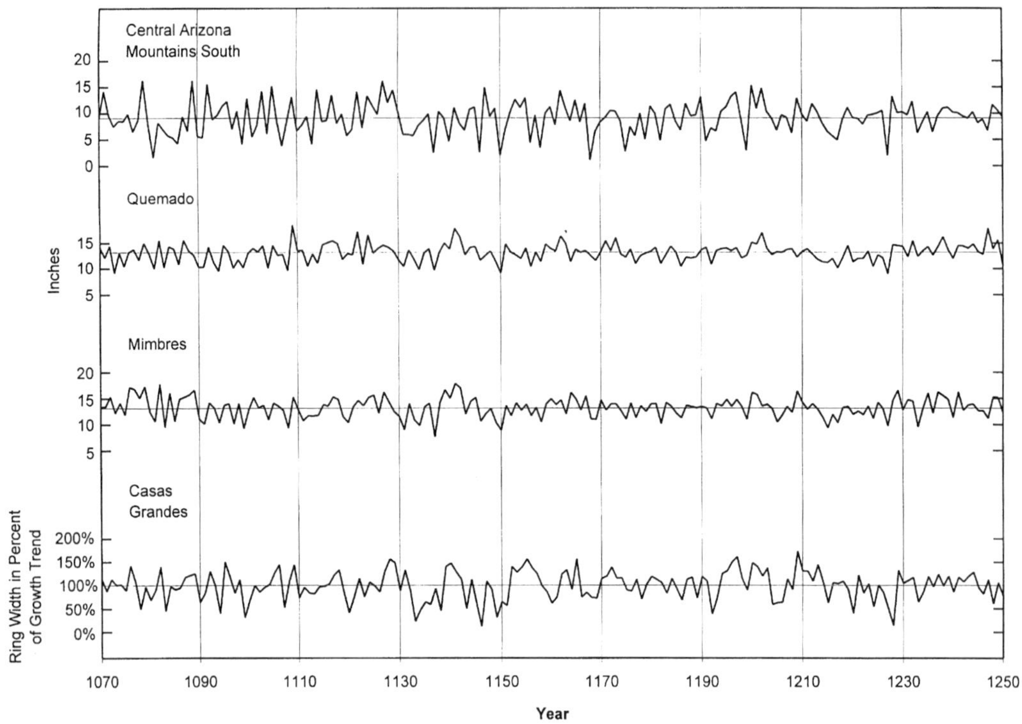

FIGURE 10.4. Reconstructed precipitation (in inches, prior August–current July) for Central Arizona Mountains South, Quemado, and Mimbres areas, A.D. 1070–1250. At bottom is a plot of ring width in percent of growth trend for the same time range in the Casas Grandes, Chihuahua, area.

reconstructed from the tree-ring record in the upper drainage, appear to have been no worse, and perhaps were even somewhat better, than in surrounding areas for which there are comparable data. As shown in Figure 10.4, areas north, west, and south of the Mimbres valley experienced dry years in the A.D. 1130s, as did the southern Rio Grande basin (Grissino-Mayer et al. 1997) and the Southwest as a whole. The Quemado, Mimbres, southern Rio Grande, and Casas Grandes tree-ring records are similar for this time period and show little spatial variability in precipitation. As a group, they differ somewhat from the Central Arizona Mountains South record during the A.D. 1130s–1140s, primarily in having experienced a less prolonged dry period.

All else being equal, the Mimbres valley seems to have been as good a place as any in the 1130s drought, at least for those with the best land. Demographic changes in this drought may have been similar to those at Oraibi, discussed by Levy (1992), in which people without access to productive land left the community during periods of scarcity. Perhaps the drought of the 1130s resulted in similar actions by the Mimbreños.

We must remember that the amount of productive, irrigable land in the Mimbres valley probably decreased as the drought continued. During a prolonged drought, the ever-dwindling base of productive land may have supported only a modest core of the former population. In addition, not only did precipitation totals decrease during this drought, but the climate regime was also more erratic and less predictable (Grissino-Mayer et al. 1997:58). As

Nelson and Anyon (1996) point out, there may well have been fallow valleys elsewhere in southwestern New Mexico that could have absorbed some people from the Mimbres valley in the A.D. 1130s, but the reality of this possibility has yet to be determined.

The marked similarity between Classic and Black Mountain phase mortuary practices also seems to me, as it did to Anyon and LeBlanc (1984), to reflect ethnic continuity, although this does not necessarily indicate continuous occupation of either specific communities or even segments of the Mimbres valley. Indeed, it does not preclude Nelson and Anyon's (1996) fallow valley hypothesis. Notable similarities in the corrugated pottery of the Classic and Black Mountain phases are further evidence of continuity (see also LeBlanc 1983:164).

Regardless of any specific explanation of the Classic–Black Mountain transition, we cannot forget that (1) cultural changes were occurring across the Southwest and (2) drought was also affecting other areas, some more drastically than others. Documented demographic and settlement pattern changes were a response, at least in part, to this drought. Regardless of their causes, the long-term demographic/settlement pattern changes surely must have affected property use rights, water rights, and other aspects of social organization in profound ways.

Obviously, much remains to be learned about the so-called Mimbres collapse and subsequent events in the area; more research is needed at Black Mountain phase sites, especially those adjacent to Classic pueblos. In addition, a thorough geomorphologic and hydrologic study of the Mimbres drainage should be made. Indeed, there is evidence for river channel changes that may well have affected irrigation capability. Such studies cannot come soon enough.

11

Tree Rings, Drought, and the Pueblo Abandonment of South-Central New Mexico in the 1670s

James A. Parks, Jeffrey S. Dean, and Julio L. Betancourt

CLIMATE-SENSITIVE TREE-RING SAMPLES FROM AN ISOLATED, LOW-elevation pinyon (*Pinus edulis*) stand in the Sevilleta National Wildlife Refuge near Socorro, New Mexico (Figure 11.1), produced high-quality dendroclimatic reconstructions that illuminate the environmental and human histories of the area over the last millennium. The stand of mostly dead trees is at an elevation of 1,585 to 1,705 m on the north slopes of washes that form the headwaters of Arroyo Milagro, east of Mesa del Yeso and south of the Los Piños Mountains. The location's aridity is exemplified by the presence of creosote bush (*Larrea tridentata*) and the northernmost population of ocotillo (*Fouquiera splendens*). The severely stressed Sevilleta pinyons and one-seed junipers (*Juniperus monosperma*) had experienced a massive dieback, perhaps in a single death event. Living trees endure only where conditions are most favorable to surviving sustained drought: in or near dry streambeds for access to runoff, near-surface underflow, and cool-air drainage. Nearly all the mature trees on the high slopes had died. The dating of the tree-ring samples showed that a high percentage of the dead trees succumbed in the 1950s during a notorious drought period across the southern United States. Ecological implications of this discovery are discussed elsewhere (Betancourt et al. 1993; Swetnam and Betancourt 1998). In this chapter, we treat the development of the Sevilleta tree-ring chronology and use this and other chronologies from the area to reconstruct precipitation in south-central New Mexico. These reconstructions illuminate the relationships between climate variability and the protohistoric and historic peoples of the area, particularly the occupants of the Piro and Salinas pueblos.

TREE-RING CHRONOLOGY DEVELOPMENT

Increment cores and cross sections were taken from 269 trees to determine recruitment ages and approximate death dates for living and dead trees. The cross sections and cores were sanded to expose the ring structure for micro-

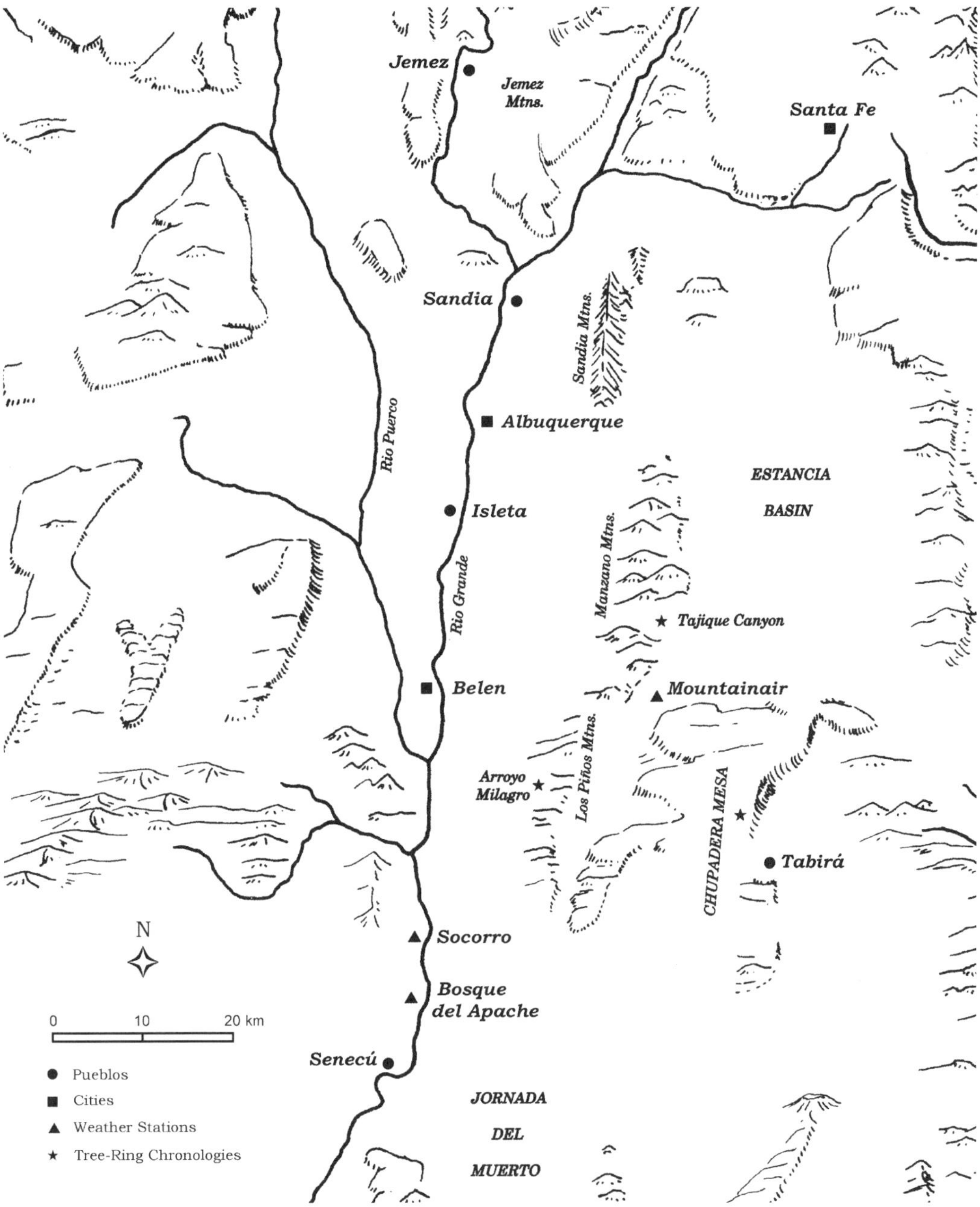

FIGURE 11.1. The study area in south-central New Mexico.

scopic examination. Standard dendrochronological procedures (Stokes and Smiley 1968) were followed to crossdate the Sevilleta ring series. Problems such as compressed ring series, erratic radial growth, and excessive numbers of false rings (intra-annular latewood bands) made crossdating impossible for all or parts of some samples. Samples with few such problems were dated from innermost to outermost ring.

The primary considerations in these analyses were the calendar dates of the pith year and the bark year (death year) of each sample. For most samples, pith rings could be only approximately dated because of the compressed growth of the early parts of the ring series. This difficulty is common in pinyons, which tend to begin growing slowly, putting on very small rings. The true outsides of many dead-tree samples were completely weathered away, and these samples could not be given exact death dates; however, most outside dates from these samples fell into the same 1940s and 1950s statistical group as the true death dates. For the dead-tree samples that retained true outsides, death dates were difficult to determine because the trees appeared to be on a steady decline before dying, which produced compressed and difficult ring series. Thus, even the most confidently determined death dates might be one or two years off.

Ring series from samples with the fewest dating problems were selected for measurement. Three subgroups are represented: living trees dating out to 1991; trees with death dates in the 1950s; and the oldest trees, dating back to the 1600s. This selection provides adequate sample depth for the range of the chronology, excluding the earliest years because of the small number of thoroughly dated samples in this period. The selected ring series were measured to the nearest 0.01 mm on a computerized measuring machine described by Robinson and Evans (1980). COFECHA, a computer program designed to assist quality control (Holmes 1983), revealed some inconsistencies in dating and measurement that were visually checked on the samples and corrected if deemed appropriate.

Each measured tree-ring series was standardized by calculating the percentage departure of each ring width from an empirically determined growth trend line, a procedure that homogenizes the variance in each ring series and removes nonclimatic growth trends (Fritts 1976:261–268). The best method of standardizing the Sevilleta pinyons was the application of a stiff cubic spline trend line (Cook and Peters 1981) to each series. In addition, each series was autoregressively modeled to remove year-to-year autocorrelation while retaining the persistence common to most of the series (Cook and Holmes 1986). Using these procedures, we derived tree-ring indices for each measured ring series and averaged them into a final index chronology for the Sevilleta site for the 1598–1991 period.

Using a segment (1600–1719) of the Sevilleta chronology to bridge a gap of eight years between the Tajique Canyon and Chupadera Mesa chronologies previously developed by Dean and Robinson (1978) produced a revised Chupadera Mesa chronology extending from 1294 to 1972. The Tajique Canyon chronology, ranging from 1656 to 1972, is from a pinyon site on the east slope of the Manzano Mountains; the Chupadera Mesa chronology spans 1294 through 1648 and includes ponderosa pine (*Pinus ponderosa*) and pinyon archaeological samples from sites in the Salinas Pueblo Missions National Monument. Although the Sevilleta and Tajique Canyon chronologies exhibit a fairly high degree of agreement (Table 11.1), the 50-year overlapping segments of the Sevilleta and Chupadera Mesa chronologies do not agree well statistically. The modest correlation between the Sevilleta and Chupadera Mesa chronologies

TABLE 11.1. Correlation Coefficients Between Pertinent Tree-Ring Chronologies and Seasonalized Precipitation Data (Prior August Through Current August of Growth Year).

	SEVILLETA CHRONOLOGY	TAJIQUE CANYON CHRONOLOGY	CHUPADERA MESA CHRONOLOGY	SOCORRO PRECIPITATION	MOUNTAINAIR PRECIPITATION	BOSQUE DEL APACHE PRECIPITATION
Sevilleta chronology	XXX					
Tajique Canyon chronology	.50	XXX				
Chupadera Mesa chronology	.31	N/A	XXX			
Socorro precipitation	.74	.59	N/A	XXX		
Mountainair precipitation	.67	.67	.67	.69	XXX	
Bosque del Apache precipitation	.56	.45	N/A	.74	.64	XXX

may be explained by the shortness of the overlap and the low sample depth during that period for both chronologies. Two other dendrochronologists confirmed the between-chronology crossdating for the overlap period. These facts, the proximity of the chronologies (Figure 11.1), and the lack of other nearby chronologies long enough to cross the gap between the Chupadera Mesa and Tajique Canyon chronologies validate the use of the Sevilleta chronology to bridge the gap.

DENDROCLIMATIC RECONSTRUCTION

Precipitation data from three New Mexico meteorological stations (Western Regional Climate Center: http://www.wrcc.dri.edu/) were chosen for comparison with the Sevilleta and revised Chupadera Mesa chronologies: Socorro at 1,400 m elevation (1892–1991), Mountainair at 1,980 m (1902–1991), and Bosque del Apache at 1,395 m (1890–1991). The stations were chosen because of their proximity to the tree-ring sites and because of their long records with no missing yearly data and relatively few missing monthly values. The missing monthly values for each station were estimated by averaging the corresponding monthly values for the two other stations.

The tree-ring chronologies were statistically compared to seasonalized precipitation data from each station. Several different seasonal periods were examined, but the best was found to be a 13-month interval covering the prior August through the current August of the growth year. Grissino-Mayer et al. (1997) found a similar conditioning of cambial growth by rainfall of the previous summer–fall and the current spring–summer, with current winter

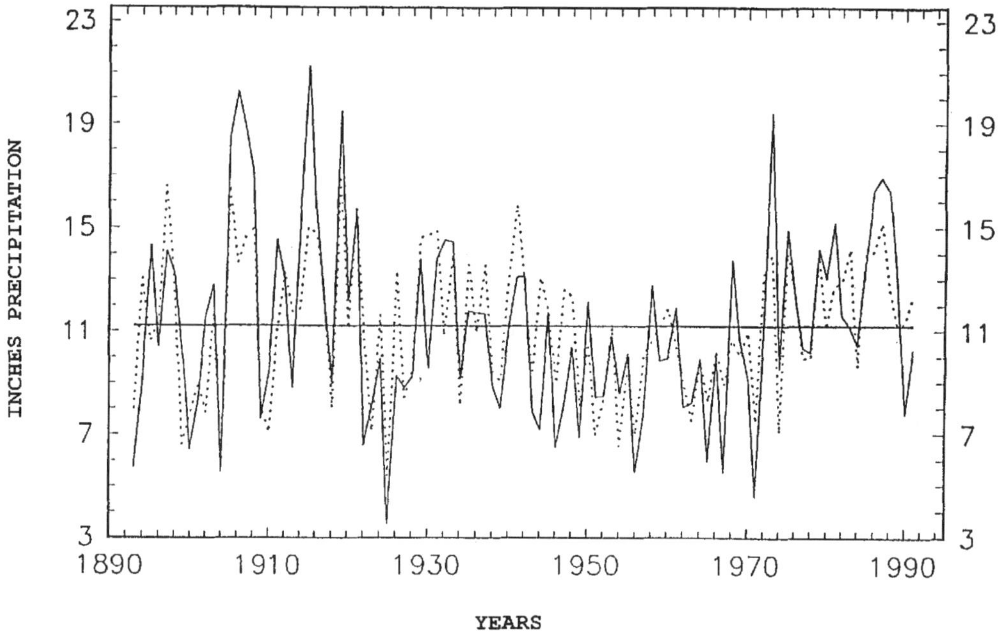

FIGURE 11.2. Actual annual Socorro, New Mexico, precipitation (prior August–current August of growth year) vs. reconstructed precipitation based on a tree-ring chronology from the Sevilleta National Wildlife Refuge. *Solid line* is actual precipitation, *dotted line* reconstructed.

(January–March) having little effect. Deficient precipitation before the growing season reduces tree growth by limiting food manufacture and storage; low precipitation during the growing season causes water stress, which limits cambial activity (Fritts 1976:207–245). The Sevilleta chronology compares most favorably with the Socorro data, explaining 55 percent of the variance (r^2) in Socorro precipitation (Table 11.1). Due to the influence on the chronology of a few old trees growing along the Arroyo Milagro wash, the Sevilleta chronology may be particularly indicative of variability in summer monsoonal precipitation and runoff in addition to winter soil-moisture conditions. The revised Chupadera Mesa chronology agrees best with the Mountainair data, explaining 45 percent of the variance in Mountainair precipitation. The percentage of variance explained by these chronologies is among the highest for previous climate reconstructions in New Mexico (Grissino-Mayer, personal communication 1995; Grissino-Mayer 1996; Grissino-Mayer et al. 1997; Rose et al. 1981).

The regression equations used to derive the relationships between the precipitation data and the corresponding tree-ring indices were transformed into transfer functions that convert the tree-ring indices into inches of precipitation. The Socorro and Mountainair reconstructions are presented in Tables 11.2 and 11.3, respectively. The similarities between the actual and reconstructed precipitation values for each station (Figures 11.2 and 11.3) and the simplicity of reconstructions made any further testing and verification of the transfer functions unnecessary.

TABLE 11.2. Annual Precipitation (Prior August Through Current August) in Inches for Socorro, New Mexico, Using a Tree-Ring Chronology from the Sevilleta National Wildlife Refuge.

	YEARS									
	0	1	2	3	4	5	6	7	8	9
1590									08.23	12.29
1600	10.19	08.79	13.20	13.10	13.54	10.15	11.76	10.00	10.57	14.55
1610	15.69	14.17	11.34	09.64	10.69	09.57	09.32	09.59	11.49	09.71
1620	12.61	11.52	11.66	09.15	08.46	07.36	11.45	13.01	09.93	12.80
1630	10.33	11.03	08.14	09.82	11.05	14.36	10.90	09.68	09.95	08.36
1640	12.18	12.37	11.43	13.30	14.32	13.61	14.03	14.67	11.69	12.51
1650	10.25	15.18	12.50	12.50	10.62	14.08	10.93	10.77	12.47	11.26
1660	12.59	14.95	12.30	11.34	10.62	14.25	12.24	09.94	10.17	09.70
1670	07.77	10.59	10.74	10.31	09.87	09.89	11.06	10.99	05.92	09.23
1680	09.37	10.09	10.68	11.88	10.27	07.40	13.18	11.24	11.48	15.54
1690	09.07	09.50	13.48	09.67	12.08	11.74	07.21	12.81	12.78	13.78
1700	14.04	12.89	11.95	07.95	10.44	13.08	12.95	09.75	13.16	12.49
1710	16.90	11.12	14.20	09.53	09.50	10.05	08.90	12.28	12.56	11.39
1720	15.48	15.38	15.81	12.78	15.60	12.61	15.84	13.03	11.24	08.66
1730	08.07	08.75	06.59	08.63	14.88	10.52	09.34	05.87	07.73	06.58
1740	09.84	09.41	07.24	10.02	09.50	10.77	13.87	14.95	06.59	13.09
1750	08.52	16.81	08.83	14.11	13.37	14.56	12.14	06.53	14.06	17.06
1760	11.91	14.72	16.08	07.02	11.58	11.79	14.06	15.91	12.77	11.03
1770	13.43	16.12	08.81	06.64	10.84	12.69	12.03	11.53	12.43	10.32
1780	08.11	08.87	09.82	12.76	11.54	07.54	06.32	08.90	10.76	08.92
1790	12.04	11.55	11.40	13.89	12.23	10.58	11.19	08.59	11.99	09.73
1800	10.85	11.39	12.31	11.39	13.72	13.32	05.93	12.50	09.11	09.76
1810	14.05	13.11	10.81	12.00	11.30	14.02	15.53	12.23	05.98	07.02
1820	13.18	12.38	07.36	09.12	09.20	11.68	07.33	10.27	12.69	12.86
1830	13.39	09.30	07.88	13.16	14.94	11.27	07.15	13.45	10.89	15.91
1840	14.06	15.24	10.45	12.45	09.88	10.34	12.64	12.02	10.18	16.66
1850	13.03	11.07	14.00	11.89	11.90	10.11	13.03	08.89	14.94	08.20
1860	11.14	09.89	06.36	10.96	06.90	10.74	10.62	13.42	14.80	15.55
1870	07.62	12.26	11.43	08.18	11.31	11.87	12.60	13.02	09.47	09.75
1880	12.38	09.85	15.24	09.85	10.01	13.89	06.34	12.90	12.66	12.83
1890	08.13	13.12	08.41	07.96	13.06	10.61	11.44	16.61	12.36	06.61
1900	07.79	08.97	07.83	11.67	05.65	16.60	13.64	14.63	14.95	08.10
1910	07.05	10.90	13.30	11.76	11.88	15.03	14.77	12.66	08.01	16.83
1920	11.25	15.81	10.66	07.11	11.65	05.37	13.32	08.42	09.53	14.59
1930	14.69	14.88	10.83	13.97	08.11	13.59	10.94	13.58	09.37	09.10
1940	13.27	15.86	12.78	09.32	13.09	11.85	08.99	12.68	12.33	08.11
1950	10.53	06.97	08.37	11.15	06.53	09.76	07.00	10.05	10.65	11.20
1960	11.90	10.50	08.95	07.52	09.82	08.26	10.25	08.92	10.68	10.03
1970	10.98	07.45	13.40	13.94	07.06	14.05	12.49	09.96	09.98	13.94
1980	11.14	12.75	12.92	14.20	09.61	14.01	14.09	15.21	12.72	11.25
1990	11.29	12.43								

TABLE 11.3. Annual Precipitation (Prior August Through Current August) in Inches for Mountainair, New Mexico, Using a Tree-Ring Chronology from the Chupadera Mesa area.

	YEARS									
	0	1	2	3	4	5	6	7	8	9
1290					14.18	15.01	13.18	15.72	15.46	20.29
1300	18.67	17.94	17.34	15.26	15.43	18.62	14.99	16.94	15.40	18.02
1310	18.62	17.37	18.68	19.55	18.70	16.54	14.67	17.58	17.35	17.75
1320	18.42	17.09	15.08	12.82	15.12	16.50	15.36	15.98	16.56	16.50
1330	17.19	16.42	19.48	19.89	18.15	16.77	14.40	16.66	14.34	16.98
1340	14.49	15.72	13.45	14.22	16.67	16.19	17.61	12.86	15.30	16.21
1350	14.90	16.82	13.58	16.26	19.14	18.09	20.64	15.31	18.78	18.78
1360	16.75	18.34	15.47	13.99	15.71	16.94	15.86	16.01	18.36	13.88
1370	17.03	16.65	19.69	17.71	18.15	14.85	15.35	17.77	20.34	16.93
1380	19.25	15.67	15.30	18.79	14.70	17.48	16.97	14.95	18.05	18.82
1390	15.18	15.53	18.75	16.71	13.32	16.26	15.19	15.02	20.32	14.98
1400	16.60	17.03	15.92	14.85	18.06	14.56	17.03	13.15	16.63	14.59
1410	18.70	14.06	18.48	16.74	19.68	14.99	19.17	20.31	16.06	14.51
1420	14.32	17.55	13.99	14.42	14.03	17.28	17.69	18.87	21.29	17.83
1430	17.75	18.08	15.77	18.55	18.28	16.45	16.19	16.34	15.14	16.85
1440	17.72	17.99	15.94	17.99	18.13	16.06	17.66	16.70	13.44	14.27
1450	13.58	16.73	18.18	20.11	17.68	15.20	16.37	17.02	16.74	14.98
1460	17.21	15.23	17.00	14.17	16.19	16.93	19.79	18.89	16.07	17.75
1470	15.59	13.41	14.59	15.77	14.97	13.94	16.98	15.64	16.88	16.02
1480	13.98	17.91	17.35	16.48	20.05	19.35	18.51	13.76	15.79	17.03
1490	18.09	16.34	19.95	18.92	18.83	13.41	15.58	14.70	17.16	19.32
1500	16.85	15.77	13.21	15.81	16.58	16.84	14.79	19.60	18.15	17.75
1510	16.97	17.89	15.77	16.63	17.25	18.00	13.27	13.65	16.35	17.74
1520	18.74	18.46	12.86	13.47	11.84	15.13	15.90	15.09	15.64	17.80
1530	18.39	19.79	17.52	16.81	20.13	15.89	16.62	18.39	16.28	18.59
1540	19.89	19.34	16.97	19.95	15.79	19.17	16.50	16.90	17.93	13.88
1550	16.05	15.60	15.42	20.06	18.53	19.44	20.33	17.31	15.73	16.48
1560	12.98	17.27	18.75	16.21	15.71	16.51	17.06	15.50	16.59	16.40
1570	15.59	15.07	14.78	13.71	15.74	15.60	14.50	17.52	18.16	15.69
1580	12.39	15.32	15.30	12.50	15.59	13.41	16.47	15.11	19.94	19.66
1590	12.44	18.90	16.84	13.57	17.63	15.98	19.47	19.85	16.45	19.78
1600	15.98	15.42	18.14	19.15	18.77	16.53	17.34	16.53	16.58	17.89
1610	20.65	19.66	17.84	15.71	16.09	15.42	16.51	16.21	17.40	16.16
1620	16.51	18.95	17.77	15.69	14.12	14.36	16.23	18.30	14.37	18.02
1630	15.78	16.12	13.88	15.71	16.54	19.06	15.73	15.57	15.43	14.68
1640	17.01	17.04	16.22	17.59	18.56	17.98	18.82	18.82	16.05	17.69
1650	15.74	19.99	17.68	17.68	16.05	19.04	15.71	15.47	16.64	16.86
1660	17.03	18.77	17.53	16.82	16.22	18.58	17.12	15.59	15.75	15.52
1670	13.45	15.56	15.96	15.76	15.54	15.90	16.55	16.87	12.99	15.23
1680	15.58	15.67	16.24	17.21	15.73	13.33	17.50	16.35	17.03	19.93
1690	15.14	15.86	18.72	15.61	17.38	16.92	13.65	17.60	17.69	18.66
1700	18.65	18.75	17.32	15.32	16.52	18.11	18.34	15.79	17.69	16.93
1710	20.37	16.06	18.71	15.71	15.61	15.19	14.31	16.92	17.70	16.50
1720	18.56	19.09	17.93	18.91	17.85	16.67	16.30	15.42	17.12	12.83
1730	15.63	17.57	16.74	14.77	19.63	15.88	16.23	14.18	15.46	14.54
1740	16.45	18.62	16.30	19.71	18.26	18.89	20.60	17.49	11.81	17.08
1750	13.69	18.71	12.69	16.90	16.65	14.32	16.01	14.19	19.30	18.19

TABLE 11.3. *(continued)*

	YEARS									
	0	1	2	3	4	5	6	7	8	9
1760	19.15	20.64	17.59	13.36	16.66	16.06	15.82	17.32	15.98	14.95
1770	15.89	19.09	15.32	14.21	15.11	14.21	16.16	14.79	15.89	14.63
1780	13.89	16.67	17.17	19.70	19.66	15.68	17.56	17.10	16.06	14.64
1790	17.45	16.53	17.62	19.81	16.90	17.13	13.87	14.60	14.78	15.65
1800	15.44	14.57	14.68	16.47	16.31	16.90	12.92	15.75	15.94	16.02
1810	17.05	16.28	16.18	18.29	17.22	18.26	19.33	15.03	13.19	13.82
1820	16.40	16.56	12.63	13.53	15.60	16.95	17.92	19.19	18.84	17.34
1830	17.87	17.49	17.97	18.78	19.25	18.58	15.56	16.51	16.47	18.49
1840	19.01	16.73	12.33	14.32	15.78	16.86	18.21	16.02	17.66	19.77
1850	18.09	15.71	18.78	18.66	17.86	17.29	16.78	19.75	20.58	13.96
1860	16.07	13.74	12.62	14.76	15.37	17.99	17.55	19.38	18.74	18.39
1870	14.59	15.07	16.14	14.16	15.62	18.05	15.91	18.20	18.90	17.78
1880	14.33	16.11	15.37	13.64	14.59	16.13	14.02	17.14	15.98	16.88
1890	15.29	16.79	14.56	15.89	13.47	16.22	12.45	17.93	16.88	13.96
1900	14.03	16.52	13.35	15.95	11.30	16.70	18.77	22.74	21.65	18.74
1910	16.86	20.39	15.67	13.95	18.37	17.54	21.24	18.93	16.71	20.21
1920	19.59	18.13	14.50	14.24	16.73	11.82	16.79	15.74	16.58	19.39
1930	17.61	18.27	19.55	23.33	14.15	20.30	18.09	19.48	14.59	16.63
1940	18.73	19.72	19.76	17.02	18.59	18.39	14.33	16.46	17.29	16.82
1950	12.03	11.63	15.38	15.39	11.58	12.19	11.56	14.72	16.84	18.73
1960	20.84	19.63	18.76	16.07	14.76	15.16	15.31	12.89	19.15	17.71
1970	19.46	14.61	16.71							

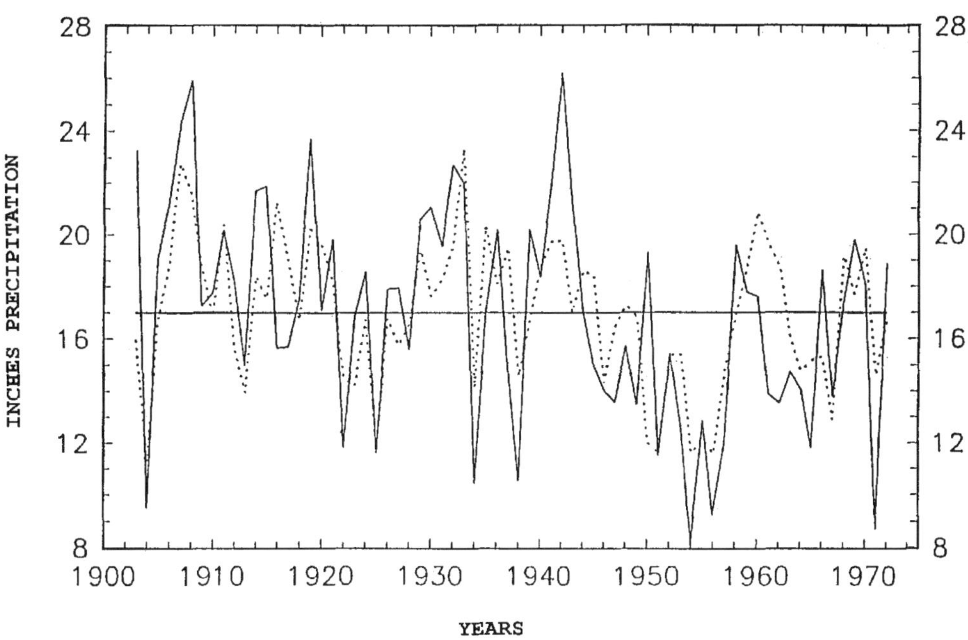

FIGURE 11.3. Actual annual Mountainair, New Mexico, precipitation (prior August–current August of growth year) vs. reconstructed precipitation based on a tree-ring chronology from the Chupadera Mesa area. *Solid line* is actual precipitation, *dotted line* reconstructed.

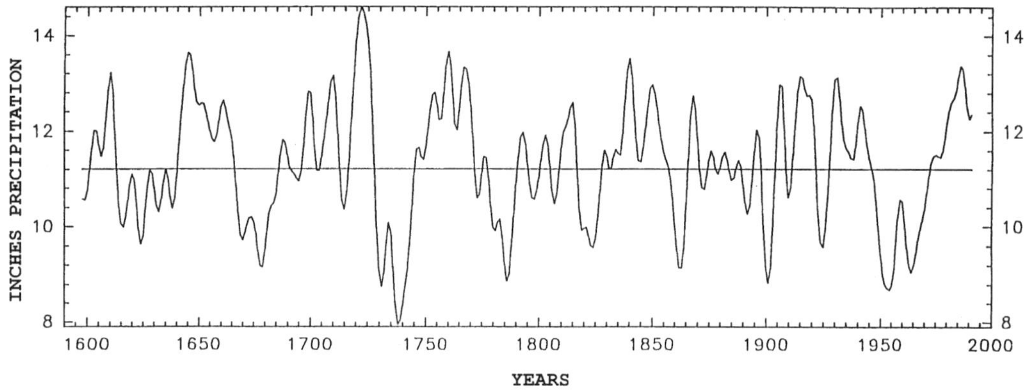

FIGURE 11.4. Reconstructed annual precipitation (prior August through current August of growth year) for Socorro, New Mexico, based on a tree-ring chronology from the Sevilleta National Wildlife Refuge.

DROUGHT HISTORY

Figure 11.4 shows the Sevilleta/Socorro reconstruction expressed in eight-year running means for the sake of clarity. The worst drought period appears to be in the 1730s and 1740s, following an exceptionally wet period in the 1720s. The mid-twentieth-century drought was not as severe but lasted longer. Many of the pinyons growing on the high slopes of the Arroyo Milagro site survived the 1730s and 1740s drought yet succumbed in the 1950s. Perhaps the duration of a drought has as much effect as severity (Grissino-Mayer et al. 1997:48–50, Table 6). In severity and duration, the 1660s and 1670s drought is the third worst of the reconstructed period, followed by the 1770s and 1780s. It is important to keep in mind that the Sevilleta chronology explains 55 percent of the variance in Socorro precipitation. Explaining the other 45 percent of the variance could make a difference in the reconstruction of drought severity.

The Mountainair reconstruction, shown in Figure 11.5 as eight-year running means, has higher average precipitation than the Socorro reconstruction, which is explained by the higher elevation of Mountainair. The most severe drought obviously took place in the 1950s, though this drought was shorter than the one in the late nineteenth century. Other droughts of significant severity and duration are in the early 1800s; the 1770s and 1780s; the 1660s and 1670s; a severe, prolonged drought in the 1570s and 1580s; a severe but relatively short drought in the 1520s; and the 1340s and 1350s. The caution concerning the absolute accuracy of the Socorro reconstruction also holds for the Mountainair reconstruction.

The Socorro and Mountainair reconstructions compare favorably with a regional dendroclimatic reconstruction for the southern Rio Grande Basin in New Mexico (Grissino-Mayer et al. 1997). In addition to samples from living trees, the southern Rio Grande Basin chronology incorporates tree-ring samples from archaeological sites distributed across the southern part of the state, including some of the sites used in the present study. Comparison of the three reconstructions indicates how aspects of Socorro and Mountainair climate resemble or differ from the regional pattern and illuminates spatial variability in climate across the region. Although the reconstructions are

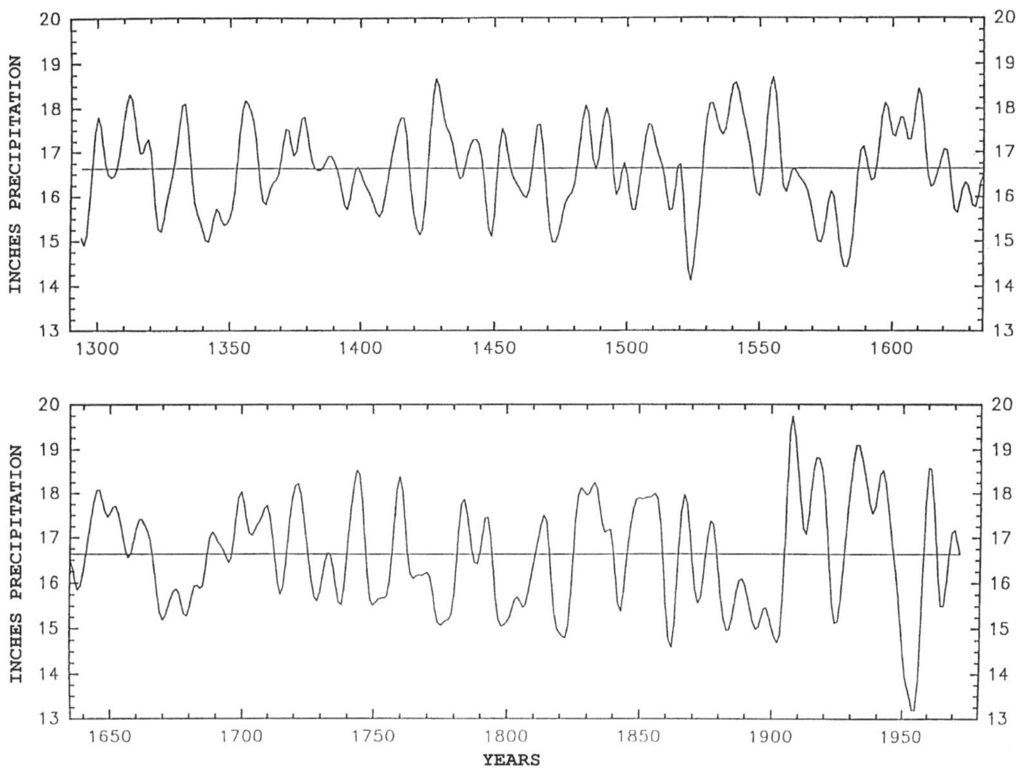

FIGURE 11.5. Reconstructed annual precipitation (prior August–current August of growth year) for Mountainair, New Mexico, based on a tree-ring chronology from the Chupadera Mesa area.

quite similar in their general characteristics (Figures 11.4 and 11.5; Grissino-Mayer et al. 1997:45–56, Tables 6 and 7, Figures 17 and 18), their differences are instructive. The Socorro droughts, in order of diminishing severity, rank fifteenth (1730s–1740s), second (1950s–1960s), third (1660s–1670s), and seventh (1770s–1780s) in Grissino-Mayer et al.'s (1997:Table 6) classification of regional droughts based on magnitude and duration. The ordered Mountainair droughts come in second (1950s), sixteenth (late 1800s), sixth (early 1800s), seventh (1770s–1780s), third (1660s–1670s), fourth (1570s–1580s), and not listed (1520s and 1340s–1350s) on the Grissino-Mayer et al. ranking. The departures from the regional pattern of the Socorro and Mountainair droughts identify local differences in climate that may have differentially affected the human populations of the area.

DROUGHT AND PUEBLO ABANDONMENT IN CENTRAL NEW MEXICO

Climate is an important thread in the fabric of history. Though rarely (some might say never) the sole cause of social change, climatic catastrophes have worked in concert with other forces to shape the destiny of many societies. If certain peoples are beset by political and social tensions, a flood or a drought (with concomitant famine and disease) can escalate those tensions to the point of societal collapse, such as the Puebloan abandonment of south-central

New Mexico that began in the 1670s. General histories have for the most part overlooked or underreported this abandonment, emphasizing instead other dramatic events during and after the abandonment.

In the middle of the seventeenth century, the Roman Catholic church and the Spanish colonial government of New Mexico were locked in battle for supremacy, each one playing the Puebloans against the other to the detriment of all concerned, especially the Puebloans (Scholes 1937). In the 1660s, drought struck New Mexico, hitting the southernmost pueblos hardest (Hackett 1937:271–272; Kessell 1979:224–225). The Socorro and Mountainair climate reconstructions (Figures 11.4 and 11.5) show that by 1670, the Piro pueblos along the Rio Grande and the Tiwa and Tompiro pueblos of the Salinas District to the east (Figure 11.1) were experiencing the worst drought since the 1580s. In fact, this drought was the worst of the Spanish colonial period up to that time and the third severest in Grissino-Mayer et al.'s (1997) ranking of droughts from 622 to 1994.

By 1669, no crops had been harvested for three years at the Tompiro pueblo of Tabirá (Figure 11.1), and more than 450 of its people starved to death (Hackett 1937:271–272; Kessell 1979:224–225; Vivian 1964). These crop failures were associated with the failure of the summer monsoon rains in 1667, 1668, and 1669. Moreover, the Spanish demand for tribute in the form of food, other commodities, slaves, and labor left little for the Puebloans. The people of the Salinas pueblos were further oppressed by the requirement to provide large quantities of salt for the Parral silver mines in southern Chihuahua (Hickerson 1994:115). Added to the burden of tribute were the outbreak of disease, a leadership crisis caused by the execution of the leaders of abortive Piro and Tompiro uprisings (John 1981:92–93), and the scourge of plundering Apaches (also reeling from the drought), which completed the recipe for catastrophe.

The Spaniards, bickering among themselves, were unable to defend the southern pueblos against Apache depredations, managing in the way of assistance only a small dole of grain to some pueblos, including Socorro and Senecú, just to keep the few clerics from leaving the missions. These grain shipments were made possible by contributions from northern missions such as Jemez and Pecos (Kessell 1979:219), where the drought was generally less severe, as can be seen in dendroclimatic reconstructions for the Santa Fe area (Rose et al. 1981).

Sometime after 1672, the Salinas pueblos began to depopulate, with some residents fleeing to the southern Tiwa pueblos of Isleta and Sandia (Katherine A. Spielmann, personal communication 1995), some to the Piro pueblos (Schroeder 1979:241), and others to El Paso (Beckett and Corbett 1992:3). Spanish attempts to resettle a few of the Salinas pueblos and defend the region from Apache raids failed. By the late 1670s or early 1680s (the documentary evidence is contradictory), the Salinas district was entirely abandoned (Forbes 1960:174–175; Hackett 1937:298; Kessell 1979:221; Scholes 1937:283).

The Piro pueblos fared only a little better. Although recipients of aid from the north, the pueblos along the Rio Grande continued to be plundered by Apaches while drought, famine, and disease marched on. When Apaches destroyed Senecú in 1675, many of the survivors sought refuge in El Paso (Beckett and Corbett 1992:3). By the time of the Pueblo Revolt of 1680, the Piros were probably too few, too weak, and too demoralized to participate. Though the Piros had staged small uprisings to protest their plight in the 1660s (John 1981:92–93; Kessell 1979:224–225), they were not involved in the planning of the great rebellion. Realizing that the leaders of the Puebloan restoration might consider them traitorous, most of the Piros cast their lot with the Spaniards and fled to El Paso.

Historians and social scientists who have closely examined the pueblo abandonment of south-central New Mexico (Barrett 2002; Forbes 1960; Kessell 1979; Scholes 1937; Vivian 1964) generally agree that it was caused by a symphony of factors whose orchestration is not entirely known. The instruments of this calamity, however, can be identified: (1) the drought of 1667–1681; (2) Apache depredations; (3) disease spreading through diminished and weakened populations living in unsanitary conditions; and (4) tribute to the Spaniards in the form of food, other commodities, slaves, and labor, which sapped the ability of the Puebloans to maintain subsistence systems finely attuned to the vicissitudes of the harsh southwestern environment.

Barrett (2002:78–79) points out that by the early 1640s, the Piro and Salinas pueblos had already been decimated by a major outbreak of smallpox. This epidemic depopulated the entire Puebloan region by 68 percent, affecting the Piros most but also significantly affecting the Salineros. The southern pueblos in general experienced the highest depopulation in the region as well as the highest rate of abandonment. During this period, eleven of the fourteen Piro pueblos and five of the eleven Salinas pueblos were abandoned. Though depopulated, few of the northern pueblos were abandoned. Perhaps the epidemic, presumed to have moved north from New Spain up the Rio Grande, had abated by the time it reached the northern pueblos; or maybe the higher populations of the northern pueblos mitigated their wholesale abandonment.

Climate played an indirect role in the pueblo abandonments of the 1640s even though precipitation during this time was above average. The southern pueblos were the smallest and were also located in the driest environments of the eastern Puebloan region. Population and environment are undoubtedly linked; the Malthusian principle of population increasing to fill the carrying capacity of its environment seems apropos here. In fact, a Malthusian corollary might be suggested: when a population check such as disease occurs in a large region, those populations with the most limited carrying capacities (in this case, drier environments) are disproportionately affected. Thus, the southern pueblos were smaller because of drier environments and, therefore, more susceptible to population checks like the smallpox epidemic of the late 1630s and early 1640s.

The Spanish attempt to exterminate Puebloan religion (Schroeder 1979:237) possibly also played a role in the failure to withstand the drought. By suppressing the native religions, the Spaniards disturbed a delicate balance between society and environment that was fostered and maintained by various beliefs, rites, and observances. Additionally, the Spaniards interfered with trade relationships that had developed in late prehistory between the Puebloans and the Plains Indians (Spielmann 1991), including the recently arrived Athapascans. This disruption, as well as slave raids by the Spaniards, turned the Plains groups against the Spaniards and, by extension, their putative allies the Puebloans (Forbes 1960:156–159).

Zubrow (1974) found a significant positive relationship between increased Puebloan contact with Spanish and American colonists and increased vulnerability of the Puebloan population to climate change, at least after 1800. Zubrow had limited population data before 1800, and hardly any data before 1700, our period of interest. Moreover, his study ignores the more southern pueblos and is thus not directly germane to our problem. Nevertheless, his findings support the overall importance of climate change in the region and the idea that Spanish colonial policies undermined the Puebloans' delicate relationship with an unpredictable environment.

Although the Spaniards treated the Puebloans more humanely and tolerantly after the Reconquest of New Mexico in 1694 (Dozier 1970:62–63), the Piro and Salinas pueblos were never reoccupied. The Rio Abajo district below Belen was not resettled until the early part of the nineteenth century (Marshall and Walt 1984:259), and fortunately so, for the Socorro climate reconstruction (Figure 11.4) shows that this area experienced a drought in the 1730s–1740s that far surpassed the one in the 1660s–1670s, though elsewhere in New Mexico the eighteenth-century drought was less severe (Grissino-Mayer et al. 1997: Table 6). A twentieth-century attempt at dry farming on Chupadera Mesa ended during the same drought that decimated the Arroyo Milagro pinyon woodland (Rautman 1994), demonstrating once again the fragile relationship between climate and society in this region.

CONCLUSIONS

Of the major factors contributing to the Puebloan abandonment of south-central New Mexico around 1670 (drought, Apache raiding, disease, and Spanish oppression), only drought is now quantifiable. Our dendroclimatic reconstructions of precipitation patterns in this tumultuous period show that the southern pueblos were experiencing the worst drought in ninety years. These people might have weathered the drought if not for a host of other adversities. They had, in fact, survived a much longer drought in the late sixteenth century (Figure 11.5; Grissino-Mayer et al. 1997:Table 6). One reason the 1667–1681 drought was more detrimental could be that the Spaniards, who were unfamiliar with the climatic extremes of this region, failed to store enough food for expected lean years during the favorable period between droughts. The

Puebloans, who under normal circumstances probably would have accumulated sufficient reserves, were prevented from doing so by the demands of the Spanish encomienda system. The Spaniards may have become accustomed to years of surplus production in the middle seventeenth century, had little experience with prolonged drought, and refused to amend their expectations of the encomienda system as the drought intensified. Perhaps the encomienda system also short-circuited any regional native solutions to shift surpluses from the more fortunate to the less fortunate pueblos.

The common denominator in most of the factors contributing to the Puebloan abandonment of central New Mexico in the 1670s and 1680s appears to be Spanish colonialism. The 1667–1681 drought was a contributing factor (one whose quantification is necessary to reach the conclusion of this study), but the primary responsibility for abandonment lies with the Spaniards and the changes they wrought. Moreover, the abandonment of the Piro and Salinas pueblos was only part of a larger pattern of pueblo abandonment and depopulation in New Mexico that had taken place since the arrival of the Spaniards (Schroeder 1979:236–254). Although not all prehispanic abandonments that characterized the region can be attributed to European colonialism, the conclusion that the presence and actions of the Spaniards created conditions that made Puebloan subsistence systems far more vulnerable to environmental perturbations than they had been before 1600 is inescapable. Political oppression, social divisiveness, and cultural suppression can turn an otherwise survivable climatic catastrophe into the collapse of a society. Furthermore, if our Malthusian corollary, that population checks disproportionately affect populations with the most limited carrying capacities, is valid, it is understandable that the Piro and Salinas pueblos were the most negatively affected by Spanish colonial policies.

ACKNOWLEDGMENTS

We thank the U.S. Fish and Wildlife Service (Sevilleta Wildlife Refuge) and the University of New Mexico (Sevilleta NSF Long-Term Ecological Research Site) for hosting our work. In particular, we thank Mark Betancourt, Carlos Mendoza, Bob and Cheryl Parmenter, Kierstan Quade, Steve Tharnstrom, and Ray Turner for wielding chain saws and extensively sampling the drought-stricken pinyon populations at Arroyo Milagro. We also thank Gary Funkhouser for valuable assistance with the dendroclimatic reconstructions and Richard Warren for performing quality control analysis of the tree-ring dating. Finally, we acknowledge funding from the USDA Forest Service, the U.S. Geological Survey, the National Science Foundation (various grants to support archaeological analysis, chronology building, and dendroclimatic reconstruction), and the National Park Service and Southwestern Parks and Monuments Association for supporting the Southwest Paleoclimate Project.

12

Sky as Environment: Solar Eclipses and Hohokam Culture Change

W. Bruce Masse and Fred Espenak

> These late Eclipses in the Sun and Moone portend no good to us: though the wisedome of Nature can reason it thus, and thus, yet Nature finds it selfe scourg'd by the sequent effects. Love cooles, friendship falls off, Brothers divide. In Cities, mutinies; in Countries, discord; in Pallaces, Treason; and the Bond crack'd, 'twixt Sonne and Father.
>
> —Shakespeare, *King Lear*

THE SKY IS AS MUCH A PART OF OUR PHYSICAL ENVIRONMENT AS ARE the more mundane features of Earth, such as topography, climate, vegetation, and water availability. Like the physical environment of Earth, the naked-eye celestial environment can be divided into objects and phenomena that are regular and predictable—the "fixed" heavens of Sun, Moon, planets, and stars—versus objects and phenomena that are transitory, unexpected, and often spectacular. These transient events include novae and supernovae, variable stars, unusual comets, solitary fireballs and bolides, great meteor storms, solar and lunar eclipses, auroras, unusual planetary conjunctions, and cosmic impacts.

The rulers and priests of all past societies viewed "signs in the sky" as omens or portents requiring interpretation. There is strong historical evidence that such portents influenced political and social decision making (e.g., Aveni 1992; Fountain and Sinclair 2005; Krupp 1997). Given the seemingly innate need for humans to manipulate, control, or at least attempt to explain objects or phenomena in their physical environment, and given the holistic nature of past cultures, we would expect the occurrence of transient celestial phenomena and events to be a part of the cognitive and behavioral framework of cultural systems. It is therefore reasonable to hypothesize that spectacular transient celestial events or clusters of events, in conjunction with other environmental events and processes, may stimulate culture change. We challenge our col-

FIGURE 12.1. General distribution of Hohokam material cultural remains and ball courts around A.D. 800 to 1200 (adapted from Crown and Judge 1991: Figure 1.1).

leagues with the proposition that transient celestial events are quantifiable and can be systematically used as a tool for identifying and explaining patterns of change observable in the archaeological and historic cultural record.

This chapter focuses on the Hohokam culture of southern Arizona (Figure 12.1). Like the Ancestral Pueblo and the Mogollon, the Hohokam first appear as a discrete archaeological entity around A.D. 200–600. They persist as a recognizable, perhaps multiethnic entity until the period of A.D. 1375–1500, during which time some Hohokam migrated to northern Mexico and Zuni, and perhaps to Hopi, with the remainder being assimilated by the linguistically related Akimel O'odham (Pima), Tohono O'odham (Papago), and River Yumans (Shaul and Hill 1998).

Attempts have been made to apply archaeoastronomy approaches to the Hohokam (e.g., Bostwick and Plum 2005; Hoskinson 1992, 2005; Mixon and White 1991; Mulloy, in Wilcox and Shenk 1977:44–45, Figure 11); however, no one has systematically considered both astronomy and traditional cosmology (but see Wilcox 1987 for a notable exception). We here compare the Hohokam archaeological and paleoenvironmental record with that of the reconstructed record of one class of historic transient celestial events, solar eclipses.[1] We also rely on oral historical and ethnographic data, especially from O'odham and Nahua, who, like the Hohokam, were members of the Uto-Aztecan language family. Although solar eclipses are just a part of the overall complex of celestial phenomena and transient natural events observed and responded to by the Hohokam, we argue that the suite of tangible physical effects that accompany solar eclipses gave them a unique role in traditional societal decision making and behavior.

These data make it possible to reasonably hypothesize that the Hohokam cognitively linked the occurrence of total solar eclipses with disruption in the universe caused by the dissatisfaction of powerful creator gods with humankind. Such disharmony in the physical environment required action and mending through substantive change in leadership, the restructuring of social institutions, settlement change, and change in religious ritual and perhaps even core religious beliefs. Evidence of this cognitive correlation is preserved in mythologies and as patterned responses observable in the archaeological record.

MYTHOLOGY AND ORAL TRADITION

Social scientists accept that mythology was vital for all traditional non-Western cultures, that myths were believed accurate and truthful by the members of each culture, and that myths contained information deemed by many cultural groups to be sacred and among their most valued possessions.[2] However, the scientific consensus is that although myths may have historical context and intrinsic psychological and social value, they are not factual history in and of themselves. Such widely discrepant views between the knowledge keepers and the scholars who study their knowledge systems should be of greater concern to Western science than has been the case. Oral tradition is viewed in a similarly skeptical manner, and its nature and validity are likewise subject to Western cultural biases. Simply stated, oral tradition represents the coherent oral transmission of various types of knowledge (e.g., songs, tales, myths, genealogies, ceremonies, histories) for more than a single generation. Donald Bahr, an expert in O'odham mythology and oral history, advises caution when using these stories to understand Hohokam history:

> As for my own opinion on what could have happened in Hohokam history, I repeat that all the stories or myths on this subject should be taken as good faith histories, that is, as stories that were offered by their tellers as true. Each myth might, could, and should be true as far as its teller was concerned; and thus I assume that no story was ever intentionally falsified, neither in tellings to Indians nor in transmission to a white recorder.
>
> Now, without going into detail, I say that all the stories on the Hohokam cannot be true. They are a collection of alternative and more or less contradictory good faith histories. Finally, I note that the stories have events that I find difficult to believe, such as the god's resurrection and the ascent of

> the Pima-Papago from the underworld. Of course, my own Christian white people have stories that are equally difficult to believe. All accounts of mystical or supernatural things are of that nature. . . . I do not suppose that the mystical part of the stories will ever be proved. (Bahr et al. 1994:7–8)

Despite our great admiration for Bahr and his work, we respectfully use these particular statements as a foil against which to paint a divergent viewpoint (see also Masse 2003a).

Evidence from Hawai'i (Masse 1995, 2003b; Masse and Tuggle 1998) strongly argues for a different view of the nature and utility of both oral tradition and mythology. Because Hawaiian mythology is chronologically embedded within royal genealogies stretching back more than 95 generations, it can be demonstrated that many genealogically ordered myths contain unmistakable descriptions of volcanic eruptions that are now radiocarbon-dated and transient celestial events historically recorded and dated in Asia and elsewhere, and that remarkable details of these events have been accurately orally transmitted for more than 1,000 years. These findings reinforce the plea made by Vansina (1985:199–201) for scholars to recognize oral tradition as a valid and particularly rich form of history.

Five findings are important here. First, cosmogonic mythology is derived from observations of major environmental events and processes taking place on the Earth and especially in the sky. In the holistic view of traditional non-Western societies, real sentient but nonhuman beings were responsible for all transient celestial and earthly environmental events and phenomena. The "supernatural" aspect of mythology arguably is derived from the observance of these visually evocative and powerful beings who appear to come and go and to move on their own volition but who lead an existence largely separate from human senses—such as in comets, which could be seen in the sky but not heard or felt. The rare times that these beings were also heard and felt (e.g., lightning strikes, volcanic eruptions, earthquakes, observed meteorite falls) served to emphasize their extraordinary and perhaps capricious power.

Second, the training and skill required to become a cultural knowledge keeper must not be underestimated. For example, Underhill's (1946:7) observations about knowledge keepers among the Tohono O'odham are applicable to all cultural groups:

> There are only a few accredited narrators. One can be found in every dialect division, though by no means in every village. Each narrator has received his lore from some relative, usually in the male line, and has learned it by the tedious process of repeatedly hearing the whole, without interruption, from beginning to end. He will hand it on in the same way to some young man, chosen for intelligence and a retentive memory. If suitable heirs are scarce, he may choose from the mother's side of the family, and even

> women have sometimes qualified. Members of the narrator's audience, too, may learn fragments of the tale and become informal narrators to small groups.

Ethnographers, as they collect oral traditions and reach conclusions about the extent of a particular culture's esoteric knowledge, often fail to distinguish whether they are basing their judgments on information derived from accredited narrators themselves or from the more general and thus more fragmentary pool of cultural knowledge.

Third, the transmission of core cosmogonic myths and other critical esoteric knowledge nearly always takes place in rigorously prescribed, typically annual, public settings, such as solstice/New Year ceremonies. The telling of the Tohono O'odham origin myth is instructive (Underhill 1946:8):

> The Papago have the usual southwestern rule that myths shall not be narrated in the summer. They wait until the biting creatures are asleep, and desert people include among these not only rattlesnakes, sidewinders, and gila monsters but also vinegaroons, tarantulas, and even the stinging ants. One informant, persuaded to tell a fragment of the tale out of season, went for weeks in fear of a flood. The official time for myth telling is during the four days of the winter solstice "when the sun stands still." Then residents of the narrator's village approach him with a handful of cornmeal, asking that he spend four nights in telling the story from beginning to end. He fasts from meat, grease, and salt during the whole time, and the old rule demanded sexual continence also. The listeners must sit upright all night, keep silent, and not doze. If the narrator sees someone sleeping, even a child, he may break off the narrative for that night.

Fourth, not only are the performances themselves entrusted to the most gifted and highly trained narrators, but also typically there is a combination of dance, chant, song, and ritual drama that serves to reinforce the accuracy of story transmission. Repetition and redundancy are also typical for particularly critical portions of cosmogonic narrations. Our ability to recognize the stability and accuracy of oral knowledge transmission has been compromised because much of the recent detailed scientific study of oral tradition has taken place in settings where the long-established systems for the transmission of oral knowledge have been affected and transformed by colonization and other acculturative forces.

Finally, the power, sanctity, and legitimization of chiefs and priests are tied to some degree to the successful performance of cosmogonic narration. Thus there is strong motivation to ensure that such narration is properly performed. In a holistic society, the quality of the performance itself is perceived as having an impact on the ability of people to communicate with their gods.

The cumulative result of these factors is that myths can and do form a coherent body of patterned natural observational data that are amenable to systematic analysis, hypothesis formulation, and testing within the framework of modern Western science. This has been exquisitely demonstrated, for example, in the identification of a previously unknown (to Western science) catastrophic volcano in Papua New Guinea (Blong 1982) and in the encoding of transient celestial events and volcanic eruptions in Hawai'i (Masse 1995, 2003b; Masse et al. 1991).

NAKED-EYE TRANSIENT CELESTIAL EVENTS

Naked-eye transient celestial events are a surprisingly common feature of the nighttime and occasionally the daytime sky. The Chinese, Koreans, Japanese, Arabs, Babylonians, and Europeans recorded more than 1,100 novae, supernovae, comets, and great meteor storms during the period 200 B.C. to A.D. 1800 (e.g., Bode and Evans 1989; Clark and Stephenson 1977; Hetherington 1996; Ho 1962; Ho and Ang 1970; Imoto and Hasegawa 1958; Kronk 1999; Rada and Stephenson 1992; Xu et al. 2000; Yeomans 1991). Most events were precisely dated by their observers and documented as to relative location against the backdrop of named constellations in the sky. Most would have been visible over much of the world, local weather permitting.

Categories of Transient Celestial Events

Naked-eye transient celestial events can be organized into several categories. Excluded from this discussion are meteorological and other physical phenomena, such as auroras and zodiacal light, that while visually impressive in the heavens are more difficult to extract and analyze from the historic record (see, however, Masse 2003b for a treatment of auroras in Hawaiian mythology).

Supernovae are cataclysmically exploding stars visible to the naked eye for months or even years of time. The most widely known (to archaeologists and other social scientists) historic supernova was first observed in the constellation Taurus in A.D. 1054. The Taurus supernova, whose remnant is the famous Crab Nebula, achieved an apparent magnitude slightly greater than that of Venus (ca. -5.0) and at its brightest was visible during daylight hours in China for 23 consecutive days. However, few scholars are aware that in A.D. 1006, fewer than fifty years before, an even brighter supernova occurred in the constellation Lupus. The Lupus supernova at its brightest was at least four times the brightness of Venus (Clark and Stephenson 1977; Schaefer 1996) and may have been visible to the naked eye for nearly four years.

Novae are stars that shed a substantial portion of their mass but retain enough to continue functioning as a star. They are often divided into long-duration, "slow novae" that may be visible for several months and are difficult to distinguish from further distant supernova events, and "fast novae" that are within our galaxy and are visible for up to 30 days depending on the energy released and proximity to the Earth.

Variable stars change apparent magnitude on a regular or irregular basis. They include geometric variables (stars whose light varies because of changes in position) and stars that change intrinsic brightness. One of the best-known geometric variables is the eclipsing binary system Algol, which regularly changes magnitude due to an eclipse cycle of 2.87 days. One extraordinary intrinsic variable is the star Eta Carina. When first classified in 1677, it had a magnitude of around 4.0, or among the dimmer stars visible to the naked eye. It slowly brightened to around 1.0 in 1838, joining a small group of the brightest stars, and increased in magnitude to around –1.0 in 1843, being surpassed only by the brightest star, Sirius. It then steadily decreased in magnitude, becoming invisible to the naked eye (less than magnitude 6.0) around 1870. Appearances of variable stars likely have occurred during the past several thousand years. Ancient Polynesians were aware of the variable nature of stars such as Algol and Delta Cephei hundreds of years before the initial documentation of this phenomenon by European astronomers (Masse 1995, 2003b).

Comets are extraordinary celestial objects that had a profound impact on early cultures. A single comet can change appearance in remarkable ways, from dull to brilliant over the course of several days or hours. Its tail can range from short (or even nonexistent) to long (or even multiple) and from straight to branched, twisting, or chain-like. Comet tails sometimes turn bright red or blue, reflecting both compositional elements within the nucleus and atmospheric observing conditions. Appearances of comets were common; more than 900 were observed between 200 B.C. and A.D. 1800.

Many of the worldwide myths about supernatural tricksters and transformers (shape changers) likely encode major comets. Postperihelion comets, with dust and ion tails moved by solar winds in front of the retreating comet's head, are often depicted in archaeological and religious iconography as "horned" or "feathered" serpents or other elongate creatures (Masse 1995, 1998, 2003b). The symbol for comet in the Codex Telleriano-Remensis is a snake emerging from a starry sky (Keber 1995:321), although at least one snake-like comet was described by the Aztecs as a "colorful worm" or caterpillar (Boone 2000:Figure 30c). Similarly, the Skidi Pawnee apparently used long-tailed comets as the inspiration for the long feathered headdresses worn by chiefs (Chamberlain 1982:142; Weltfish 1977:225, 377). Indeed, the name "feather headdress star" was given to a bright comet in 1882 (Murie 1981:Table 2). A painted kiva (Kiva 8) at Pottery Mound in New Mexico (Vivian 1994:89–90) depicts a number of coiled "snake warriors" (serpent bodies with humanoid heads), including one that is nested inside a feathered Sun shield (Vivian 1994:Plate 10). Of particular note are snake warrior figures that, instead of having human heads, are depicted with "soul heads." Such heads are painted in the shape of four-pointed stars with fearsome faces, each figure wearing a long headdress of possible eagle feathers (Vivian 1994:Plate 12).

Meteors, meteor showers, and meteor storms occur when fast-moving extraterrestrial particles combust because of friction in the Earth's atmosphere.

Meteor showers are the product of comet debris—objects ranging in size from dust to pebbles—known as meteor streams, which follow the orbits of their progenitor comets and typically intersect the Earth's orbit on an annual basis. Some recent streams, such as that of the Leonid, have widely dispersed large clusters of debris that intersect the Earth's orbit only a few times each century but form the basis for dramatic meteor displays. Meteor-shower phenomena are roughly sorted into three classes for our analysis: showers (fewer than 100 naked-eye-visible meteors per hour), substorms (between 100 and 999 meteors per hour), and storms (1,000 meteors or more per hour).

Meteor storms are visible for a few hours over less than half the world for any given event, although a few spectacular storms are particularly widespread and sporadically persist for two or even three days. The 1966 Leonid storm at its height may have exceeded 45 meteor streaks per second (Littmann 1998:216). The spectacular Leonid meteor storm of 1833 caused consternation throughout North America (Littmann 1998:1–9), including the Southwest, and is widely preserved in Native American art, calendar sticks, and oral traditions.

Fireballs (meteors brighter than magnitude –5.0, that is, brighter than the planet Venus), and to a lesser degree *bolides* (exploding fireballs), often accompany larger meteor storms, such as the 1966 Leonid storm, but can occur in isolation as well.

Planetary conjunctions are predictable (for modern astronomy) occurrences of two or more planets appearing within a few degrees of one another. Five planets are visible to the naked eye (Venus, Jupiter, Saturn, Mars, Mercury), and all are the brightest objects in the fixed heavens other than the Sun and Moon. Occasionally, the brightest stars, in particular Sirius, will approach or exceed the apparent magnitude of all planets but Venus. Most cultures understood that planetary conjunctions were cyclical in nature. Indeed, conjunctions of two or three planets are common (Table 12.1) and for that reason were used in astrological forecasting. Conjunctions of four or five planets are rare and were therefore considered portentous.

Witnessed meteorite falls and larger cosmic impacts together may represent the most significant class of transient event for overall effects on human civilization (Masse 1998, in press). The 10–20-megaton atmospheric impact of a probable stony meteorite in 1908 leveled 2,000 km^2 of forest in the Tunguska region of Siberia and has been modeled to occur on the average of every 100–200 years (Chyba et al. 1993; Morrison 1997).

Detailed examination of southwestern Native American oral traditions suggests that no major catastrophic cosmic impacts have taken place in or near the region in the past few thousand years. Nevertheless, it is certain that many groups in the region observed significant meteorite falls and could recognize meteorites when encountered on the ground.

For example, Navajo observers witnessed a meteorite fall two centuries ago and preserved this knowledge—including the location of the impact

Table 12.1. Approximate Average Numbers of Transient Celestial Events in Selected Categories and Relative Significance by 20-Year Human Generation for a Large Cultural Region such as the Hohokam.

TIME PERIOD	CATEGORY 1 MINOR EVENT	CATEGORY 2 MODERATE EVENT	CATEGORY 3 POSSIBLY MAJOR EVENT	CATEGORY 4 MAJOR EVENT	CATEGORY 5 MULTISENSORY EVENT
20 years 1 generation	60 (2°) 2-planet conjunctions 40 meteor showers 3 comets 1 nova	7 partial lunar eclipses 3 (2°) 3-planet conjunctions 1 comet 1 meteor substorm	8 total lunar eclipses 4 fireballs 2 comets 2 meteor substorms		
100 years 5 generations	300 (2°) 2-planet conjunctions 200 meteor showers 16 comets 4 novae	35 partial lunar eclipses 13 (2°) 3-planet conjunctions 7 comets 5 meteor substorms	40 total lunar eclipses 20 fireballs 7 comets 4 meteor substorms 1 nova 1 (5°) 4-planet conjunction	2 bolides 2 comets 1 meteor storm	
200 years 10 generations	600 (2°) 2-planet conjunctions 400 meteor showers 32 comets 7 novae	70 partial lunar eclipses 26 (2°) 3-planet conjunctions 14 comets 10 meteor substorms 1 nova	80 total lunar eclipses 40 fireballs 15 comets 8 meteor substorms 3 novae 2 (5°) 4-planet conjunctions 1–2 annular (94–97%) solar eclipses	4 bolides 3 comets 2 meteor storms 1–2 annular (98–99%) solar eclipses 1 nova	1–2 total solar eclipses
500 years 25 generations	1500 (2°) 2-planet conjunctions 1,000 meteor showers 81 comets 18 novae	175 partial lunar eclipses 66 (2°) 3-planet conjunctions 35 comets 25 meteor substorms 2 novae	200 total lunar eclipses 100 fireballs 37 comets 19 meteor substorms 7 novae 5 (5°) 4-planet conjunctions 3-4 annular (94–97%) solar eclipses	11 comets 10 bolides 5 meteor storms 3–4 annular (98–99%) solar eclipses 2 novae	3–4 total solar eclipses 1 witnessed meteorite fall

Note: Based on counts of actual comets, novae, and meteor substorms/storms for the period A.D. 500–1500 (see Masse 2003b). The other figures are extrapolated from general observations and astronomical theory.

crater—in their oral traditions. The remnant impact crater (23 by 34 m) was recently documented near Chinle on Navajo Nation lands (Shoemaker et al. 1995). Astronomer William Hartmann (1991:Figure 13-18) notes the discovery (in 1928) of fragments of a stony meteorite in a likely twelfth- or thirteenth-century crypt in the vicinity of Elden Pueblo, near Winona, Arizona, which he speculates represent a witnessed fall. A Tohono O'odham myth about the origin of fire encodes a description of fire being produced by either a lightning strike or a meteorite impact (Wright 1929:37–42). Likewise, O'odham myths about the nature and death of the "man-eagle" (Bahr et al. 1994:157–177) are similar, in part, to the stories surrounding a Tsawi Pawnee warrior bundle which tell of a meteor that came down to Earth in the form of a man wearing a bald eagle on his back (Murie 1981:2:190–195).

A 1.7-ton (1,545-kg) meteorite, now at the Smithsonian Institution, was discovered in the 1860s at Casas Grandes (Paquimé) in Chihuahua, Mexico (DiPeso 1974:3). The stone, "carefully and curiously wrapped with a kind of coarse linen, similar to that with which the Egyptians inclose [*sic*] or wrap their mummies," was found in "a kind of tomb made of adobe-brick" situated in the middle of a large room (Pierson 1874). This placement of the cotton-wrapped meteorite indicates an association with the Medio period at Casas Grandes, currently dated to about A.D. 1150–1425 (Schaafsma and Riley 1999: Table 1). Another notable southwestern meteorite find occurred in 1915 when an amateur collector was excavating a slab-lined cist inside a tenth–eleventh-century Sinagua house near Camp Verde, Arizona (Hudson 1973). Inside the cist, carefully wrapped in a feather blanket, was a 135.5-pound (61.5-kg) meteorite from the Canyon Diablo impact that created Meteor Crater during the late Pleistocene. The transport of these heavy meteorites over substantial distances and the care with which they were wrapped and placed strongly suggest that both were viewed as something special, perhaps as the earthly remnant of a deceased celestial being.

A myth described in 1634 by Alonso de Benavides from earlier observations by Zárate Salmerón links the Mexica and Tewa with a demon that "carried on her head an enormous and terrible boulder of iron" (Hodge et al. 1945:40–41; Zárate Salmerón 1966:91). This boulder was historically described by Spanish observers as weighing between 200 and 800 quintals (2,000–8,000 kg) and was used as a trail and boundary marker near Santa Bárbara in Chihuahua (460 km south of Casas Grandes). The details of the story suggest that this object was part of a witnessed meteorite fall and was carried by hand several hundred leagues south of its original location. The Santa Bárbara and Casas Grandes stones could conceivably derive from a single witnessed impact event; it would be of considerable interest to relocate the now lost Santa Bárbara stone for quantitative comparison.

The Skidi and other Pawnee tribes were aware of meteorites and kept them in medicine bundles along with other odd stones such as fossils and pyrite, which were also apparently classified as meteorites (Chamberlain 1982:142–

153). One tribe even has a story about a girl killed by a meteorite. In Mesoamerica (Taube 2000), meteorites and obsidian were considered "star feces," and the Aztec gods Huitzilopochtli and Obsidian Butterfly were associated with meteors, meteorites, and possibly comets. Taube (2000) suggests that the Xiuhcoatl, or fiery serpent weapon of Huitzilopochtli, represents a meteor or shower of meteors.

Eclipses of the Sun and Moon form the final category of transient celestial event, and the basis of this chapter. Eclipses are predictable (for modern astronomy) orbital phenomena that occur when the Moon is in a position to block the path of direct light from the Sun to the surface of the Earth (solar eclipse) or the Earth blocks the path of direct light to the surface of the Moon (lunar eclipse).

About 150 lunar eclipses occur in an average century: 80 total (i.e., the Earth's shadow, or umbra, covers the entire Moon) and 70 partial (Allen and Allen 1987). Each eclipse is visible, weather permitting, from slightly more than half the Earth's surface. Lunar eclipses vary in brightness (from roughly equal to Jupiter to nearly invisible) and color (shades of red, orange, or brown), depending on magnitude and atmospheric conditions, with a theoretical maximum duration of 1 hour 44 minutes for totality (Vanin 1999:133). Although lunar eclipses were important to southwestern cultures, including the O'odham, they lacked the overwhelming cultural significance of solar eclipses (see also Xu et al. 2000:61). Space limitations preclude detailed consideration of lunar eclipses here, but a future study would likely prove interesting and profitable.

Based on the 4,530-year period 2004 B.C. to A.D. 2526, solar eclipses average 238 per century (Littmann et al. 1999; Mucke and Meeus 1983). On average, about 84 are partial (the Moon's umbra fails to touch the Earth); 79 are annular (the Moon is too far from Earth to completely cover the Sun's disk, thus giving a ring-like appearance); 64 are total (i.e., the Moon completely covers the disk of the Sun); and 11 are a hybrid annular/total (eclipse path changes from annular to total; the apparent diameters of the Moon and Sun are nearly equal). Our ability to accurately plot the longitude of solar eclipse paths diminishes as we move back in time. For example, whereas we have total confidence in the calculated path of an eighteenth-century eclipse, by A.D. 1400 the uncertainty increases to approximately ±0.2° of longitude (~22 km at the equator); by A.D. 1100 it is ±0.3°; by A.D. 700 it is ±0.5°; and in A.D. 500 the uncertainty is about ±0.8°. Reliable earlier historic records can help to reduce this margin.

Unlike a lunar eclipse, which can encompass the entire disk of the Moon, the umbral shadow of a solar eclipse covers only a small swath of the Earth. For a nonpolar total eclipse, this path averages about 12,000 km long and 130 km wide, but it can range in width from less than a kilometer to around 270 km. Annular eclipse paths can be a hundred kilometers wider still. During a total solar eclipse at the general latitude of southern Arizona, the Moon's umbra races east along the ground at an apparent speed of 30 km or more per minute.

The theoretical maximum length of time for solar eclipse totality (near the equator) has been calculated at 7 minutes 31 seconds, while totality for a typical total eclipse averages only 2 or 3 minutes. The theoretical maximum duration for an annular eclipse (near the equator) is 12 minutes 30 seconds. The duration of an eclipse increases as one moves from the edges of the path inward to the central line of the path of totality. The increase in duration is especially dramatic within the first 10 km near the edges due to the grazing geometry on the Moon's limb with respect to the Sun.

The magnitude of a solar eclipse is the fraction of the apparent diameter of the solar disk covered by the Moon at eclipse maximum for a given location, and always equals or exceeds 1.00 in a total eclipse. This differs from (and always exceeds) the degree of obscuration, which is the percentage of the area of the Sun's disk that is covered. A trained observer might be able to detect the presence of a partial eclipse as low as 0.80 magnitude on a clear day. However, because of the brightness of the Sun, it is only when a partial or annular eclipse reaches a magnitude of around 0.94 to 0.95 on a relatively clear day that the effects become pronounced enough for everyone to realize that something is happening to the Sun (eclipses with much smaller magnitudes may be detected for rare sunrise and sunset eclipses due to the filtering effect of Earth's atmosphere near the horizon). Even a distant eclipse path can have a pronounced visible effect, as was the case with the narrow total eclipse of 1318 along the northern edge of the Mogollon Rim, which produced a partial eclipse magnitude of 0.961 at the Hohokam Casa Grande Great House some 136 km from the edge of the path of totality (see Figure 12.8).

From any given spot, a total solar eclipse will be visible, weather permitting, on the average of only once every 375 years (Meeus 1982). However, because of the complex nature of the orbital dynamics of the Moon around the Earth, and that of the Earth around the Sun, it is possible to have total solar eclipses occur in consecutive years at a given location (very rare), as was the case in Papua New Guinea in 1983 and 1984, or to occur more than 1,100 years apart, as has been computed for Jerusalem from A.D. 1133 to 2241 (Littmann et al. 1999).

We are thus presented with a seeming conundrum: how can an event that lasts for only a few minutes—often with centuries of time between each occurrence—have the large cognitive and behavioral impact evident in historic cultural records? The answer to this puzzle comes partly from the suite of physical effects that accompany a total or near-total solar eclipse and partly from the nature of the cultural cosmological systems that attempt to grapple with the meaning of eclipses.

Although the total phase of a solar eclipse lasts only a few minutes, other accompanying phenomena last considerably longer and impress even professional astronomers (e.g., Allen and Allen 1987:86; Kauffman 1993:31–32; Littmann et al. 1999:1–4; Vanin 1999:115–116; Zirker 1995). There is considerable variation between total eclipses, but a "typical" moderate-duration midday eclipse on a relatively clear day might proceed as follows.

The partial eclipse begins about one hour before the total phase but is largely unnoticed. About 10 minutes before totality, the blue sky slowly grows darker and the landscape begins to take on a steel-gray metallic cast. An astute observer under the speckled light of a leafy tree or brush-covered ramada would notice that the usual circular or blob-like pinpricks of light hitting the ground were taking on a crescent-like shape. About 5 minutes before totality, the western sky is darker than the east, and by now everyone suspects that something strange is happening. Darkness begins to close in around the dimming Sun, and the sky deepens to blue-gray or violet. Many diurnal animals become silent, wary, and confused (particularly birds) and take on patterns typical for twilight and nightfall (Murdin 2001; Tramer 2000). Flowers begin to close. Venus appears in the darkening sky. Temperature drops a few degrees centigrade, and the cooling atmosphere may cause the wind to stir.

Eclipse-generated wind changes are contingent on factors such as environment, season, and time of day (Jay Anderson, personal communication 2003). In normal surface-based convection, thermal bubbles are created as the ground surface is heated. The bubbles rise into the air and by mid-afternoon may extend thousands of meters into the atmosphere. Air descending from aloft to replace rising thermals brings a gustiness to the ground as the stronger winds above the surface are carried earthward. An eclipse cools the ground surface and causes an inversion in which the surface winds are decoupled from the higher winds. In the earlier stages of cooling, the gustiness may increase as the friction of the air against the ground is reduced, but as the eclipse progresses toward totality, the gustiness (and steady wind) gradually diminishes. The combination of sunny skies, late afternoon, dry soils and surfaces, and high surface temperatures would be most conducive for the generation of noticeable eclipse-driven wind gusts. The contrast between the early gusty winds and the light breezes of mid-eclipse would be very noticeable.

At 2 minutes prior to totality, a dark amorphous form begins to spread out along the western horizon, silent and ominous. In the remaining seconds, the Sun narrows to a blazing white crescent. Daylight-active birds begin to roost, pigs lie down, daytime insects alter their activity (such as bees flying en masse back to their hives and butterflies settling into the grass), and cows have been observed to begin marching in line back to their barns or in one curious case formed a circle with all animals facing outward. In the final minute before (and after) totality, atmospheric conditions often produce "shadow bands." This light phenomenon appears as irregular, rapidly moving light and dark bands a few tens of centimeters wide that crowd closely together as totality approaches. This phenomenon, if present, would be easily observable on the plastered walls of adobe buildings. The colors of objects (and people) seem to flicker and change. Jupiter may now be visible. Temperature is 5 or 6 degrees centigrade lower than before the eclipse, and the wind may blow in fitful gusts.

As totality approaches, the dark orb of the Moon takes shape opposite the waning crescent of light. A common experience in the final few seconds before and after totality is for the crescent Sun to appear to coalesce into one final brilliant diamond. If you are looking down at the ground from a high vantage point, it might be possible to detect the shadow of the Moon racing toward you, particularly in the broad, open basin valleys of southern Arizona.

At totality in the darker eclipses, first- (and rarely second-) magnitude stars join the planets in visibility, as do any bright comets near their perihelion passage around the Sun. The sky appears similar to that of normal twilight some 30 to 45 minutes after sunset. The temperature during totality in a summertime eclipse might be 10 or more degrees centigrade lower than before and after the eclipse. After several minutes of deathly animal stillness prior to totality, evening insects, such as cicadas, and some roosting birds such as sparrows, bust into a cacophony of song; nocturnal moths, bats, and owls begin to emerge, as do snakes and earthworms.

One extraordinary phenomenon is the sudden appearance of the Sun's corona, or glowing outer atmosphere, which begins a few seconds before totality and ends a few seconds after. No two eclipse coronas are identical; each has its own unique appearance. During sunspot minimum the shape of the corona is elongated at the Sun's equator, whereas during maximum it tends to be more symmetrical. However, to the traditional naked-eye human observer of the past, the corona gave the appearance of a jet of flame, feathers, or even, disturbingly, the white of a human eyeball surrounding its black iris—the darkened Moon (Robin Edgar, personal communication 2002).

Because of the importance of the Sun to past cultures, and because of the suite of physical effects that accompany a total solar eclipse (weather permitting), historical annals worldwide are filled with instances in which eclipses had profound impacts on the behavior of cultural systems. For example, the rise of at least two major revitalistic movements in North American Indian tribes can be directly linked with the occurrence of solar eclipses. The Shawnee prophet Tenskwatawa, who allegedly had advance knowledge of an 1806 eclipse, used the event to create a following and a major problem for American general (and later president) William Henry Harrison. Similarly, the coincidence of a severe illness and solar eclipse in 1889 provided Paiute prophet Wovoka with some of the inspiration for the famous Ghost Dance religion (Mooney 1973:674, 720, 771–774).

Chronology, Tempo, and Mode of Transient Celestial Events

Based on the historical record, a minimum of 321 likely comets, 87 meteor substorms (including a few large meteor showers), 10 meteor storms, 3 meteor storms with meteorite impacts, 55 probable fast novae, and 3 supernovae were observed and recorded between A.D. 500 and 1500 (Masse 2003b). In terms of human generations, if we assume that these figures, primarily from observations made in Asia (China, Korea, and Japan), are roughly representative for any region on Earth, every generation of 20 years has the potential to witness nine or more of these naked-eye transient celestial events. This figure does not include annual meteor showers, solar and lunar eclipses, bolides, fireballs, mid-latitude auroras, and planetary conjunctions.

Taken as a set, the visual properties of transient events are stochastic and are similar to any physical phenomena that follow a power law distribution. The events range in character from the commonplace to rare spectacles that leave a lasting powerful impression. To study the potential effects of transient events on culture, we devised a numeric scale ranging between 1 and 5 based on frequency, duration, and visual character, as derived from historical records and astronomical modeling. Table 12.1 represents a first approximation of a linked temporal and modal scale for categories of transient celestial events. These data are derived from compendia of historically recorded observations (Masse 1995, 2003b), reconstructions based on astronomy computer software programs,[2] and modern astronomical theory.

Categories 1 and 2 (minor and moderate events) are those that commonly occur in a 20-year human generation and are not particularly visually impressive; examples include comets that have small tails or lack tails entirely and are visible for a few weeks at most. Other examples are annual meteor showers (some of which may rarely approach the magnitude of meteor substorms), partial lunar eclipses, and nova events that do not exceed the brightness of second- or third-magnitude stars.

Category 3 (possibly major events) occur less frequently and are visually more impressive. Examples include comets that have tails longer than 10 degrees (20 times the apparent diameter of the full moon) and that are visible for 30 days or longer, moderate meteor substorms, novae and supernovae that achieve the brightness of first-magnitude stars, fireballs, total lunar eclipses, and annular solar eclipses with a magnitude of approximately 94–97 percent. Category 4 (major events) include rare novae and supernovae exceeding the apparent magnitude of the brightest planets, bolides, annular solar eclipses with a magnitude of 98–99 percent, great meteor storms of substantive duration and intensity, and near-Earth long-tailed comets of unusual shape, color, and duration. Category 5 events (multisensory) are extraordinary and include total solar eclipses and witnessed cosmic impacts, both of which are accompanied by a broad range of perceptible physical effects beyond that of simple sight and sound.

The record of comets, novae, meteor storms and substorms, and solar eclipses between A.D. 500 and 1500 reveals their stochastic nature (Table 12.2). Although not all specific events would have been visible to the Hohokam—and although the Hohokam would have witnessed some events not observed and recorded in Asia, Europe, and the Near East—the pattern and chronological context would have been similar worldwide. Minor and moderate events dominate this 1,000-year period. Their frequency led to their being closely woven into cultural cosmologies, religious and political systems, astrological prognostication, and science-based conceptions of the normal workings of nature. However, major and especially multisensory events were treated differently, as were periods of exceptionally high (and perhaps low) numbers of events.

The "scores" listed in Table 12.2 can be viewed as gross indicators of the potential impact of transient celestial events on cultures during each 20-year generation period. Scores from 11 to 29 fall within one standard deviation of the mean score (20.2 ± 9.7) and can be considered the norm for the appearance of transient celestial events. Scores falling outside this range (less than 11 or more than 29), particularly at the high end of the scale, likely represent periods in which transient celestial events figured prominently in cultural behavior. Notable are the high scores and large numbers of high-magnitude events around the periods A.D. 821–840, 881–940, 1061–1100, and 1361–1440, and the paucity of events between 621–680, 781–820, and 1241–1360. Such situations demanded official explanations, which often resulted in heightened ritual behavior and pronounced social, political, and ideological change.

Each category of transient event and indeed the totality of all events behave in much the same manner as climatic variables, such as the streamflow reconstructions developed by Graybill and colleagues elsewhere in this volume. We suggest that Category 1, Category 2, and perhaps Category 3 events in Tables 12.1 and 12.2 behave in part like "high-frequency process," or HFP, variables (Dean et al. 1985, 1994), as defined for other aspects of environment in the ancient Southwest. These specific types of transient celestial events are a common and expected part of the environment, and their cycles and fluctuations are woven into the ideological fabric of each culture. However, because of their location in the sky, such events are imbued with supernatural overtones. The formal observation and interpretation of these HFP transient celestial events are limited to a small group of knowledge keepers, typically chiefs and priests/shamans and their novitiates. Category 2 and Category 3 events, when occurring within a few years of other major environmental and cultural events (e.g., floods, major droughts, earthquakes, deaths of chiefs), are likely to be cognitively linked with these other events.

Category 4 events (and perhaps some Category 3 events), though not quite identical with "low-frequency process," or LFP, environmental variables (Dean et al. 1985; Dean et al. 1994), are spectacles that leave lasting impressions on knowledge keepers and on virtually all members of the society who observed the event. Therefore, the appearance of such an event, particularly a cluster of two or more events within a single generation, triggers social behavior in roughly the same way as LFP environmental processes. This observational memory is almost always passed on to succeeding generations (Masse 1995, 1997, 1998, 2003b).

Category 5 events are not only rare but produce a suite of physical effects that stress social systems, especially priests/shamans, who mediate with deities to foster and preserve harmony in the universe. Total solar eclipses and cosmic impacts require changes in or enhancements of the basic fabric of a cultural

TABLE 12.2. Transient Celestial Events Potentially Seen by the Hohokam Between A.D. 500 and A.D. 1500, by 20-Year Human Generation Period.

DATE (A.D.)	CATEGORY 1 (MINOR)	CATEGORY 2 (MODERATE)	CATEGORY 3 (POSSIBLY MAJOR)	CATEGORY 4 (MAJOR)	CATEGORY 5 (MULTISENSORY)	SCORE
501–520	C		CCE			10
521–540	CCN		CCMM	M		19
541–560	CN		MMME			14
561–580	CCN	CCM	CCCC			21
581–600	C	CCMM	CMMM			21
601–620	CCCC		CCC			13
621–640	C	CCC				7*
641–660	CN	CMM				8*
661–680	CCCCCN		C			9*
681–700	CCC	CMM	CNNE	M		25
701–720	CCCCC	CM	MMMM			21
721–740	CCCCN		M		E	13
741–760	C	CMM	CCC			16
761–780	CCCCCCC	MM	CMMM			23
781–800			M		E	8*
801–820	CCC	CMM	C			10*
821–840	CCCCCCCNN	CNMMM	CCNMM	CC		42**
841–860	CCC	CM	CM			13
861–880	CCCCCN	C	CM			14
881–900	CCCCNN	CMMM	CCCCCNM	CC		43**
901–920	CCN	MMM	CCNMM	C		23
921–940	CCCCCCN	CMM	NMMM	MM	M	38*
941–960	CCC	CCC	CC	M		19
961–980	CCC	CCC	CME	C		25
981–1000	CCC	CCM				9*
1001–1020	CCN	CM	CCCM	S		35*
1021–1040	C	CNMMM	CMM	CM		28
1041–1060	CCC	CCCCCM		S		19
1061–1080	CCCCNNN	CCM	CNNM	C	E	34*
1081–1100		MMM	CCCMM	M	M	30*
1101–1120	CCCC	CCM	E	C		17
1121–1140	CCCNN	CCMM	CCC			22
1141–1160	CC		CCCE			14
1161–1180	CCN	C	CNM	L		15
1181–1200	CC	CM	CSE			15

TABLE 12.2. *(continued)*

DATE (A.D.)	CATEGORY 1 (MINOR)	CATEGORY 2 (MODERATE)	CATEGORY 3 (POSSIBLY MAJOR)	CATEGORY 4 (MAJOR)	CATEGORY 5 (MULTISENSORY)	SCORE
1201–1220	CCC	CNMM	CCN		M	25
1221–1240	CCCCNN	CCCM		CCM		26
1241–1260	C	CM	M	N		12
1261–1280	CCCCCC	C		C		12
1281–1300	C	CCC	C			10*
1301–1320	CCCCCCC	C	CE	C		19
1321–1340	CC	CC	C			9*
1341–1360	CCCCNN	CC				10*
1361–1380[a]	CCCCCCC	CCCCM	CCCCCM	C	E	44**
1381–1400	CCNN	CCCCM	CCE	M		27
1401–1420	CCCCCCCCCN	N	CNNME	C		31*
1421–1440	CCNNN	CC	CCCCCNM			30*
1441–1460	CCCCCCNN	CC	CCC			21
1461–1480	CCCCCNNNN	CCMM	CC	C		27
1481–1500	CCCCN	CCCM	N			16

Notes: Data from Masse 2003b. C = comet; N = nova; S = supernova; E = solar eclipse; M = meteor shower/substorm/storm; L = possible lunar impact. The "score" represents the assigned category number (1–5) times the number of events for each category, summed for each 20-year generation period. The mean score is 20.2 ± 9.7. Numbers with * are one standard deviation from the mean; with **, two standard deviations.

[a] Period of major auroral activity.

group's cosmology and world view. Because of the holistic nature of past civilizations and traditional non-Western cultures, culture histories worldwide are replete with the direct correlation of Category 5 transient celestial events with changes in leadership, religious ritual, settlement patterns, and social processes (Masse 1998, 2003b).

The historic record of transient celestial events is analogous to a master tree-ring sequence. In much the same manner that sequences of tree-ring widths provide a mechanism for absolute dating, the chronological spacing and nature of individual historically recorded transient celestial events, event clusters, and hiatuses provide a yardstick against which cultural/archaeological traditions can be systematically scrutinized and fitted (e.g., Table 12.2).

Our focus on one unique LFP-like class of events, solar eclipses, serves as a model for the broader set of all transient celestial events. Table 12.3 lists all total and annular solar eclipses between A.D. 1 and 2000 that would have been visible to southern and central Arizona Hohokam and later populations.[3] This list includes eclipses in vicinity of Yuma, in the Trincheras and Río Sonora culture regions of northern Sonora, Mexico, and in the Casas Grandes region of Chihuahua, Mexico (Figure 12.1), given the important economic and

TABLE 12.3. Catalog of Total and Annular Solar Eclipses Passing Through and Near Southern and Central Arizona Between A.D. 1 and A.D. 2000.

DATE[a]	LOCAL TIME OF MAXIMUM ECLIPSE[b]	ALTITUDE OF SUN AT MAXIMUM ECLIPSE	ECLIPSE TYPE (ANNULAR OR TOTAL)	ECLIPSE DURATION (IN MINUTES AND SECONDS)	ECLIPSE MAGNITUDE[c]	ECLIPSE OBSCURATION[d]	ECLIPSE CENTRAL LATITUDE[e]	MAGNITUDE AT CASA GRANDE[f]
0052 Sep 11	10:27	53°	Annular	4 min 14 sec	0.956	0.915	29°37' N	0.893
0085 Jun 10	15:50	43°	Annular	3 min 47 sec	0.959	0.919	34°04' N	0.949
0098 Mar 21	06:41	2°	Annular	1 min 34 sec	0.975	0.951	35°41' N	0.937
0102 Jan 06	14:27	31°	Total	2 min 28 sec	1.029	1.000	30°27' N	0.961
0106 Oct 14	09:46	39°	Annular	4 min 07 sec	0.954	0.909	28°23' N	0.866
0128 Aug 13	09:49	50°	Annular	2 min 42 sec	0.971	0.943	32°39' N	0.975 A
0310 May 15	06:27	12°	Total	2 min 42 sec	1.052	1.000	34°46' N	0.999
0323 Feb 21	16:38	20°	Annular	0 min 45 sec	0.990	0.980	31°30' N	0.952
0501 Feb 03	18:02	0°	Total	1 min 41 sec	1.030	1.000	31°15' N	0.960
0505 Nov 11	12:07	41°	Annular	4 min 36 sec	0.958	0.918	30°42' N	0.936
0559 Dec 14	14:49	25°	Annular	3 min 40 sec	0.961	0.923	31°55' N	0.958
0619 Mar 21	08:15	23°	Total	1 min 45 sec	1.026	1.000	29°07' N	0.934
0621 Aug 22	17:10	22°	Total	3 min 23 sec	1.057	1.000	29°38' N	0.958
0684 Sep 14	17:00	18°	Total	2 min 34 sec	1.042	1.000	29°18' N	0.967
0736 Jun 13	15:17	51°	Total	3 min 08 sec	1.040	1.000	32°44' N	1.012 T
0797 Aug 26	11:19	61°	Total	4 min 06 sec	1.052	1.000	33°24' N	1.015 T
0800 June 25	18:51	9°	Annular	2 min 55 sec	0.953	0.908	36°21' N	0.922
0804 Apr 13	14:49	49°	Total	5 min 08 sec	1.069	0.915	36°17' N	0.928
0830 May 25	12:34	82°	Total	4 min 31 sec	1.051	1.000	29°32' N	0.919
0979 May 28	06:46	16°	Annular	1 min 41 sec	0.973	0.948	31°58' N	0.970
0985 Jul 20	05:32	1°	Total	0 min 01 sec	1.000	1.000	33°53' N	0.948
1011 Sep 11	07:25	17°	Total	2 min 54 sec	1.048	1.000	28°56' N	0.892
1062 Dec 03	11:33	32°	Annular	9 min 17 sec	0.922	0.850	34°18' N	0.940 A
1076 Mar 07	16:50	20°	Total	1 min 08 sec	1.017	1.000	33°20' N	0.998
1102 Oct 13	15:33	26°	Annular	1 min 39 sec	0.980	0.961	31°33' N	0.962

TABLE 12.3. *(continued)*

DATE[a]	LOCAL TIME OF MAXIMUM ECLIPSE[b]	ALTITUDE OF SUN AT MAXIMUM ECLIPSE	ECLIPSE TYPE (ANNULAR OR TOTAL)	ECLIPSE DURATION (IN MINUTES AND SECONDS)	ECLIPSE MAGNITUDE[c]	ECLIPSE OBSCURATION[d]	ECLIPSE CENTRAL LATITUDE[e]	MAGNITUDE AT CASA GRANDE[f]
1156 Nov 14	16:45	5°	Annular	3 min 02 sec	0.960	0.921	35°09' N	0.933
1196 Sep 23	13:08	54°	Total	4 min 06 sec	1.049	1.000	30°27' N	0.967
1234 Mar 01	07:08	4°	Annular	2 min 39 sec	0.961	0.924	35°05' N	0.940
1318 Aug 26	10:13	49°	Total	0 min 44 sec	1.009	1.000	34°32' N	0.961
1379 May 16	08:10	33°	Total	3 min 48 sec	1.061	1.000	32°53' N	1.027 T
1397 May 26	18:17	14°	Total	3 min 05 sec	1.057	1.000	35°57' N	0.959
1402 Aug 28	13:45	59°	Annular	0 min 33 sec	0.994	0.988	30°33' N	0.943
1496 Aug 08	14:06	63°	Total	5 min 15 sec	1.067	1.000	28°47' N	0.922
1531 Mar 18	13:18	61°	Total	0 min 17 sec	1.003	1.000	30°08' N	0.926
1557 Apr 28	16:21	34°	Total	4 min 07 sec	1.061	1.000	34°49' N	0.970
1578 Sep 01	13:06	62°	Total	3 min 16 sec	1.041	1.000	30°30' N	0.975
1623 Oct 23	14:53	31°	Total	2 min 13 sec	1.028	1.000	32°57' N	1.013 T
1651 Apr 19	16:42	28°	Annular	0 min 54 sec	0.988	0.976	31°08' N	0.936
1740 Dec 18	17:01	4°	Annular	1 min 20 sec	0.981	0.963	31°41' N	0.961
1745 Sep 25	10:27	53°	Total	4 min 14 sec	0.956	0.915	29°37' N	0.879
1806 Jun 16	08:02	32°	Total	3 min 21 sec	1.053	1.000	33°16' N	1.019 T
1821 Aug 27	06:25	5°	Annular	3 min 14 sec	0.953	0.907	31°22' N	0.928
1923 Sep 10	14:16	54°	Total	3 min 27 sec	1.042	1.000	28°55' N	0.920

Note: Table lists all eclipses of magnitude 0.94 or greater. Eclipse circumstances are those at the location where the eclipse's central line crosses longitude 111°30' W and between latitude 28°30' and 36°30' N. The one exception is the total eclipse of 0985, for which the central line values are given for longitude 110° W.

[a] Julian calendar is used before 15 October 1582.

[b] Local time is Greenwich Mean Time minus 7 hours.

[c] Fraction of the Sun's diameter obscured by the Moon. For total eclipses, the diameter ratio of Moon/Sun is given.

[d] Fraction of the Sun's area obscured by the Moon.

[e] Latitude of the eclipse's central line at longitude 111°30' N, the approximate longitude at Casa Grande National Monument.

[f] Maximum magnitude of the eclipse as calculated for Casa Grande National Monument (latitude 32°59.8' N, longitude 111°31.9' N). T (total) and A (annular) designate cases in which Casa Grande itself was within the eclipse path.

political role these regions potentially played in portions of Hohokam culture history. In discussing these eclipses, we refer to other categories of transient celestial events and other significant environmental events (floods, droughts, earthquakes) that occurred near the time of solar eclipses that may have been holistically connected by the Hohokam with the solar eclipses.

Alfredo López Austin (1996:63–64) eloquently summarizes this perspective:

> Fear can be the source of great creations. Since ancient times fear of eclipses has motivated humans to observe the sky. . . .
>
> Fear of cataclysms, wars, diminished crops, deaths of rulers, malformations, cross-eyed children with harelips—at first, native beliefs may give the impression that, in traditional thought, disparate ideas can accumulate with no rhyme or reason and, being readily accepted by the people, lead to a heap of superstition. The matter was not that simple. The secular persistence of these ideas should make us suspect that there were fundamental, if unconscious, reasons for coherence among believers in a particular cosmovision. Even if they could not perceive the structural basis of their culture, they expected the supernatural to be explicable. Mere faith was not enough; they looked for rational reasons for their beliefs.

THE SOUTHWESTERN SKYWORLD

Traditional Astronomy and Cosmology

The astronomy of traditional southwestern Native American cultures is poorly known, partly because formal sky watching was usually restricted to a small group of priests and shamans who zealously guarded their knowledge. For example, a typical historic O'odham village of 500–600 individuals would have had only about five *makai* ("magicians"), or medicine men, who had special powers over crops, weather, and wars, and perhaps star lore (Russell 1975:21, 256–259). The *makai*, along with *si'atcokam*, or medicine men who dealt with disease, were considered the real leaders of the villages and exercised more influence than the chiefs. As with other southwestern groups (see, e.g., Griffin-Pierce 1992:39; Mails and Evehema 1995; Parsons 1996:82–83; Waters 1963), star lore among the O'odham was tied into cosmology and world harmony and was not shared with outsiders or even one's village, clan, gender, or other small segments of the society. This makes such esoteric knowledge vulnerable to loss (Lesser 1933:50).

Our lack of understanding about traditional astronomy also reflects the fact that few southwestern ethnologists are trained to collect and evaluate such data, a situation clearly evident from southwestern ethnographies. For example, Underhill's (1946) masterful overview of Tohono O'odham religion barely mentions astronomical phenomena. Densmore (1929:15) provides the following tantalizing statement but does not elaborate on the topic in her excellent study of Tohono O'odham music: "The Papago watch the stars and when they see the Pleiades cross the sky in one night they say that the proper time for story telling has arrived."

Despite this dearth of ethnographic detail, generalizations can be made about Hohokam astronomy and cosmology by focusing on the ethnohistoric record of neighboring tribes and especially presumed linguistically related groups, including both fellow Tepiman speakers and other members of the Uto-Aztecan language family (Shaul and Hill 1998).

Southwestern cultures, like virtually all past societies, viewed the heavens as a separate but connected otherworld, one filled with sentient and mysterious beings (Masse 1995, 1997, 2003b). That celestial objects were viewed as living supernatural beings by pre-European southwestern cultures is evident from the previously mentioned treatment of meteorites found in archaeological contexts.

The O'odham thought of the stars, planets, Sun, and Moon as living beings (Russell 1975:250–252). Venus, in its aspect as the morning star, was considered the daughter of a magician. Sun and Moon were husband and wife, and Coyote was their son. Coyote helped to create the Milky Way and the other stars in the sky.

O'odham religion was described as "a belief in the supernatural or magic power of animals, and especially in the omnipotence of the Sun. . . . The disk was not regarded as the 'shield' or 'headdress', but as the veritable person of the god" (Russell 1975:250). The O'odham shared this solar emphasis with other past cultures, including their linguistic cousins, the Mexica (Aztecs) and other Nahuatl speakers (collectively termed *Nahua*). In fact, because the Nahua believed that four preceding world ages had ended in the death of the Sun and cataclysm, they performed elaborate rituals and ceremonies, involving bloodletting and human sacrifice, so that the Sun would continue to function properly (León-Portilla 1963, 1992).

The O'odham recognized two minor deities who lived on the eastern horizon at the points of the rising Sun on summer and winter solstice, as well as four minor deities who occupied positions along the path of the Sun (Russell 1975:251). In addition to the Sun and Coyote, the O'odham recognized two major deities who lived in the east and who shared the duties of the universe. One is Jewed Makai, or Earth Doctor, and the other is Siuuhu/I'itoi, or Elder Brother/Drinker (Bahr et al. 1994:41–43, 137–141, 179–181). Earth Doctor is a beneficial god associated with wind and rain, and Elder Brother, in the form of the Drinker, is a perilous god associated with the Sun and with drought (Bahr et al. 1994:38, 317 n. 7). As discussed below, Elder Brother is likely associated with solar eclipses and with the appearance of comets.

While the lack of ethnographically preserved astronomical detail is lamented for the Southwest, in particular for the O'odham, studies of groups such as the Pawnee (Chamberlain 1982; Murie 1981; Weltfish 1977) and the Chumash (Hudson and Underhay 1978) provide perspectives on how traditional Native American societies wove the events and processes of the heavens into rituals and daily lives. That each culture had its own cosmological view of the universe does not detract from the fact that the celestial realm, including transient events, served a critical role in the functioning of every traditional culture.

Native American cultures also expressed a strong sense of cosmology in terms of both village structure and regional spatial geography. The observations by Harrington (1916:41–45) and Ortiz (1969) for the Tewa and Elsie Clews Parsons (1996:1:98–103, 365–368) for Pueblo Indian religion regarding the importance of directions and sacred numbers can be extended to other groups, including the Hohokam. Parsons (1996:365) notes: "The ritual directions are six—North, West, South, East, Above, and Below. . . . And in passing from one direction to the other the order as given for the cardinal points is followed, so that the ceremonial circuit is sinistral or anti-sunwise. Associated with the directions are colors, clouds, lightning and rain, the prey animals, birds, ants, corn and other plants, trees, stones or shells, and many of the Spirits."

The importance of direction extended to the placement of sleeping mats (always with the head away from the orientation of the head in extended burials), travel into and out of villages, arts such as basket coiling and pottery design, and dancing. It is uncertain whether rules existed regarding the selection of locations for new villages. Significantly, every Tewa village had a specialist (e.g., Winter Chief) who observed solstice occurrences against the backdrop of appropriate positions on distant mountain ranges (Ortiz 1969:102). It can be reasonably assumed that southwestern villages, including those of the Hohokam, had similar arrangements for viewing solstices, equinoxes, and other calendrical midpoints. Our lack of documented examples of calendrical features in the archaeological record reflects the dearth of training by archaeologists to understand and look for such features (see Fountain 2005; Preston and Preston 2005).

Portents, Omens, and Astrology

One curious and poignant aspect of the conquest of Mexico in A.D. 1519–1521 by a few hundred Spanish soldiers was the initial belief by Montezuma and his priests that Hernán Cortés was the long-lost benevolent Toltec king/god Topiltzin Quetzalcoatl. Legend indicated that Topiltzin Quetzalcoatl had left Mexico centuries earlier with a promise to return and help his people. Belief in this myth (Carrasco 1992) and the coincidence of iconographic parallels between Aztec deities and the regalia of Cortés and his priests helped lead to the unlikely capture of the large city of Tenochtitlan by the tiny Spanish force, and ultimately led to the demise of the Aztec state.

The Mexica had an intricate system of personal astrology (*tonalpohualli*) based on horoscopes (*tonalámatl*), referred to as the "book of the days and destinies" (León-Portilla 1963), as well as a cyclic view of human history, one punctuated by episodes of worldwide cataclysm (Carrasco 1992; León-Portilla 1963, 1995; Milbrath 1995). The Aztecs believed that fate—both individual and cultural—was not ascribed and unavoidable and that human and cultural intervention could change history or delay catastrophes (León-Portilla 1963). For this reason, omens and portents, including celestial signs, were of grave importance and required continuous monitoring and evaluation.

Portents and omens were a part of the fabric of Nahua life and political decision making. Because of the human need to understand, predict, and control our environment, unusual physical events, including celestial events, were interpreted as signs of change or impending (or even preceding) disasters (Schaefer 2005). For example, the famous Florentine Codex, prepared by Fray Bernardino de Sahagún from Nahuatl informants in the latter half of the sixteenth century (Sahagún 1953:13), stated that comets were considered omens for rulers. The appearance of a comet portended the death or imprisonment of a ruler or other nobility, warfare, or famine. For major "disasters" such as the Spanish Conquest, celestial events (comets and a fireball) and other phenomena witnessed a decade earlier were perceived as part of the package of signs that forecast this catastrophic outcome (Carrasco 1992:187–191).

A comet, solar eclipse, and an earthquake over a seven-month span in 1695–1696 alarmed the Jesuit missionaries in western Chihuahua, Mexico, not so much because of the events themselves but for how they would be interpreted and acted on by the already rebellious Tarahumara Indians. Father Joseph Neumann wrote: "Orally or in writing, the missionaries described these portents to the Father Visitor, and he, too, felt that they presaged some sad and terrible misfortune for the missions. Therefore, he instructed the Fathers to pay close heed to all that the Indians did and to learn as much as possible about their plans" (Brown and Sinclair 1998:123). Neumann was not mistaken in his perception of the critical importance of these portents for Tarahumara decision making: in 1697, he and his fellow priests faced the largest and last of the Tarahumara uprisings.

Omens and portents were a fact of life for the O'odham, Hopi, and Maricopa as well. A clear example, described by Russell, is recorded on a calendar stick covering the period 1833–1902. Russell's (1975:38) description begins with reference to the spectacular Leonid meteor storm of 1833: "In the early winter the meteoric shower took place. This event was followed by heavy rains that caused floods in the Salt and Gila rivers. The spectacle of falling stars was to the Pimas an augury of disaster, and the succeeding floods [winter 1833–1834] were regarded as a punishment for sins which they had committed." Although some Maricopa calendar stick records also began with the two-day Leonid storm of 1833, "the time the stars fell," no portents or omens were recorded in association with it (Spier 1978:138–142). Spier (1978:295) notes, however, that the appearance of a comet for the Maricopa "betokened the death of a prominent person or a general disaster."

An O'odham myth links celestial phenomena with omens and portents (Shaw 1968:111). A young maiden (Morning Star) and her uncle (Meteor) are carried to the sky by Whirlwind to provide care and protection for the people. The uncle specifically states: "When they see me shooting across the evening sky, they will know the enemy plans to wage war against them. The men will get their bows and arrows ready for the enemy." The name provided to Russell (1975:47) for a meteor or fireball observed in 1859 and recorded on calendar sticks was *pai-ikam ho-o*, "angry star." A similar conception of meteors was held by the Tewa, and meteors were likewise viewed at Isleta pueblo as a portent of war (Hewett and Dutton 1945:28).

The Hopi have legends of clan migrations that were triggered or accompanied by the appearance of fireballs (Courlander 1971:72), meteor showers (Nequatewa 1967:29), comets (Courlander 1971:76), eclipses (Courlander 1971:99), and nova events (Nequatewa 1967:29, 35, 38; Waters 1963:22, 39). One story tells of the destruction by flood and earthquake of Palotquopi (Nequatewa 1967:97), which some scholars (e.g., Teague 1993) believe was a Hohokam village.

Solar Eclipses as Portents

Studies of Mexica and Maya art and chronicles (Milbrath 1995, 2000) underscore the importance of solar and lunar eclipses as major cultural events. In Nahua chronicles, such as the Codex Telleriano-Remensis (Keber 1995) and the Codex Chimalpopoca (Bierhorst 1992), only a few environmental and cultural events are deemed important each year, primarily major battles, the accession or death of a ruler, floods, earthquakes, and solar eclipses. Among the codices there are minor discrepancies in dates and in the events being portrayed (Boone 2000:223–228), but solar eclipses are among the more commonly noted events. Solar eclipses at or near the time of a death of a ruler were linked in some manner to that death. During lunar eclipses pregnant women put obsidian in their mouths or on their bellies to protect themselves and to prevent disfigurement and disease in their unborn children (Parsons 1996:181; Sahagún 1953).

Solar eclipses were more traumatic than lunar ones because of the belief that the four preceding cosmological ages had each ended with cataclysm and the death of the Sun (Bierhorst 1992; León-Portilla 1963, 1995; Masse 1995, 1997). Each new eclipse could thus portend the end of the present age. López Austin's (1996:62) translation of Sahagún's sixteenth-century record of the Nahua reaction to total solar eclipses is instructive:

> When this occurs [the Sun] turns red. It is no longer quiet, no longer tranquil. It is only balancing itself. There is immediate turmoil. People are restless, there is an uproar, fear and weeping. People wail, give cries of alarm, scream, clamor, rattle hawk bells. Albinos are sacrificed. Captives are sacrificed. People shed their own blood, piercing their ears with reeds, and flowery songs are sung in the temples. The noise continues, the wailing persists. It was said, "If it comes to an end, if the Sun is eaten, everything will become dark forever. The [feminine monsters called] *tzitzimime* will come down to devour the people."

The din raised by people during a solar eclipse was to scare away the *tzitzimime* demons believed to be devouring the Sun (Milbrath 1995; Miller and Taube 1993) and to prevent them from diving headlong from the heavens to attack people on earth. *Tzitzimime* were visualized, in part, as constellations, individual stars, and planets transformed during the eclipse into star-like, clawed demons. This change would be frightening since the Mexica viewed the normal appearance of planets, stars, and constellations as aspects of powerful gods or supernatural beings. Like star demons, *tzitzimime* are represented in the rock art of the Jornada and northern Rio Grande regions of New Mexico during the late fourteenth century, along the paths of total solar eclipses that transited the area in 1379 and 1397 (Masse and Soklow 2005).

The documentary evidence for the manner in which traditional southwestern cultures reacted to eclipses is intriguing. Spier (1978:149) notes that eclipses are a time when the Sun and Moon "died." Memory of one eclipse (16 June 1806) was passed down through 124 years of oral transmission. The O'odham considered eclipses as a time when the Sun or Moon "dies" (Masse 2003a; Russell 1975:61, 63). Russell (1975:61) was puzzled that only two lunar eclipses were listed in the 1833–1901 O'odham annals but rationalized this situation with the following important statement: "As it is known from American testimony that the Pimas were profoundly impressed by such phenomena [lunar eclipses], the failure of the annalists to note them can be accounted for only by their aversion to even a mention of supernatural events supposed to be threatening in character."

As related by my Hia C'ed O'odham friend and colleague Lorraine Eiler (personal communication 1997), a fear that exposure to lunar eclipses would cause problems for pregnancy was prevalent into the mid-twentieth century. A fear of solar eclipses was noted among groups in central and northern Mexico as recently as the total solar eclipse of July 1991 (Miller and Taube 1993:84), when "native peoples in many parts of the country made a commotion to stave off any baleful effects."

The Hopi viewed eclipses as ominous signs. Like the Pima, they believed that the Sun and Moon died during an eclipse (Parsons 1996:86, 180). Disease and death were associated with eclipses, and purifying ceremonies were required after an eclipse. It was believed that if a solar eclipse occurred immediately after sunrise, Zuni and Rio Grande pueblo peoples would die; if in the middle of morning, Hopi people would die; and if at noon, people to the southwest would die. The Tewa considered an eclipse to be the death of the Sun or the Moon (Hewett and Dutton 1945:27).

SOLAR ECLIPSES AND THE HOHOKAM

Hohokam culture history between A.D. 500 and 1500 is divided into five temporal periods and approximately twelve subperiods (phases) based on perceived changes in architecture, social structure, and material culture, especially decorated ceramics, with period boundaries presumably marking the greatest amount of change (Crown and Judge 1991; Gumerman 1991; Haury 1976).

Chronology is a hotly debated topic because of the general absence of tree-ring-datable wood at Hohokam sites (Dean et al. 1996), although scholars may have reached a seeming general consensus (Ciolek-Torrello 1995; Dean 1991; Deaver and Ciolek-Torrello 1995; Elson 1996; Elson and Gregory 1995; Wallace et al. 1995). Even this "consensus" yields an average range of 25–50 years for the inception of each period and phase, depending on the views of individual scholars. Table 12.4 presents a recent consensus dating for the Phoenix Basin (Salt-Gila) and the Tucson Basin, including the chronological uncertainties. The table provides current estimates for the inception of specific significant, rapid (within one or two generations) cultural processes/events that generally coincide with period and phase boundaries. These chronological sequences and events are plotted against the record of known and reconstructible eclipses in the Hohokam area (Table 12.3).

It is notable that the inceptions of four of the five Hohokam cultural periods (Pioneer, Colonial, Sedentary, and Postclassic) appear to coincide with total solar eclipses in the area. A total solar eclipse along the southwestern edge of the Hohokam region highlights the inception of the fifth period, the Classic. These correlations take on added significance with the remarkable fact that total and high-magnitude annular solar eclipses closely coincide with nearly all the most pronounced rapid changes in the Hohokam historical record (Table 12.4) and yet are virtually absent from the middle portions of Hohokam periods, with the single exception being the last half of the Sedentary period, when a total solar eclipse and annular eclipse coincide with the beginning of the Santan and Late Rincon phases around A.D. 1075–1100. Some scholars feel that the cultural changes that occurred ca. A.D. 1075 are equal in magnitude to those occurring at the inception of the five previously defined periods (e.g., Doyel 2000).

The Year of the Diminished Sun and the Dawn of the Hohokam Culture

Around A.D. 550, the Southwest experienced broad cultural changes. Among them are changes in pottery, architecture, and settlement patterns in the Mogollon area, warranting the creation of named local phases (Cordell 1997:205). Similar material culture changes in the Ancestral Pueblo area are tied to possible changes in ritual and belief systems, including the appearance of masked images along with the humpbacked flute player, Kokopelli (Cordell 1997:247), which may be linked with solar eclipses (Masse and Soklow 2005).

The Hohokam region also displays substantive changes. The period around A.D. 500 reflects a time of cultural divergence of populations in the Tucson Basin from those in the Phoenix Basin (Wallace et al. 1995:596). According to Wallace and Lindeman (2000), it is currently estimated that plazas made their appearance at the sites of Snaketown and perhaps Grewe along the Gila River and at Valencia Vieja in the Tucson Basin around A.D. 550.

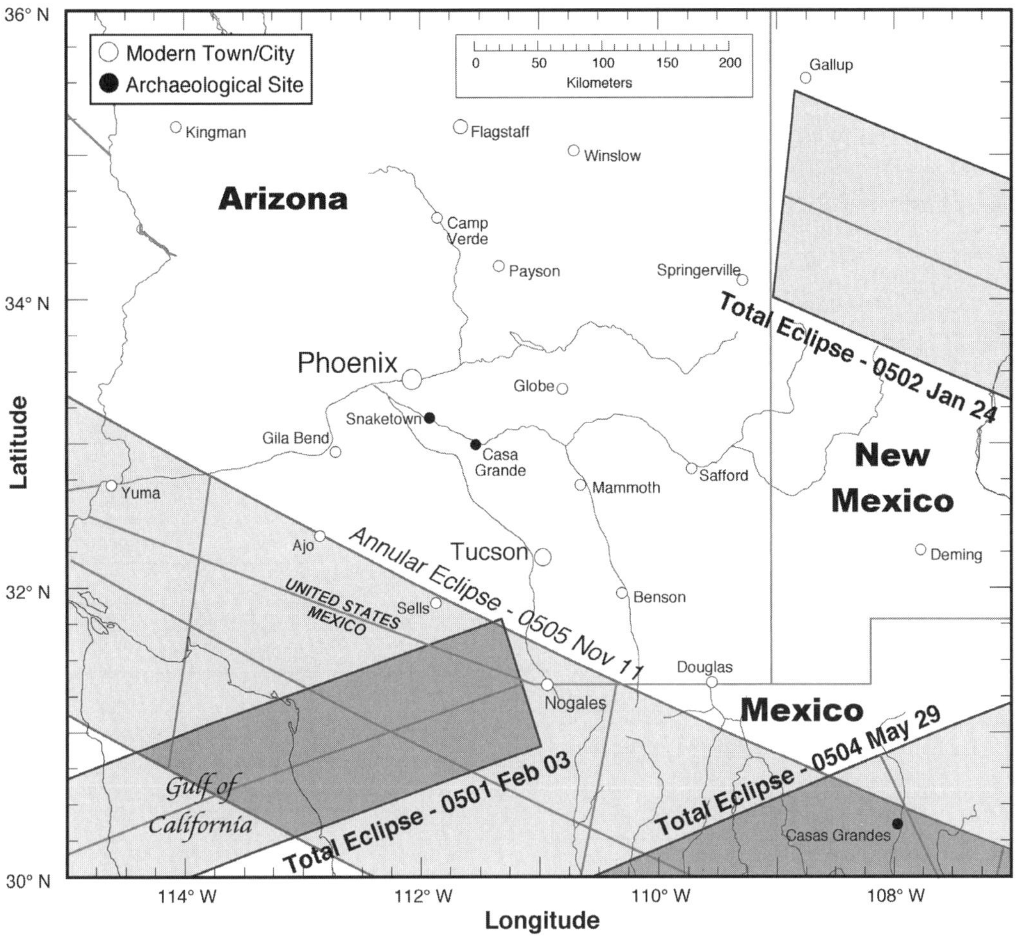

FIGURE 12.2.
Paths of total and annular solar eclipses between A.D. 501 and 505.

A remarkable cluster of five eclipses occurred in fewer than five years between A.D. 501 and 505 and would have been most pronounced in and south of the Tucson Basin (Figure 12.2). This cluster began with a rare sunset total eclipse (TSE-501, Feb 03; 0.960 magnitude at Casa Grande), followed by an equally rare sunrise total eclipse northeast of the Hohokam region (TSE-502, Jan 24; 0.863 magnitude at Casa Grande), followed by a sunset partial eclipse (PSE-503, Jun 09; 0.318 magnitude at Casa Grande), followed by an annular eclipse (ASE-504, May 29; 0.899 magnitude at Casa Grande), followed by a final annular eclipse (TSE-505, Nov 11; 0.936 at Casa Grande). As previously noted, the Hohokam could have perceived a low-magnitude sunrise/sunset eclipse, such as that of A.D. 503.

During the period A.D. 536–537, the cultural records of Asia, Europe, and the Middle East note that normal sunlight was greatly diminished for more than a year, an observation matched by tree rings worldwide that indicate little or no growth (Baillie 1999; Gunn 2000). Examples of the concern created by this phenomenon include the eyewitness account of the Byzantine historian Procopius (Schove and Fletcher 1984:96): "during this year a most dread

TABLE 12.4. Correlation of Solar Eclipses with Hohokam Chronology and Culture History.

DATE (A.D.)	ECLIPSE (MAGNITUDE)[a]	PERIOD (? = DATE RANGE)	PHASE SALT-GILA/TUCSON	MAJOR CULTURAL EVENTS AND CULTURAL PROCESSES
490–509	**TSE-501** (0.960, sunrise); ASE-505 (0.936)	?? ?? Pioneer ??	Vahki/Tortolita	Reorganization: Beginning of Tucson basin divergence from Phoenix basin
510–529				
530–549	"year of diminished sun"			Reorganization: First use of plazas and community houses
550–569	ASE-559 (0.958)			
570–589				
590–609				
610–629	PSE-621 (0.958)			
630–649			?	
650–669			? Estrella/Estrella ? Sweetwater/Sweetwater	First decorated pottery
670–689	PSE-684 (0.967)			
690–709			? Snaketown/Snaketown	Major change in pottery clay and designs
710–729				
730–749	**TSE-736** (1.012)		?? Late Snaketown/Late ?? Snaketown	Reorganization: Mexican iconography and material culture; change in ritual systems
750–769				
770–789				
790–809	**TSE-797** (1.015)	?? Colonial ??	Gila Butte/Canada del Oro	Reorganization: First use of ball courts and the creation of the regional ball-court system
810–829				
830–849				
850–869			?? Santa Cruz/Rillito ??	
870–889				
890–909				
910–929				
930–949				
950–969				
970–989	ASE-979 (0.970); PSE-985 (0.948, sunrise)	?? ?? Sedentary ??	Sacaton/Rincon	Maximum extent of Hohokam regional ball-court system; ritual intensification
990–1009				
1010–1029				
1030–1049				
1050–1069	ASE-1062 (0.940)			

Table 12.4. *(continued)*

Date (A.D.)	Eclipse (Magnitude)[a]	Period (? = Date Range)	Phase Salt-Gila/Tucson	Major Cultural Events and Cultural Processes
1070–1089	**TSE-1076** (0.998)		?? Santan/Late Rincon ??	Reorganization: Abandonment of cores of large villages, including Snaketown
1090–1109	ASE-1102 (0.962)			
1110–1129				
1130–1149				
1150–1169	PSE-1156 (0.933)			
1170–1189				
1190–1209	PSE-1196 (0.967)	?? Classic ??	Soho/Tanque Verde	Reorganization: end of regional ball-court system; *cerros de trincheras;* platform mounds
1210–1229				
1230–1249	PSE-1234 (0.940)			
1250–1269				
1270–1289				
1290–1309			? Civano/Tucson ?	Construction of great houses/towers
1310–1329	PSE-1318 (0.961)			
1330–1349				
1350–1369		?? ?? Postclassic ??	Polvoron/Tucson	Reorganization: War with O'odham; end of great houses/towers; migrations/population loss; significant changes in material culture
1370–1389	**TSE-1379** (1.027)			
1390–1409	PSE-1397 (0.959); PSE-1402 (0.943)			
1410–1429				
1430–1449				
1450–1469		?? ?? Protohistoric ??		Reorganization: End of material aspects of archaeological Hohokam culture
1470–1489				
1490–1509	PSE-1496 (0.922)			

[a] TSE = total solar eclipse; ASE = annular solar eclipse; PSE = partial solar eclipse (eclipse path not directly within the Hohokam cultural region). Magnitude is that at Casa Grande National Monument.

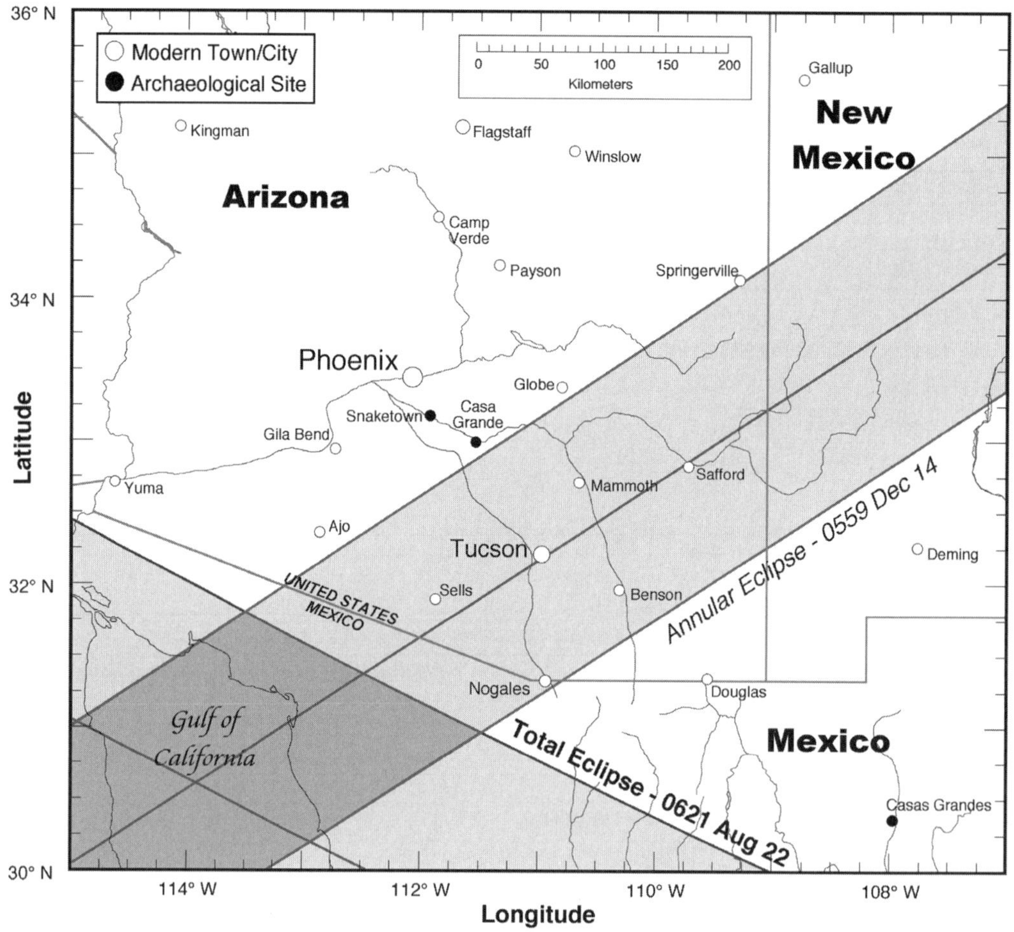

FIGURE 12.3. Paths of the annular solar eclipse of A.D. 559 and the total solar eclipse of A.D. 621.

portent took place. For the Sun gave forth its light without brightness, like the Moon, during this whole year, and it seemed exceedingly like the Sun in eclipse." The cause of this event is still debated (Baillie in press). Nevertheless, the presence of prolonged diminished sunlight would have been ominous for the Hohokam and would have fostered fears about the well-being of the Sun and harmony in the universe.

In A.D. 559, a scant generation after the year of diminished sunlight, an afternoon annular eclipse centered on the Tucson Basin (Figure 12.3) had a high enough magnitude (0.961) and was close enough to the winter solstice to have been an alarming portent. Two total solar eclipses south of the Hohokam region, in A.D. 621 (Figure 12.3) and A.D. 684 (Table 12.3), with nearly identical paths, resulted in partial eclipses at Casa Grande of a magnitude around 96 percent, similar to that of ASE-559.

The Total Eclipse of A.D. 736 as Precursor to the Hohokam Regional System

The first major Hohokam total solar eclipse occurred in mid-afternoon on 13 June 736. TSE-736 transited through the heart of the Hohokam region, and

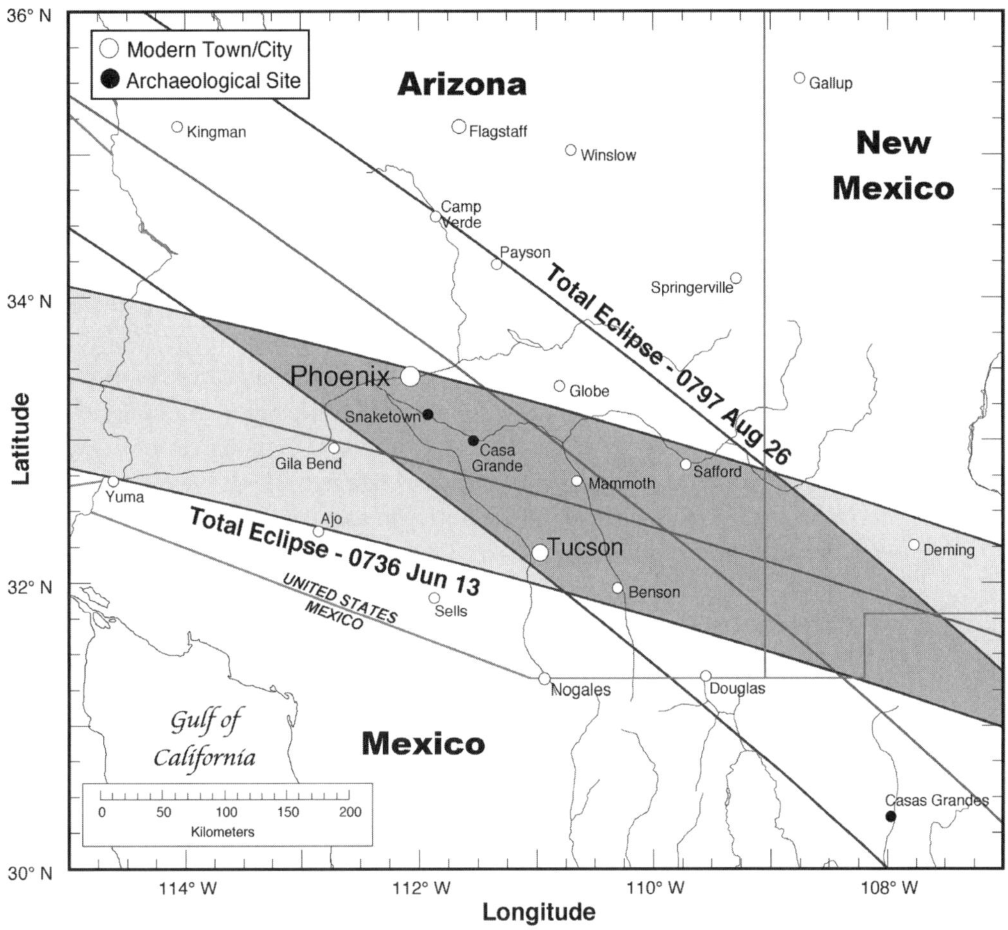

FIGURE 12.4. Paths of the total solar eclipses of A.D. 736 and 797.

totality lasted more than three minutes (Figure 12.4). The appearance of this eclipse just prior to summer solstice would have been ominous, as would the location in the constellation Gemini, surrounded by many of the brightest constellations and stars. The fact that Mars and Saturn were in conjunction near the eclipse and that Mercury and Venus were also visible would have heightened the eclipse's overall effect.

TSE-736 marks the approximate middle of the Snaketown phase, according to current dating estimates (Table 12.4). This is a time of major change in the material culture of the Hohokam that some scholars attribute to direct Mesoamerican influence (e.g., Doyel 1991a:247–248). It is possible that a spectacular eclipse in A.D. 750 along the latitude of the "turning point of the Sun" north of the Valley of Mexico helped to promulgate these traits. Masse (1995, 1997) elsewhere notes that the A.D. 750 eclipse may have been a factor in the fall of the sacred city of Teotihuacan and the dispersal of peoples and cosmology throughout northern Mexico. The cultural ripple effects of the A.D. 750 Mexican eclipse and TSE-736 may have set the stage for the Hohokam reaction to the solar eclipse of A.D. 797.

The Total Eclipse of A.D. 797 and the Spread of the Hohokam Regional Ball-Court System

Just before noon on 26 August 797, a total solar eclipse transited the entire Hohokam culture area, coinciding with the path of the TSE-736 eclipse (Figure 12.4). Totality lasted more than four minutes along the center of the eclipse path. The fact that TSE-797 occurred at the height of summertime, possibly after the end of the summer monsoons, would have enhanced its effects, including changes in ambient temperature and winds. The eclipse occurred in the constellation Leo when Saturn and Venus were visible in conjunction and Jupiter likewise was visible.

High-magnitude streamflow occurred along the Salt River (Table 12.5) less than a decade after TSE-797. Reconstructed discharges for A.D. 803–805 are each among the 15 highest for the 1,417-year record of the Salt River (Graybill et al., this volume). Although we cannot be certain that high-magnitude streamflow years always resulted in substantial flooding of the river valleys, this particular sequence of multiple large discharges strongly suggests that Phoenix Basin Hohokam irrigation systems—and perhaps even village features on the first terrace above the river—would have been devastated (Gregory 1991; Masse 1991). TSE-797 would have been viewed as a powerful portent for the floods. A similar sequence of high-magnitude streamflow events occurred about the time of TSE-736 and TSE-1076 (Table 12.5), a coincidence to which we return below in the discussion of TSE-1196.

Around A.D. 800 (Wallace et al. 1995:609), the regional system first became established, expanding into the very regions impacted by the eclipse. The underpinning of the Hohokam regional system was the construction of large earthen ball courts, likely related to the widespread Mesoamerican ball game (Haury 1976; Scarborough and Wilcox 1991; Wilcox 1991a, 1991b; Wilcox and Sternberg 1983). The Hohokam manufactured and used many ritual objects, perhaps in connection with the ball game or even an associated cult (Doyel 1994), including palettes, stone censers, figurines, and serrated projectile points (Doyel 1991b; Haury 1976; Neitzel 1991). Between A.D. 800 and 1200, the Hohokam acquired objects that were likely manufactured in Mexico, such as rubber balls for use in the ball courts, copper bells, and mosaic plaques or mirrors.

The Mesoamerican ball game was associated with solar symbolism and potentially with the movement of stars and constellations (Wilcox 1987, 1991b) and was viewed as a reenactment of the struggle between day and night, light and darkness (Gillespie 1991:319–321; Schele and Freidel 1990:126, 243). Ball games on the dates of the equinoxes and solstices reenacted the descent (death) and ascent (birth) of the Sun into and from the underworld (Leyenaar 1992). Although Hohokam ball courts lacked some of the features of Mesoamerican courts (such as stone targets in the shape of a ring or hoop), Carrasco's (1990:35) evocative discussion of the Aztec ball game is pertinent for insights into the game played by the Hohokam:

TABLE 12.5. "Cognitive Correlation" of Solar Eclipses with High-Magnitude Streamflow in the Salt and Gila Rivers.

DATE (A.D.)	ECLIPSE (TOTAL, ANNULAR, PARTIAL)[a]	GILA RIVER HIGH STREAMFLOW YEAR (AND OVERALL RANK)	SALT RIVER HIGH STREAMFLOW YEAR (AND OVERALL RANK)
490–509	TSE-501; ASE-505		
510–529			
530–549	"year of diminished sun"	Start of record	
550–569	ASE-559		
570–589			Start of record
590–609			
610–629	PSE-621		622 (7)
630–649		649 (13)	
650–669			
670–689	PSE-684		
690–709			691 (20)
710–729		728 (23)	
730–749	**TSE-736**	**731** (18), **747** (20), **749** (1)	**730** (14), **731** (4)
750–769			
770–789			
790–809	**TSE-797**		**803** (13), **804** (15), **805** (8)
810–829			822 (21)
830–849		833 (22), 834 (8), 848 (24)	
859–869			
870–889			25
890–909		897 (17), 898 (2)	899 (5)
910–929			
930–949			
950–969			
970–989	**ASE-979; PSE-985**	**983** (25), 988 (3), **989** (4)	
990–1009			
1010–1029			
1030–1049			
1050–1069	ASE-1062		
1070–1089	**TSE-1076**	**1077** (12), **1089** (19)	
1090–1109	ASE-1102		
1110–1129			
1130–1149		1141 (9), 1142 (10)	
1150–1169	PSE-1156		
1170–1189			
1190–1209	PSE-1196		

TABLE 12.5. *(continued)*

DATE (A.D.)	ECLIPSE (TOTAL, ANNULAR, PARTIAL)[a]	GILA RIVER HIGH STREAMFLOW YEAR (AND OVERALL RANK)	SALT RIVER HIGH STREAMFLOW YEAR (AND OVERALL RANK)
1210–1229			
1230–1249	PSE-1234		
1250–1269			
1270–1289			
1290–1309			
1310–1329	PSE-1318	1313 (6)	
1330–1349			
1350–1369			
1370–1389	**TSE-1379**		**1382** (6), **1384** (18)
1390–1409	PSE-1397; PSE-1402		
1410–1429		1420 (14), 1426 (7)	
1430–1449			
1450–1469			
1470–1489			1482 (10)
1490–1509	PSE-1496		
1510–1997		1 Event	12 Events

Note: Bold entries represent potential episodes of putative cognitive correlation between eclipses and streamflow events.
[a] Partial solar eclipses are annular and total solar eclipses whose paths did not extend within the boundaries of the Hohokam regional system.

> It appears that the playing court represented the narrow passageway of the underworld through which the sun traveled at night. The game represented a cosmic struggle between competing factions to see which group could bring the sun out of the underworld by hitting the ball through one of the two perforated rings on the sides of the court. The ball court, then, is a kind of temple in which the solar drama is acted out in human time and space.

A second symbolic theme for ball courts, closely linked with solar symbolism, is that of agricultural fertility and seasonal renewal, also connected with the growing of maize (Cohodas 1991:255; Gillespie 1991; Schele and Freidel 1991). The rapid construction and expansion of the regional Hohokam ball-court system at around A.D. 800 (Masse 1991) may thus represent the ritualized attempt by the Hohokam to prevent further occurrences of both dreaded solar eclipses and disastrous floods, as occurred during the period A.D. 797–805.

The Eclipses of A.D. 979 and 985 and the Height of the Hohokam Regional System

The period A.D. 979–985 witnessed a high-magnitude annular solar eclipse and a sunrise total solar eclipse (Figure 12.5). ASE-979 and TSE-985 come

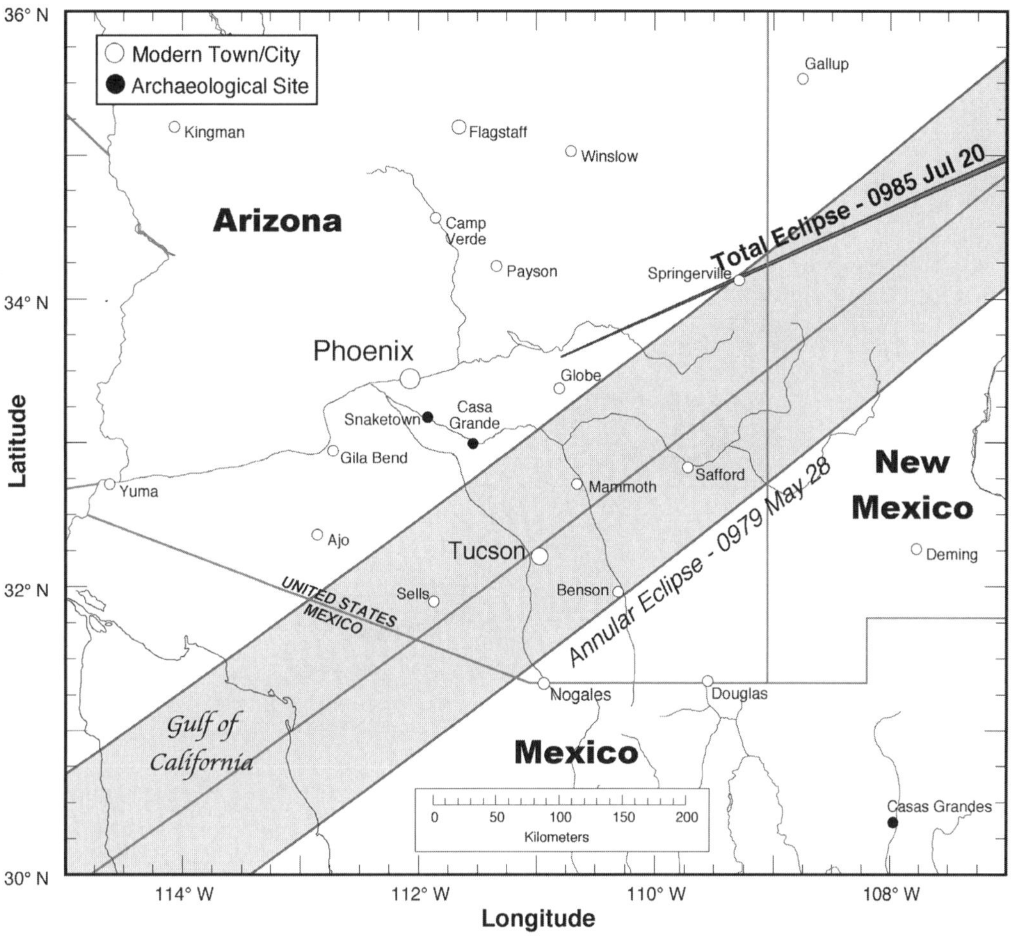

FIGURE 12.5.
Paths of the annular solar eclipse of A.D. 979 and the total solar eclipse of A.D. 985.

near the beginning of the maximum expansion of the Hohokam regional ball-court system and seemingly at a time of intensification and elaboration in ritual activities (see Crown and Judge 1991; Gumerman 1991). However, the period was not marked by obvious reorganization of religious ritual or changes in settlement patterns, as was true for the beginnings of other designated periods (Table 12.4). In fact, despite peaks of population at sites such as Snaketown (Haury 1976), the Sedentary period is remarkable for its seamless continuity with the earlier Colonial period.

It is possible that the early morning timing and magnitudes of ASE-979 and TSE-985 were high enough to cause major alarm, requiring ritual intensification and cultural elaboration to preserve "solar harmony," but the absence of totality in the Hohokam region precluded the need for major culture change. Perhaps more important for Hohokam ideology during the first half of the eleventh century were the A.D. 1006 Lupus supernova and the A.D. 1054 Taurus supernova, which, as argued later in this chapter, may be reflected in pottery designs.

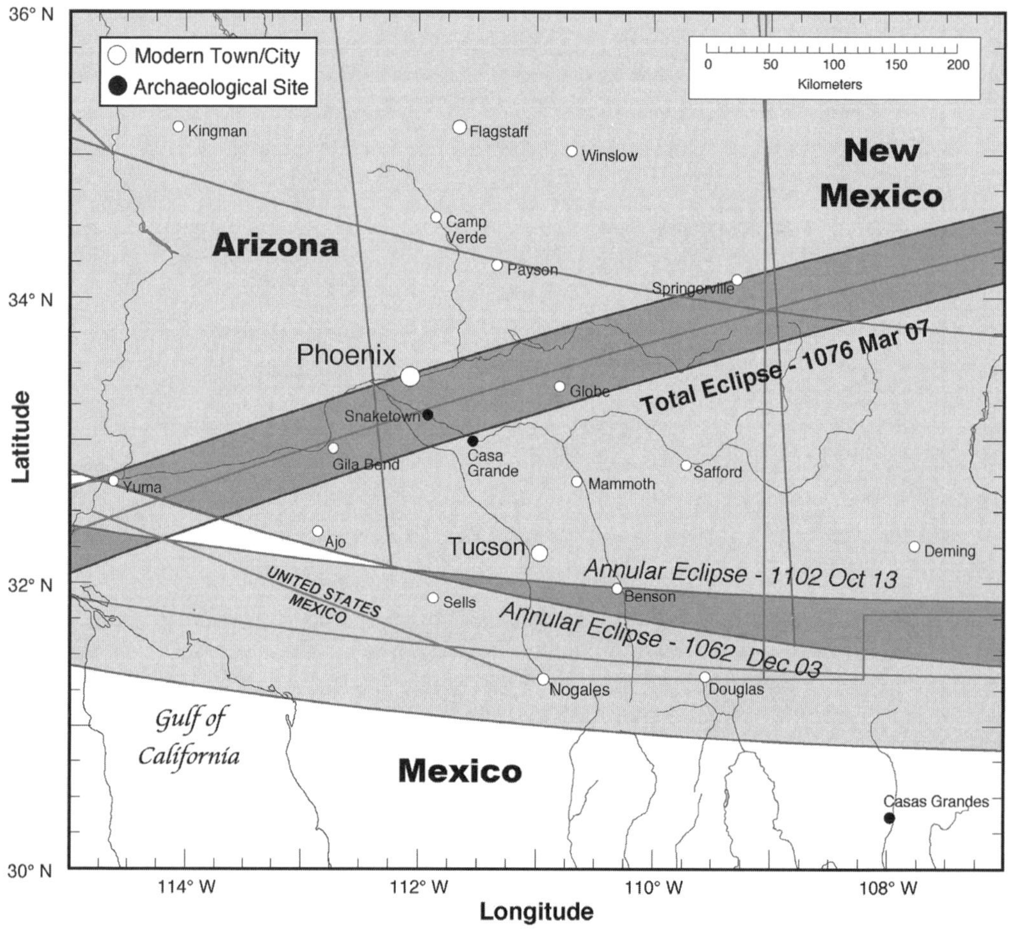

FIGURE 12.6. Paths of the annular solar eclipses of A.D. 1062 and 1102 and the total solar eclipse of A.D. 1076.

The Total Eclipse of A.D. 1076 and the Reorganization of the Hohokam Regional System

In December A.D. 1062, a high-magnitude annular solar eclipse at least nine minutes long transited virtually the entire Hohokam region (Figure 12.6). Occurring several weeks before winter solstice, this sign prepared the way for the most dramatic transient celestial event of the eleventh century in the Hohokam region, a total eclipse that occurred in the late afternoon on 7 March 1076, two weeks before the vernal equinox. The narrow path of TSE-1076 centered on Snaketown and Gila Bend. The eclipse, whose totality lasted about one minute, was located in the constellation Pisces, with Mercury in proximity and both Venus and Jupiter visible. Slightly more than a generation later, in A.D. 1102, the southern edge of the Hohokam region was transited by a high-magnitude annular eclipse (Figure 12.6). Other events witnessed by the previous generation would also have been portentous (Masse 1995, 1997), in particular the A.D. 1054 supernova and the mid- to late-eleventh-century eruption of Sunset Crater near present-day Flagstaff after many centuries of quiescence (Elson et al. 2002). The appearance of Halley's comet in A.D. 1066 and a series of bright nova events between A.D. 1065 and 1074 would have heightened the effect of TSE-1076.

Late in the eleventh century, the Hohokam regional system underwent dramatic rapid changes that included the depopulation of preeminent pit-house villages such as Snaketown, a change in religious/ritual orientation and material correlates, and a restructuring of the ball-court system, including size and orientation (Crown and Judge 1991; Gumerman 1991; Wilcox 1991a, 1991b). There was possibly a dramatic increase in the use of obsidian (Shackley 1995), although this increase may instead date to the beginning of the early Classic period, around A.D. 1200 (Doyel 1996a). The current dating for the start of this cultural reorganization period is around A.D. 1075 (Cordell et al. 1994; Doyel 1991b, 1996a, 1996b, 2000; Wilcox 1991a).

The Total Eclipse of A.D. 1196 and the Rise of the Hohokam Classic Period

The beginning of the Soho phase of the Hohokam Classic period at around A.D. 1200 (Dean 1991) is marked by significant changes in Hohokam settlement patterns and material culture. Notable are the use of adobe-walled compounds and the erection of large, multicelled platform mounds (Doyel 1991a). The ball-court system and the associated production of ritual objects ceases at around 1200, along with the importation of many objects from Mexico. The Hohokam occupation of the Gila Bend area on the southwestern frontier of the Hohokam region was greatly diminished around 1200 (Doyel 2004).

A high-magnitude annular solar eclipse occurred immediately northwest of the Hohokam region in A.D. 1156 (Figure 12.7). This was followed forty years later at autumnal equinox by a total solar eclipse of long duration along the southwestern edge of the Hohokam region. This in turn was followed nearly forty years later by a high-magnitude early morning annular eclipse in A.D. 1234 (Figure 12.7). The path of TSE-1196 largely coincides with the locations of numerous hillside stone terraces termed *cerros de trincheras*, many of which were constructed around A.D. 1200 in southern Arizona and in northern Sonora, Mexico (Downum et al. 1994; Greenleaf 1975; McGuire and Villalpando 1993; Stacy 1977). In some cases, *cerros de trincheras* constructed hundreds of years earlier were reoccupied between A.D. 1100 to 1300. Such structures were long thought to be defensive features related to warfare during the A.D. 1200s (Wilcox 1979; Wilcox and Haas 1994). However, recent studies suggest instead that they represent villages and agricultural areas moved upslope from previous riverine settings to the flanks and tops of nearby volcanic hills (Downum 1993; Downum et al. 1994).

It is possible that TSE-1196 contributed to general unease between Sonoran cultures and the Hohokam and a concomitant need for defensive posturing. Equally intriguing is the possibility of a perceived relationship between eclipses and flooding stemming from the coincidence of the two during earlier total and high-magnitude annular eclipses (Table 12.5). The appearance of TSE-1196 may have portended flooding. In response, the occupants in the area may

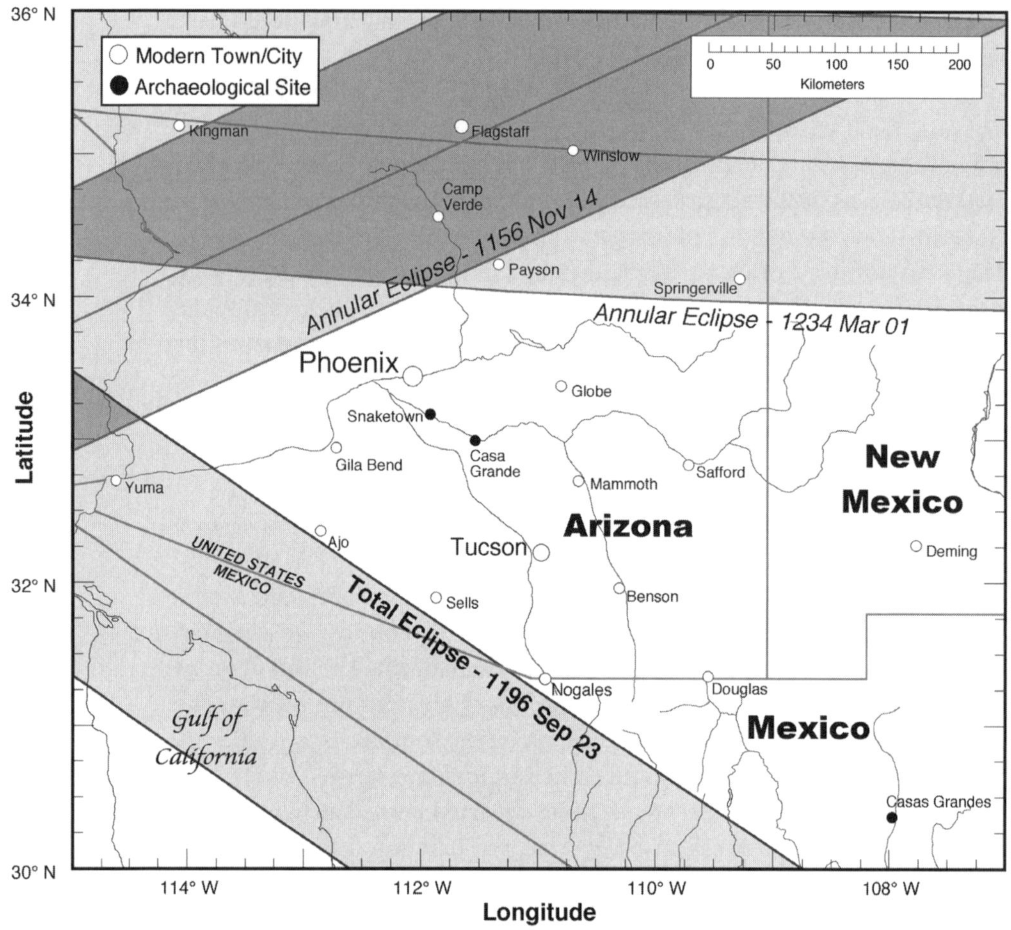

FIGURE 12.7. Paths of the annular solar eclipses of A.D. 1156 and 1234 and the total solar eclipse of A.D. 1196.

have deliberately moved homes and fields to higher ground, constructing (or reoccupying earlier) *cerros de trincheras*, platform mounds, and surrounding compounds, or moved to higher terraces near existing villages. Such behavior would have been enhanced by a widespread belief throughout Mesoamerica (and presumably among the Hohokam) that the Sun and Moon contain a spillable "liquid" (light) that is released during eclipses (López Austin 1996: 66–67).

However, if the Hohokam did embrace such a belief, a disturbing situation would have transpired after TSE-1196 eclipse. For the first time in traditional memory, a total or near-total solar eclipse was not accompanied by high streamflow discharges on either the Salt or the Gila River (Table 12.5; Graybill et al., this volume). Given this hypothesized effort to rapidly mobilize settlements in anticipation of floods, it is conceivable that Hohokam shaman-priests and chiefs lost face, as did their whole solar cosmography, when floods failed to materialize after TSE-1196. Such a scenario may help to explain the demise of the ball-court system and the general change in ritual among the Hohokam shortly after A.D. 1200.

The Eclipse of A.D. 1379 and the Fall of the Hohokam

By A.D. 1300 to 1325, the Hohokam added impressive multistory adobe structures and towers (collectively termed by the Spanish *casas grandes*, "great houses") to their platform mound communities (Gregory 1991; Wilcox 1991a; Wilcox and Shenk 1977). At least nine multistory structures and three towers were present along the lower Salt River, and four multistory structures and four towers along the middle Gila River (Doelle et al. 1995). The four-story tower at Casa Grande National Monument was sufficiently well preserved that in 1969 John Molloy proposed using it as an astronomical observatory for marking solstices and equinoxes (Raymond White, personal communication 1996; Wilcox and Shenk 1977:44–45).

One of the great mysteries in southwestern archaeology is the disappearance of the Hohokam from the archaeological record around A.D. 1450–1500. This process actually begins with dramatic changes between A.D. 1375 and 1400, during which time the Hohokam towers and multistoried structures, such as at Casa Grande, were abandoned (Wilcox and Shenk 1977), as were most rooms inside the platform mound compounds. These structures were replaced by small pithouse rooms excavated into the earlier Classic compounds and platform mounds, some of which penetrate into earlier Classic period adobe-walled rooms, and by sporadic remodeling and reuse of earlier Classic period adobe-walled rooms. This trend was accompanied by a decline in overall population and perhaps health, by the apparent demise of large-scale irrigation, and by changes in material culture.

These dramatic changes are now considered by Hohokam scholars to signal the Hohokam Postclassic period, represented by a single phase, Polvorón, the last in the Hohokam culture sequence (Bostwick et al. 1996; Crown 1991; Sires 1984; Teague 1989). We do not dispute Abbott's (2003:226–227) claim that the Polvorón phase was the product of long-term processes of deterioration within Hohokam society and the surrounding environment, but we believe he and other Hohokam scholars fail to adequately consider the evidence embedded in O'odham oral history.

David Gregory (personal communication 1987), Teague (1989, 1993), and Bahr et al. (1994) conjecture that O'odham oral history about the "conquest" of the Hohokam by the O'odham represents the real history of the Polvorón phase (see also Shaul and Hill 1998). Teague (1993:443–444) links the overthrow of the great houses, and thus the beginning of the Polvorón phase, with a drought and subsequent catastrophic flood (e.g., Mason 1921). The drought presumably coincided with a period of low-magnitude streamflow between 1335 and 1353 (Graybill et al., this volume), and the flood correlates with the reconstructed high-magnitude streamflow event of 1358 (Nials et al. 1989).

Teague (1993) and Abbott and his colleagues (2003) use a date of A.D. 1350 for the start of the Polvorón phase, the latter reference curiously contrasting with the date of A.D. 1375 used by Abbott in another study (2000:34). There is no compelling archaeological evidence to suggest that the Polvorón phase itself began any earlier than the late A.D. 1370s or early 1380s (Gregory 1991; Masse 1991). Streamflow analysis (Graybill et al., this volume) indicates that the discharge for A.D. 1358 was not as great as previously believed and points instead to discharges in A.D. 1381–1384 as having even greater potential impact on Hohokam irrigation systems, particularly in A.D. 1382, the largest since A.D. 899. Significantly, this date is supported by O'odham oral traditions that describe two absolutely datable transient celestial events—an aurora and a total solar eclipse—that occurred a few months or years before the conquest and the attack on the Casa Grande Great House.

These events are part of an O'odham origin story told by Thin Leather to Frank Russell in 1901–1902, to J. William Lloyd in 1908, and to Jesse Fewkes in 1911 (Fewkes 1997; Lloyd 1911:133–165; Russell 1975:224–230); by Juan Dolores to J. Alden Jones around 1919 (Saxton and Saxton 1973:150–168); and by William Blackwater and Thomas Vanyiko to Ruth Benedict in 1927 (Bahr 2001:31–53). In the last-named version, the solar eclipse is referred to in a song sung by the solar deity, Elder Brother (Bahr 2001:45): "With the powers of all the medicine men we have made the sun and the moon stand still." Thin Leather described the episode as follows (Russell 1975:225):

> Vulture was living in this va'aki [Classic period Great House] when the people came to him with their complaints concerning Elder Brother. They asked if he could do anything to help them. Vulture said he had never used his magic power, but he would test it. He asked the people to come to his va'aki and he would make the trial in their presence.
>
> After the people had gathered in the house and the doors had been closed he brought on darkness with his magic power while it was yet daytime. The darkness was so heavy that the people could see nothing. A beam of light arose which grew stronger and stronger until during the second night of their sojourn in the house it became as brilliant as sunlight. There were four colors, four threads of light, that extended upward until they reached the sun.

One tradition collected by Fewkes (1997:47) from Thin Leather combines the details of the eclipse with flooding: "But when the chief of Casa Grande heard that Parrot had been sent to steal his turquoises, he was greatly vexed and caused a violent rain to fall that extinguished all fires in the East. His magic power over the Rain god was so great that he was even able to extinguish the light of the Sun, making it very cold."

The most evocative and detailed eclipse description was that given by Juan Smith in 1935 to Julian Hayden, who was assisting in the excavation of Snaketown (Bahr et al. 1994:187):

> When the Children started out to play one day, they went and saw Siuuhu [Elder Brother] sitting [in front of the Casa Grande Great House] with his face all painted black and with a sharp stick sticking in his back.... The

children also saw that he was fixing his canteen. They went back to their father and told him that the man was alive again, they saw him with his face painted black and a pointed stick in his back. The fathers of these children told them that something terrible was going to happen. While Siuuhu was sitting there, sometimes he thought that he would punish them. He was crying for the great wrong they had done, and he sang:

"I am going to do this to my people
Which will be very sad.
The sun has died halfway."

When he finished this song, the sun came up, but it didn't shine like it used to and darkness fell over the earth. When the people saw this, they didn't understand for they had not seen anything like it before. Then, while sitting, he sang another song:

"I am going to do this,
To spoil everything
That was made so good
When the moon comes up
It has died halfway."

A Tohono O'odham version of this story collected in 1972 by Bahr adds the detail that this event occurred at around "saguaro cactus ripe time" (Donald Bahr, personal communication 2000). The saguaro (*Carnegiea gigantea*) typically flowers in May and fruits in mid-June. This dating is supported by the presence of a lunar month in the O'odham calendar called "Saguaro Ripe Moon" (Underhill et al. 1997:17–19), which corresponds to May or June.

The Smith version collected by Hayden is a single flowing text, and as noted by Bahr (Bahr et al. 1994:281), "the events have a temporal order; the mythology is a linear chronicle." Over a relatively short period of stated time within this chronicle are three pertinent consecutive events. First are the mysterious multiple deaths and resurrections of Elder Brother. Next is the initial punishment of the Hohokam through the death or holding still of the Sun and Moon. And finally comes the formation of the O'odham army and the penultimate punishment of the Hohokam by the overthrow of their great houses. The one O'odham myth deviating from this pattern is that provided by Vanyiko and Blackwater, in which the holding still of the Sun and Moon takes place during the assault on Casa Grande, the third (in their version) of the great houses to be attacked (Bahr 2001:45–46); however, even in this case the event is presaged in a speech by Elder Brother during the earlier period of his death and resurrection (Bahr 2001:33).

These stories unquestionably encode the total solar eclipse of 16 May 1379, whose path encompassed the entire Hohokam region, with Casa Grande along its center line (Figure 12.8). This event—the first total eclipse in the central

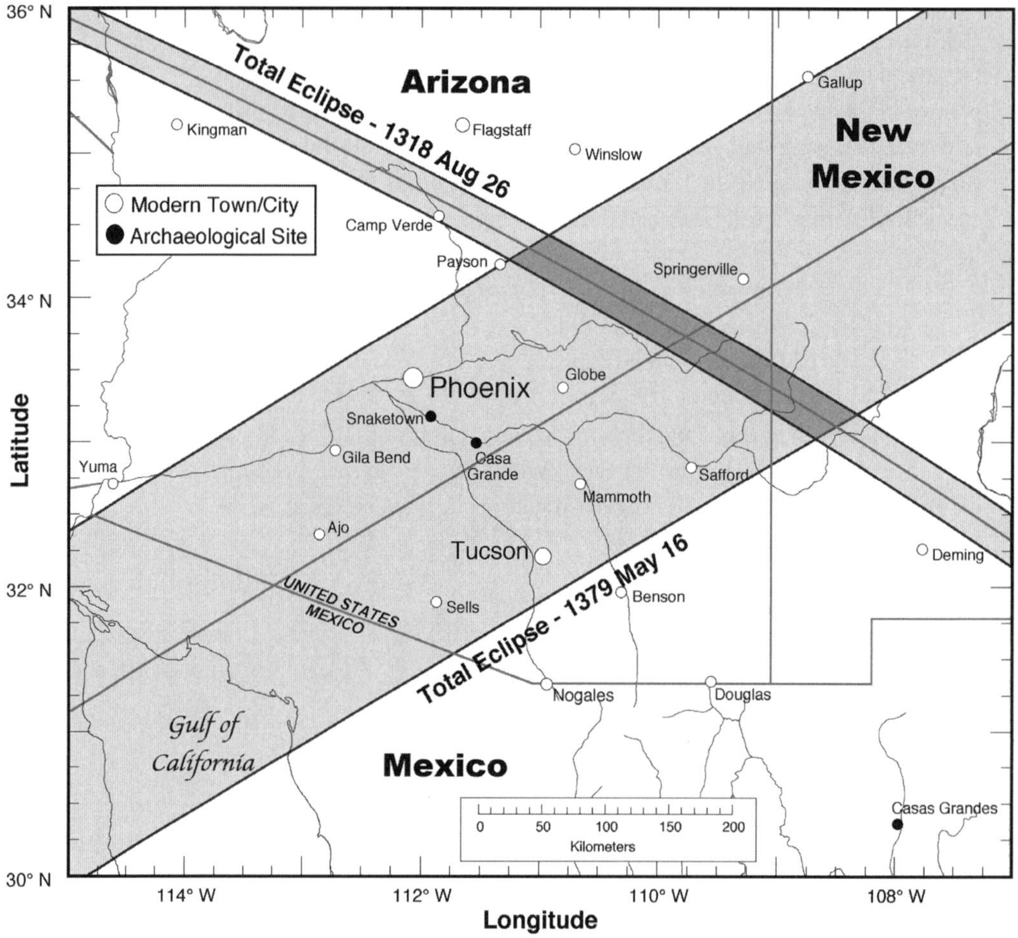

FIGURE 12.8. Paths of the total solar eclipses of A.D. 1318 and 1379.

Hohokam region since TSE-1076 and the first eclipse of any kind since the 0.961-magnitude partial eclipse in A.D. 1318—matches closely the O'odham oral histories, the archaeological dating of the demise of the Casa Grande Great House and the beginning of Polvorón phase traits, and the dendroclimatological reconstructions of streamflow events. The eclipse began shortly after 0700 and reached totality shortly after 0800, some 2.5 hours after sunrise, with totality lasting nearly four minutes. The eclipse appeared in the constellation Taurus surrounded by a number of significant stars and asterisms, including the Pleiades and Orion. It was made more spectacular by a triple conjunction of Mercury, Venus, and Mars in the neighboring constellation Aries, with Saturn also visible nearby in the constellation Pisces.

The period A.D. 1360 to 1370 also witnessed by far the largest number of auroras (12) in any decade in the historic record (Xu et al. 2000), and the following decade yielded by far the largest number of sunspot observances (26). The Hohokam no doubt witnessed some of this auroral behavior. The stories surrounding the death and resurrection of Elder Brother and the formation of his army (e.g., Bahr 1975; Bahr et al. 1994:212–233; Lloyd 1911:154–161) contain intriguing references to a "light" and an interesting use of color symbolism, separate from the solar eclipse imagery, that are compatible with observations of auroral phenomena. Lloyd (1911:164) notes: "The uniting of the bands by the 'Light' is very curious. My Indians could not tell me what this was, only something occult and mysterious by which they had clairvoyant ken of each other's needs." Remarkably, the stories surrounding the death and resurrection of Elder Brother and the formation of his army are nearly identical in story-line content to the Hawaiian auroral myth of the supernatural prophetess Wa'aia and the subsequent battles of the royal chief Kalaunui'ohua (Masse 2003b), the latter linked with an auroral substorm observed in Japan in A.D. 1370. This congruence is best explained by invoking shared observations of the same auroral phenomena and by similar cognitive processes that transformed the auroral spectacle into a portent associated with warfare.

There were other striking environmental portents in addition to the drought, auroras, solar eclipse, and subsequent floods. After more than 100 years of inactive skies, they literally erupted between A.D. 1361 and 1380 (Table 12.2). In the space of a single generation, there were 17 comets, at least 6 of which were remarkable. Seven had tails at least 10 degrees in length (20 times the apparent diameter of the Moon), some of which were bent, chain-like, or otherwise unusual. Especially ominous would have been the simultaneous appearance in different parts of the sky in March 1364 of as many as 4 comets, one of which was bright red. Also of note was a spectacular conjunction of Jupiter, Venus, and Saturn at dawn in July 1384.

The nature and actions of Siuuhu (Elder Brother) in the O'odham traditions (Bahr et al. 1994) raise interesting questions. Particularly notable is his ability to transform and reshape himself, his power to come back to life from death, his frequent travels or changes in abode, and his periodic disappearance for months or years at a time. These traits, coupled with personal attributes such as his having a hat with a gold hatband, suggest that Elder Brother was associated with the apparitions of comets and perhaps auroras. Elder Brother's characteristics are similar to those of other deities with cometary associations, such as the war god Ku in Hawai'i and the creator god Quetzalcoatl in Mexico (Masse 1995, 1997, 2003b).

Of particular interest is the alternative name, I'itoi ("Drinker" or "Drink-it-all-up"), given to Elder Brother, as recorded by Pedro Font in 1775 (Fewkes 1997:43–44). Bahr (Bahr et al. 1994:298–299) expressed puzzlement at this name. Referring to the story noted above about the parrot stealing turquoises from Morning Green, chief of the Casa Grande Great House, Bahr provides the following analysis (Bahr et al. 1994:139):

> Concerning the Font text, I speculated that Siuuhu's name "Drink-it-all-up" is solar, and I take this parrot episode as support of that speculation. The sun "drinks" water; when water (or wetness or dew) evaporates, the sun "drinks" it. The parrot does similarly, with the flood-turned-to-stone

> [turquoise]. That this parrot is green does not cancel its solarity, I submit. There could be a green sun as a kind of limiting or bridging condition between hot light and rain, and indeed I suspect that this is the meaning of Morning Green's name.

Bahr's analysis raises possibilities regarding the source of the name Morning Green. As noted by Parsons (1996:46–47) for the modern Pueblos and Russell for the O'odham, nicknames are common and may supplant the birth name; Russell (1975:188) specifically notes that "any unusual event or physical peculiarity may impose a name upon an individual." The Tewa of northern New Mexico considered the color of the summer Sun to be green, in opposition to the yellow Sun of winter (Harrington 1916:46), and Lloyd (1911:17) reported that the O'odham considered the cosmological color of the sky to be green. Assuming that the Hohokam had similar conceptions and naming practices, the name Morning Green could be an allusion to the morning summertime appearance of TSE-1379 itself.

More certain is the linkage of the name "Drinker" with comets. Bahr (Bahr et al. 1994:298) hypothesizes that "the Drinker name links him to the Aztec sun who 'drank' human blood." This in turn may be related to the Nahua concept that the Sun contains a liquid that can be spilled during eclipses, and is thus linked to bloodletting ceremonies associated with the perpetuation of the Sun. Skilled, trained celestial observers, including Aztec priests and O'odham/Hohokam *makai*, would have understood that comets have an intimate relation to the Sun. Comets race toward the Sun, only to disappear completely or reappear on the other side of the Sun several weeks later. It would be natural to assume that on disappearance the comet "drinks" the Sun's liquid, which would then also explain why the comet "shines" like the Sun.

This interpretation also explains the curious description of the "pointed stick" in the back of Elder Brother during TSE-1379. A tailed comet was observed from Japan in April 1379, and it is certainly possible that the comet could have been close to the Sun (looking like a "pointed stick") on 16 May at the time of eclipse. Historically, other naked-eye-visible comets have been observed adjacent to the Sun during total eclipses (Brown 1973:102). The sudden appearance of a comet at the edge of the Sun during an eclipse reinforces the idea that it is drinking the solar liquid and helps explain the curious detail preserved in the quoted myth about Elder Brother "fixing his canteen."

The language, context, and story line surrounding TSE-1379 in O'odham mythology all suggest that the eclipse was more than a "passive" portent coinciding with the O'odham conquest. Rather, TSE-1379 was the initial step in the punishment of the Hohokam and served as a clarion call to assemble the armies to overthrow the Hohokam. As such, TSE-1379 signals the start of the Polvorón phase and was an active element in the decision making leading to this significant act of Hohokam culture change.

THE ECLIPSE OF THE HOHOKAM

The Hohokam great-house communities of the Phoenix Basin were effectively abandoned and/or dismantled between A.D. 1380 and 1400, and the primary chiefs and priests were killed, driven away, or otherwise subjugated. However, despite its impoverished archaeological record, the Polvorón phase indicates that some Hohokam clung tenuously to their territories, trading networks, and aspects of their social systems and religious beliefs through about A.D. 1450–1500. This situation may accord with the O'odham conquest legends that suggest that after the initial destruction of the Hohokam great houses, the O'odham continued to the coast of California, returning to southern Arizona only after an indeterminate and perhaps lengthy period of time (Bahr et al. 1994:268–276). The oral traditions indicate that on their return (ca. A.D. 1450?), the O'odham intermarried with the remaining Hohokam and molded the framework for the modern distribution of O'odham communities.

The celestial story of the Hohokam and the O'odham likewise does not end in A.D. 1385. A partial eclipse in A.D. 1397 reached a magnitude of 0.959 at Casa Grande (Figure 12.9). Indeed, a large number of extraordinary comets, novae, and planetary conjunctions occurred between A.D. 1402 and the return of Halley's comet around summer solstice in A.D. 1456, and continued throughout the remainder of the fifteenth century (Table 12.2). In A.D. 1402, two spectacularly bright comets appeared, one in February and the other in June, the latter coinciding at maximum brilliance in late August with a 0.994-magnitude annular solar eclipse that transited immediately south of the Arizona border (Figure 12.9). In the early afternoon of 8 August 1496, a major total solar eclipse, lasting more than five minutes, transited parallel to and immediately south of the path of ASE-1402 (Figure 12.9).

Research on the effects of solar eclipses on the Hohokam and on Puebloan populations (Masse and Soklow 2005) suggests that some cultural groups responded to total solar eclipses by abandoning villages in the path of totality and moving into surrounding regions where the Sun had not completely disappeared. The cognitive link between floods and eclipses noted above is just one of several factors that may have influenced such an abandonment process. Perhaps the mysterious two *despoblado*, or unoccupied, lands encountered by Coronado in his A.D. 1540 march to find the "seven cities of Cibola" (Bolton 1964:31–32) are those very portions of northern Mexico, southeastern Arizona, and southwestern New Mexico that were transited by the path of TSE-1496 and TSE-1531.

More significant is the fact that TSE-1496 and the next total eclipse, in A.D. 1531, mark the brief period between the last radiocarbon dates known for the Hohokam in the late fifteenth century (Bostwick et al. 1996; Dean 1991; Mitchell 1995) and the earliest known dates (around A.D. 1520–1570) of the protohistoric O'odham (Dart 1994) in southern Arizona. The material culture of the protohistoric and historic O'odham in southern Arizona is different

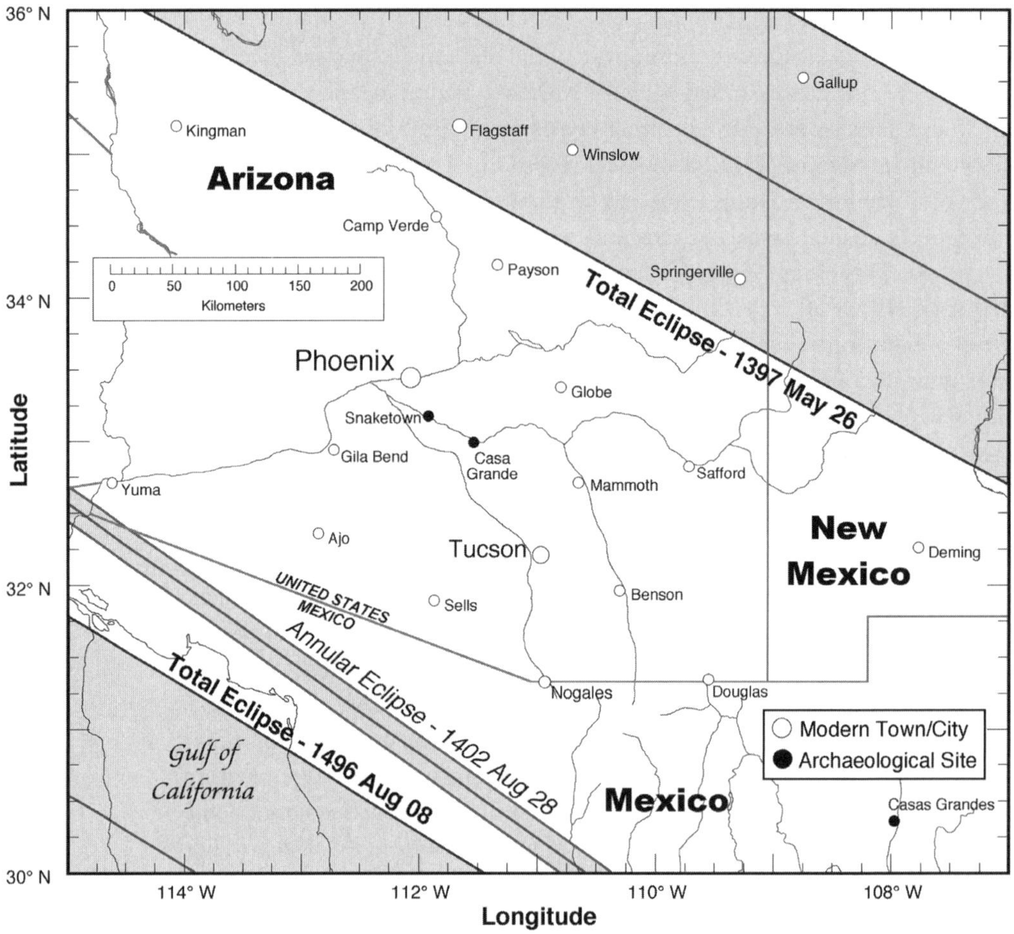

FIGURE 12.9. Paths of the total solar eclipses of A.D. 1397 and 1496 and the annular solar eclipse of A.D. 1402.

from that of the most recent Hohokam remains and appears to be more similar to the known material culture of the Lower Pima (Pima Bajo) in the Río Sonora and Chihuahua regions (Dart 1994:380; Masse 1981, 1982). Analysis of O'odham dialects and evidence from oral histories support the notion that the Aji-Totoguañ dialect currently spoken throughout much of the Tohono Indian Nation represents a post–A.D. 1450 spread of southern peoples into the area once occupied by the Hohokam (Dart 1994:380–382; see also Shaul and Hill 1998).

It is possible that TSE-1496 (and perhaps ASE-1402) disrupted Tepiman-speaking populations in Río Sonora and Chihuahua and that as a consequence they emigrated to the Papaguería and the Santa Cruz and San Pedro valleys of southern Arizona to become the branches of the historic O'odham. Another potential trigger for the move north could have been increasing incursions by Aztec-related groups (or by groups fleeing from Aztec slaving expeditions) resulting from the drought and famine that devastated the Valley of Mexico in A.D. 1454–1456 (Bierhorst 1992:107; Boone 2000:224; León-Portilla 1992:106).

The few Polvorón phase Hohokam who were attempting to hang onto their traditional lifeways following the floods, warfare, and diasporas of the late fourteenth and fifteenth centuries were no match for their southern cousins, who were likely augmented by the O'odham returning from California. Thus the mysterious "disappearance" of the Polvorón phase Hohokam after around A.D. 1450–1500 represents their rapid assimilation into O'odham society.

TSE-1531 (Figure 12.10), occurring in the early afternoon just before the vernal equinox, surely portended the arrival of the Spanish (Núñez Cabeza de Vaca, Fray Marco de Niza, and the huge expedition of Francisco Vázquez de Coronado) a few years later (Bolton 1964). But equally or even more portentous would have been the four-minute-long total eclipse of 28 April 1557 (Figure 12.10), transiting the entire region devastated by Coronado's army only sixteen years earlier.

CONCLUSIONS

Based on the striking correlation between the appearance of solar eclipses in southern Arizona and the current dating of most periods of rapid culture change for the Hohokam, and given our present understanding of southwestern and Nahua documentary and oral history, it is likely that solar eclipses stimulated the Hohokam and their neighbors to make important social and political decisions, which were reflected in settlement patterns, population movements, iconography, and religious and ritual behavior. These eclipses thus become an objective chronological referent for segmenting and understanding Hohokam culture history.

Despite the compelling evidence for the direct linkage of total solar eclipses to Hohokam culture change, it should be evident that these eclipses are just a part of a much larger series of celestial events and processes. In this regard, pottery decoration may represent a window on unusual clusters of comets, meteor storms, and novae.

For example, the use of small repeated elements on red-on-buff pottery, including many tiny star-like motifs, is prevalent at the end of the Gila Butte phase and the first half of the Santa Cruz phase, as is the use of densely packed parallel hachure without fringing lines (e.g., Gladwin et al. 1965:Figure 77; Haury 1976:Figures 12.27, 12.105). The peak use of these design elements coincides with a period of uniquely prevalent meteor storms and substorms between A.D. 880 and 940 (Table 12.2), and the motifs may represent an intentional evocative encoding of the meteor storms themselves. The distinctive "flying bird" motif (Haury 1976:243–245), which peaks during the Santa Cruz phase, is also suggestive of meteor storms.

Similarly, the imagery of the Hohokam Sedentary period, with its sudden plethora of large star- and Sun-like motifs (e.g., Gladwin et al. 1965:Plates 136, 143, 148; Haury 1976:Figures 12.23, 12.99), may reflect the remarkable Lupus supernova of A.D. 1006 and the equally remarkable Crab Nebula supernova

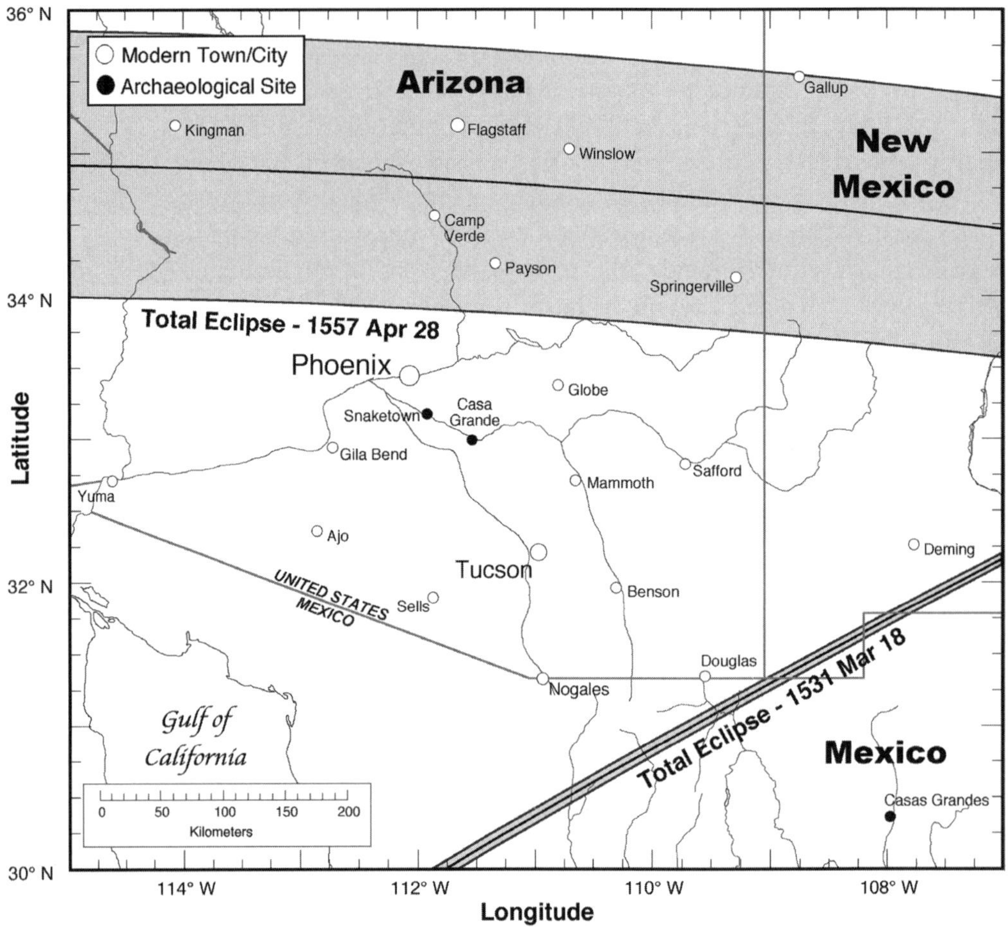

FIGURE 12.10. Paths of the total solar eclipses of A.D. 1531 and 1557.

of A.D. 1054, along with other substantive nova events (Table 12.2). The development of Mound 16 at Snaketown (Haury 1976), a possible astronomical sighting platform, may have been stimulated by the appearance of these stellar spectacles.

The spread of thirteen–fifteenth-century Salado redware pottery, in particular Gila Polychrome (Crown 1994), may represent the ideology of a period that alternated between generations that observed numerous spectacular comets (depicted on pottery as birds and feathered serpents), along with other transient celestial phenomena, and generations lacking in known transient celestial events. Indeed, such events may have played a role in the development and spread of Uto-Aztecan "flower world" imagery (Hays-Gilpin and Hill 1999; Hill 1992). In a holistic world where sky and earth are part of a single continuum, the appearance of novae, supernovae, comets, and meteor storms may have been viewed as a kind of celestial "blossom," a cosmological notion enhanced by the annual cycle of prominent meteor showers. The association of the Sun with flowers was undoubtedly enhanced by the observation of flowering plants closing their blossoms during a total eclipse and by the flower- or feather-like appearance of the eclipse corona itself. This hypothesized relationship is certainly plausible given the prevalence of earth-sky dualism in Nahua literature.

The Sun was also associated with the underworld, where it stayed or traveled during roughly half of each 24-hour day. Some caves thought to have been used as hunting or emergence shrines may have instead (or in addition) been used to placate the gods after a high-magnitude partial or annular solar eclipse. For example, the layout, natural clear water basin, and artifact assemblage of Red Cave (Ferg and Mead 1993), with its plethora of sherd discs (perhaps in multiple sets of two discs each), coupled with beads of apparent cosmological material types (e.g., turquoise and shell) and directional colors, tobacco, saltpeter, cane arrow, and so on, could fit known material culture associations with Elder Brother and the Sun. The nature and dating of the ceramic artifacts in Red Cave fit well with the high-magnitude (0.980) annular eclipse of A.D. 1102, whose path (Figure 12.6) included the cave and nearby village areas.

Not all transient environmental events are celestial. The innovative study of Picacho Mountains petroglyphs by Wallace and Holmlund (1986) indicates that a major earthquake was associated with the beginning of the "Gila style" of art, perhaps dating to around A.D. 750–800, with another earthquake taking place in the late Sedentary or early Classic period, perhaps dating to around A.D. 1200. If these earthquakes occurred at around the time of TSE-797 and TSE-1196, it would have further enhanced the portentous nature of the eclipses. In fact, the earthquake of ca. A.D. 1200 may allude to the Hopi legend of the destruction of Palotquopi (Courlander 1971:56–72; Nequatewa 1967:85–102), an event that allegedly occurred three or four generations before the settlement of the Homolovi villages, the latter being archaeologically dated at about A.D. 1260 (Adams 2002). Oral histories (Bahr et al. 1994:252–253) also indicate that an earthquake took place toward the end of the initial O'odham conquest of the Hohokam, destroying La Ciudad de los Pueblitos (Pueblo Grande). Frank Hamilton Cushing's belief that earthquakes were responsible for the demise of the Hohokam, based on his 1887–1888 excavations of Los Muertos and other Phoenix Basin sites (Haury 1945:210), may thus have some validity. These same oral traditions also note that shortly after the La Ciudad earthquake a large dust devil or tornado frightened the attackers.

We should also pay greater attention to the spatial geography of pre-European southwestern settlement and its cosmological correlates. Notable in this regard are attempts by Wilcox (1987) and Lekson (1999) to interpret the overall structure of settlement through time for the Hohokam and Chacoan regional systems. There certainly is no reason to assume that the use of cosmology in village contexts and in the structure of regional settlement was restricted to the most complex societies. The worldwide evidence (e.g., Coyne and Sinclair 1995; Fountain and Sinclair 2005) suggests that many past cultures embodied a strong cosmological perspective in site structure and in regional settlement as well.

Haury (1976:9), pondering the existence and nature of Snaketown, notes: "The choice of the site of Snaketown is difficult to assess. The land on which it resides, about 1 km^2 in area, looks like any other comparable area of desert for a long distance up and down the valley. Why should this precise spot have been selected?" In 1997, the senior author fortuitously discovered that the central plaza at Snaketown, with its associated Mesoamerican-inspired ball courts and capped trash mounds, is situated in such a manner as to directly view winter solstice sunrise and sunset over the tops of the two most prominent isolated buttes in the lower Gila Valley.[4] Such a circumstance indicates that the cosmological spatial geography of Snaketown was critical for its preeminent status. In a real sense, Snaketown was born as the product of the path of the Sun, and its demise was its great misfortune to be located along the center line of the A.D. 1076 total solar eclipse. Doyel (1993a) has suggested that Snaketown was depopulated if not abandoned by A.D. 1075, which closely accords with the eclipse data.

The Hohokam and other traditional peoples were astute observers of the natural environment, of which the sky was a critically important part. Thus the study of the relationship between the physical environment and human behavior must be considered incomplete without proper attention to events in the sky.

Transient environmental events, especially the known and reconstructed events in the heavens, provide us with a marvelous potential chronological anchor by which to study and understand both culture history and culture process. More important, by paying close attention to how these transient environmental events are organized and portrayed in the oral histories and the archaeological record, we not only can better appreciate the cosmology and ideology of past cultures but can relearn critical aspects of our surrounding environment and indeed of history itself that Western culture and science should not have forgotten.

ACKNOWLEDGMENTS

We thank Dave Doyel for his audacity in soliciting this chapter for this volume and for his own insights into things Hohokam and cosmological. An earlier version of part of this chapter was presented by the senior author at the 1996 Oxford V Archaeoastronomy Conference in Santa Fe, New Mexico. The original Oxford V paper treated eclipses throughout the Southwest; however, space limitations in the conference volume necessitated limiting that paper to just those eclipses occurring in the Puebloan culture area (Masse and Soklow 2005). The Hohokam portion of the original paper has been greatly revised and expanded to form the basis of the present chapter. The section of our chapter treating the chronology, tempo, and mode of transient celestial events was initially prepared for a symposium at the February 2000 annual meeting of the American Association for the Advancement of Science. We thank Rolf

Sinclair, Raymond White, George Gumerman, John Fountain, and Robin Gould for supporting the senior author's attendance at the Oxford V conference and for discussions of things celestial; Robert Soklow for stimulating discussions as the Oxford V paper was prepared for publication; Rolf Sinclair for inviting the senior author to attend the AAAS 2000 meetings; and George Kaplan of the U.S. Naval Observatory for suggesting Fred Espenak as a collaborator on this chapter. We have benefited from discussions or correspondence with a number of individuals, including Jay Anderson, Don Bahr, Ben Brown, Vine Deloria, Jr., Robin Edgar, Lorraine Eiler, John Fountain, Dave Gregory, Joel Gunn, James Gutmann, Bill Hartmann, Randy McGuire, Curt Schaafsma, Polly Schaafsma, Christy Turner, Dorothy Washburn, and Dave Wilcox. Brown and Curt Schaafsma pointed out the Chihuahua meteorite literature, and Deloria reminded us of the important role of the solar eclipse in the Ghost Dance religion. The senior author valued the opportunity to discuss some of his celestial ideas with the elders of the Hopi cultural resource advisory task team and with O'odham and Tewa friends and colleagues. Their kind interest has helped to sustain his research.

NOTES

1. Shortly before going to press with this study, we discovered that in 1997, Cameron Wade, then recently retired from the National Radio Astronomy Observatory, made an eloquent plea for the rigorous study of reconstructed total solar eclipse paths in the Ancestral Pueblo portion of the Southwest (Wade 1997). Wade's prescient concern with eclipses presaged much of what is contained here and especially in Masse and Soklow 2005. The fact that Wade's study seemingly did not spark a more general interest in the topic among southwestern archaeologists prior to our work is a reflection of the barriers that exist between disciplines and the general lack of training in and sensitivity to things astronomical on the part of most archaeologists.
2. We here make a distinction between mythology and folklore (Dundes 1984:1). Unlike folklore, myth is considered sacred narrative and served a different and more significant role in traditional cultures than folklore.
3. This chapter uses three astronomy software programs to reconstruct aspects of the historic southern Arizona sky: *RedShift Multimedia Astronomy 3.0*, *The Sky* (version 5), and *Voyager III* 3.52. The paths and table of solar eclipses were computed by Fred Espenak. The reader is referred to his website for details regarding the nature of and methods employed in such reconstruction: http://sunearth.gsfc.nasa.gov/eclipse/eclipse.html.
4. It is likely no accident that the twin peaks of Gila Butte are prominently displayed in artist Peter Bianchi's dramatic reconstruction of the Hohokam playing a ball game in the large ball court at Snaketown (see Haury 1967:678–679). From the plaza and ball court, the winter solstice morning

Sun rises at the center of the base of the saddle between the twin peaks of Gila Butte. The winter solstice evening Sun sets at the top of Pima Butte to the west. Discussions in 1998 with Elaine Notah of the Gila River Indian Community determined that she and perhaps others had anticipated this relationship before 1997 but had not then yet had the opportunity to field test her ideas.

13

The Resurgence of the Co-evolution of Human and Natural Landscape Studies in the American Southwest

George J. Gumerman

Environmental and ecological studies in anthropology are again rising from an era of dormancy and even denigration. The reasons why cultural ecological studies fell into disrepute can be summarized by the rationale for the 1995 Durango Conference. The call for conference papers published in the *Archaeological Record* stated that for the past quarter century, Americanist archaeology, and that of the U.S. Southwest in particular, has been dominated by a school of thought known as the new archaeology, or "processual" archaeology. The authors identified themselves as southwestern archaeologists who questioned a number of the aspects of that paradigm and who, despite differences in both method and theory, shared perspectives about how the past might best be understood. The theories and practices of processual archaeology, they recognized, are many and varied but are exemplified by an approach they termed "ecological functionalism." This approach was based on the assumption that societies are assumed to be systems, in the sense that groups (or subsystems) within a society are functionally interrelated and do not vary with any degree of independence. As a consequence, social interaction within a society is not viewed as a meaningful source of historical change. Instead, change is explained by the interaction between the entire cultural "system" and external factors such the environment, relatively independent demographic growth, and other cultural systems.

In contrast to what the organizers considered the existing paradigm, they assumed that prehistoric southwestern societies consisted of a variety of social subgroups within the larger society, groups that acted and interacted with varying degrees of dependence and independence and ranged in scale from individual and household to community, locality, and region. From their perspective, change emerged not only from the external relations between societies and the environment and demography but also from interaction between and within social groups. The organizers believed that the social relationships in a society are seldom in a state of equilibrium and that therefore social order is inherently dynamic and unstable. Thus, the internal dynamics of a society are not solely at the mercy of external forces but are themselves a source of change and need to be considered an engine of change in archaeological studies of change.

A list of the participants in the Durango Conference indicates that most of the contemporary practitioners who drive southwestern archaeological theory by publishing on a regular basis were represented on the program (Phillips and Sebastian 2001). Who, then, were those processual archaeologists who were practicing "ecological functionalism"? I believe that the organizers were operating with an outdated perception of the theoretical perspective of most southwestern archaeologists. Most archaeological practitioners who are interested in the broad theme of culture change (at least those who publish their theoretical perspectives) no longer work within the conceptual framework of a functional, balanced, and systemic view of cultural ecology. The chapters in this book are a testament to this changing perspective. Few archaeologists continue to view human societies as existing in a delicate homeostatic balance between culture and the environment, a notion that was indeed current in the 1970s and 1980s. This perspective was common not only among Americanist archaeologists but among biologically oriented ecologists, few of whom even considered humans an integral part of the environment.

The tendency to reject concepts that are somewhat flawed but still provide valuable insights has been the rule not only in archaeology but in most scholarly disciplines. Murray Gell-Mann, the Nobel laureate who has been described as one of the great minds of the last century, provides an illustration. I told him that I believed one of the keys to his intellectual success was his ability to master and retain the details of a subject while not allowing those details to blind him to the bigger picture and theoretical implications of a concept. He agreed with me but added something he felt was of equal importance: his understanding that, especially in the social and historical sciences, seldom was a theoretical perspective entirely incorrect and his willingness to correct what is perceived as wrong without rejecting the entire theory, perspective, or orientation out of hand. My retelling of our conversation is not an appeal to authority but rather an example of how archaeologists might avoid the too-quick abandonment of a flawed but nevertheless useful approach to understanding the past. Archaeologists' dismissal of traditionally oriented cultural ecological studies is just such an example of the tendency to abandon a potentially useful but somewhat flawed concept rather than make adjustments to correct the problems.

Interestingly, cultural anthropologists had modified the older imperfect notions of cultural ecology that southwestern archaeologists continued to champion, often into the early 1990s. Until a decade or so ago, many southwestern archaeologists still focused on the cultural ecological model based on a framework of Julian Steward and Leslie White with more than a dash of Louis Binford and system science thrown in. As early as 1971, Roy Rappaport, the progenitor of contemporary cultural ecological studies, declared that even in its reincarnations cultural ecology did not provide causal explanations for culture change. More often, he suggested, it provided insights and perspectives (Rappaport 1971). In fact, rather than think of cultural ecology as a theoretical construct, it became common to think of it as providing a perspective that provokes questions and defines problems.

In a retrospective of Rappaport's work and an overview of the evolution of the "new ecology" of the 1960s, Biersack (1999) defined several emerging "ecologies": symbolic, historical, and political. These emerging ecologies are being used to delimit problems according to the nature of the data and the questions being addressed.

The chapters in this volume reflect these emerging ecologies, notably historical and political ecologies, depending on the nature of the data and the questions defined by the researcher. Symbolic ecology is the least relevant to archaeological research of the three ecologies (or, at least, the most difficult to address in the archaeological record). The agenda of symbolic ecology is an archaeology of the human mind and its relationship to the landscape on which the play is staged, but it is not easy to find archaeological remnants of that play. Early efforts in cultural anthropology tried to distinguish between the "emic," or possibly flawed cultural understanding of the environment, and the "etic," or objective or actual representation of the environment (Harris 1979). More recent studies have dealt with the social construction of nature (Lansing 1991). In a rarity of symbolic ecology in archaeology, at least in the American Southwest, Masse and Espenak's chapter in this volume is an outstanding example of archaeology's ability to explore the nonmaterial aspects of culture. Unfortunately, too often these aspects are dismissed by archaeologists as overly difficult or impossible to reconstruct.

Historical ecology evolved from a relatively static perspective of human cultural adaptation to the environment to a more dynamic concept of the environment as the result of a historical interaction with culture. Kirch and Hunt (1997), for example, have explored the co-evolution of cultures and island environments in the Pacific. In many areas of the world, the landscape can no longer be understood in terms of natural processes but must include the historical cultural factors that shaped it. Crumley (1994) argues that an ecology is a historical reflection of decisions, ideas, and practices that are based on mental activities that remain encoded in the landscape. As dramatic examples, one needs only to scan the horizon of the terraced rice-growing hillsides of the Philippines or the "fortified" terraces on the islands of Palau in the Pacific to understand that these landscapes are more cultural than natural. Crumley's concept maintains a tenet of historical ecology, that it is inherently anthropocentric, not ecocentric (Biersack 1999).

Political ecology explores the consequences of power relationships on a large spatial scale in understanding how humans use the environment. The role of agency has begun to take on considerable importance in political ecological studies and, in fact, may be in danger of becoming as overused to its own detriment as "cultural ecology" was in the recent past. Most of the political ecological research efforts involving the concept of agency have concentrated on historical periods (Wolf 1982), but archaeological studies are beginning to appear in greater numbers, mostly using more complex societies as an example, especially in the Near East and in Nuclear America (Stein 1999).

The emerging ecologies have not completely displaced the traditional cultural ecological studies, especially in social and behavioral research by non-anthropologists. In fact, there seems to be a resurgence of studies that can be grouped under the rubric of environmental determinism (deMonocal 2001; Weiss and Bradley 2001). The most prominent example is the widely read and influential work *Guns, Germs, and Steel: The Fates of Human Societies* (Diamond 1997), which was awarded the Pulitzer Prize. Too often the explanation offered for culture change in these studies is a simple correlation of rapid behavioral change with dramatic environmental change (Haug et al. 2003). Contemporary southwestern archaeologists, at least, tend to require more than simple correlation for causal explanation and try to understand the dynamic relationship between the natural and cultural landscapes.

Many of the studies that make up this book reflect the dynamic nature of the new ecologies and are a refinement of traditional cultural ecological perspective. Social sciences, however, have not in general kept pace with the non-equilibrium theory and mathematical foundations of biological ecology for understanding systems (Scoones 1999). The chapters by Waters, Graybill et al., Dean, and Sullivan and Ruter in this volume are important steps toward a more quantitatively based understanding of the past. Archaeologists now often have the data and tools to offer more convincing explanations about the past and to test theories that are grounded in quantitative data analysis, rather than the more qualitative narrative approach common even today. Furthermore, these studies view the human–nature co-evolution with a greater appreciation for the complex dynamics of nonlinear interactions among culture, sub-elements of society, and the environment, although they tend to still lack an appreciation of human agency in environmental transformation.

The resurgence of ecological studies in understanding human behavior is signaled by new programs at the National Science Foundation with titles such as "Human and Social Dynamics" and with a strong emphasis on computer modeling of the natural and cultural environment combined with both theoretical and empirical components. A second NSF program is "Biocomplexity Research: Emergence of Cooperation from Human–Environmental Interactions." There are three large and important ongoing archaeological projects funded under this program, one focused in the Southwest with T. Kohler as principal investigator, one in Oceania directed by P. Kirch, and one in Mesopotamia directed by J. Christiansen. These efforts promise a more detailed understanding of past culture–environment interactions. The Santa Fe Institute has two new programs with the titles "Robustness of Coupled Natural and Human Systems" and "Robustness in Natural, Engineering, and Human Systems." The words *resilient, surprise, persistence, sensitivity, robustness,* and *fragility* are used frequently in these programs and reflect the increasing incorporation of non-equilibrium theory, the acknowledgment of the importance of initial conditions, and a realization of the difficulty of understanding long-term predictability in ecological and human behavioral studies.

These new programs, though still in their infancy, are not without their critics. Some researchers have deplored the lack of rigor in the definition of these programs and the confusion in terminology. When researchers from the natural sciences and the social sciences attempt to collaborate, as they must in an ecological archaeology, they often use the same words but speak different languages. When a team of physicists tried to apply the concept of scaling laws to biological systems (and are now attempting to do the same thing with social systems), it took a year of weekly meetings between physicists and biologists simply to learn the other's language (Brown and West 2000). Patience is not only a virtue but a requirement. A second criticism, especially in ecological archaeology, is the tendency to treat supporting studies as those done by technique-driven specialists rather than by contributing partners in theory and explanation.

The model of cultural adaptation that the authors of this volume were asked to consider, *The Anasazi in a Changing Environment* (Gumerman 1988), was the result of a twenty-year effort that included surficial geologists, palynologists, dendroclimatologists, and zooarchaeologists, as well as archaeologists. This was truly a team effort. Often for years at a time, researchers in other disciplines were allowed to pursue their interests, and were sometimes funded, although that aspect of their work may not have informed the overall goals of the archaeological research. For example, at times the surficial geologists explored the temporal periods well before humans were in the New World. By supporting research that didn't directly relate to archaeological problems, the project attracted researchers from other disciplines who were theoretically sophisticated and made major contributions to the intellectual thrust of the project.

So where does ecological archaeology go from here? Although predicting future intellectual directions is difficult, I suggest that the use of modeling and more rigorous quantification for understanding patterns and randomness are most likely to advance our understanding of the past in the near future.

Computer modeling using large environmental databases has made major strides in understanding the relative role of environmental and cultural attributes that affect culture change. The most sophisticated and detailed studies have been undertaken in the American Southwest, largely because of the availability of extremely fine-grained environmental and chronological data (Axtell et al. 2002; Dean et al. 2000; Gumerman et al. 2003; Kohler 2000; Van West 1994a).

The great advantage of computer modeling in archaeology research is that it permits researchers to practice a form of experimentation. Archaeology is a historical science, and as in other historical sciences, such as paleontology and cosmology, it is impossible to experiment on past events and processes. The traditional solution for archaeologists has been to take a cross-cultural approach to processes and patterns for understanding evolutionary trajectories. Many researchers have found this approach generally unsatisfactory, for numerous well-known reasons, such as the effects of historical contingencies and the issue of comparability between cultures. For these reasons, the cross-cultural approach has fallen into disfavor. Computer modeling, however, permits an infinite number of "experiments" that rely on environmental and cultural data that can be varied in numerous ways. The simulations can then be compared with the archaeological record. Sensitivity analyses can help determine what cultural and environmental factors are most likely to have produced the archaeological record.

An advantage of ecological studies in archaeology is their relative adaptability to quantification. If studies of social resilience, robustness, fragility, and other such terms are to have credibility, they will have to be quantitatively based. Narrative-based explanations of past behavior will be perceived as too subjective and the conclusions amenable to multiple interpretations. True, most archaeology in the last fifty years has been "quantitative" in the sense that sherd types are counted, projectile points are measured and compared, vessel volume is calculated, and the minimum number of individuals of *Capromeryx arizonensi* are compared at different sites. This form of "quantification" permits accurate description and comparison but does not offer explanation for the patterns that are observable in the record of the past. The need for greater quantification in ecological archaeology studies does not mean that archaeology has to go the way of neo-economics, which in its attempt to be mathematically rigorous has lost much of its ability to explain economic behavior in rational fashion. Instead, ecological archaeology might look to the example of mathematical ecology, where the seminal publication is Robert May's *Stability and Complexity in Model Ecosystems* (1973). The book inspired much current ecological research, especially the modeling of food webs, and it has fueled ongoing debates about the relationship between species richness and ecosystem function as well as the relationship between diversity or complexity and ecosystem stability and change, concepts that have obvious relevance to ecological archaeology. Mathematical ecology has become so dominant in the discipline that the term is often used as a synonym for theoretical ecology. The danger of a mathematical ecological archaeology is that environmental data are relatively easy to quantify and therefore that there might be a tendency to rely on ecological characteristics at the expense of less easily quantifiable sociological or ideological attributes to explain cultural stasis or change. However, most contemporary archaeologists, as many of the chapters in this volume demonstrate, no longer treat environment as an independent variable (Dincauze 2000).

The evolution of ecological anthropological studies into the emerging ecologies described by Biersack should not be considered a classification that is exclusionary and immutable. Just as the Pecos Classification has justifiably suffered critics for generations and yet provides a useful rubric for discussion, so the different emerging ecologies provide a framework for a general classification of theoretical perspectives. For example, the chapters by Alan Simmons and Michael Waters in this volume can be considered historical ecological studies, and the chapter by Suzanne and Paul Fish a political ecological effort. Parks, Dean, and Betancourt's study of environmental change and its relationship to political and social change in prehistoric and historic south-central New Mexico is both a historical and a political ecological study, with implications for symbolic ecology as well. In sum, while a classification of ecological studies can help researchers comprehend the types of relationship between cultural behavior and the environment, the typology should not restrict our schemata or blind us to ecologies yet to be defined. It is gratifying to know, however, that after a period of little progress in intellectually stimulating archaeological research that integrates ecological and behavioral studies, the future again looks bright.

References

Abbott, David R.

2000 *Ceramics and Community Organization Among the Hohokam.* University of Arizona Press, Tucson.

Abbott, David R. (editor)

2003 *Centuries of Decline During the Hohokam Classic Period at Pueblo Grande.* University of Arizona Press, Tucson.

Ackerly, N. W., J. B. Howard, and R. H. McGuire

1987 *La Ciudad Canals: A Study of Hohokam Irrigation Systems at the Community Level.* Anthropological Field Studies No. 17. Arizona State University, Tempe.

Adams, E. Charles

2002 *Homo'olovi: An Ancient Hopi Settlement Cluster in Northeastern Arizona.* University of Arizona Press, Tucson.

Adams, K. R.

1980 *Pollen, Parched Seeds, and Prehistory: A Pilot Investigation of Prehistoric Plant Remains from Salmon Ruin, A Chacoan Pueblo in Northwestern New Mexico.* Contributions in Anthropology Vol. 9. Eastern New Mexico University, Portales.

Adams, K. R., and V. L. Bohrer

1998 Archaeobotanical Indicators of Seasonality: Examples from Arid Southwestern United States. In *Seasonality and Sedentism: Archaeological Perspectives from Old and New World Sites,* edited by T. R. Rocek and O. Bar-Yosef, pp. 129–141. Peabody Museum of Archaeology and Ethnology. Harvard University, Cambridge.

Allen, David, and Carol Allen

1987 *Eclipse.* Allen & Unwin, Sydney.

Altschul, J. H., and C. R. Van West

1992 Agricultural Productivity Estimates for the Tonto Basin, A.D. 740–1370. In *Proceedings of the Second Salado Conference, Globe, AZ,* edited by R. C. Lange and S. Germick, pp. 172–182. Occasional Paper 1992. Arizona Archaeological Society, Phoenix.

Ambler, J. Richard

1985 *Navajo National Monument: An Archaeological Assessment.* Northern Arizona University Archaeological Series No. 1. Flagstaff.

Anyon, Roger, Patricia Gilman, and Steven A. LeBlanc

1981 A Reevaluation of the Mogollon-Mimbres Archaeological Sequence. *The Kiva* 46: 209–225.

Anyon, Roger, and Steven A. LeBlanc

1984 *The Galaz Ruin, a Prehistoric Mimbres Village in Southwestern New Mexico.* Maxwell Museum of Anthropology and University of New Mexico Press, Albuquerque.

Aveni, Anthony

1992 *Conversing with the Planets.* Times Books, New York.

Axtell, Robert L., Joshua M. Epstein, Jeffrey S. Dean, George J. Gumerman, Alan C. Swedlund, Jason Harburger, Shubha Chakravarty, Ross Hammond, Jon Parker, and Miles Parker

2002 Population Growth and Collapse in a Multiagent Model of the Kayenta Anasazi in Long House Valley. In *Adaptive Agents, Intelligence, and Emergent Human*

Organization: Capturing Complexity Through Agent-Based Modeling, edited by Brian J. L. Berry, Douglas Kiel, and Euel Elliott. *Proceedings of the National Academy of Sciences of the United States of America* 99 (Suppl. 3): 7275–7279.

Babenroth, D. L., and A. N. Strahler

1945 Geomorphology and Structure of the East Kaibab Monocline, Arizona and Utah. *Bulletin of the Geological Society of America* 56:107–150.

Bahr, Donald M.

1975 *Pima and Papago Ritual Oratory: A Study of Three Texts.* Indian Historian Press, San Francisco.

Bahr, Donald (editor)

2001 *O'odham Creation and Related Events, as Told to Ruth Benedict in 1927 in Prose, Oratory, and Song by the Pimas William Blackwater, Thomas Vanyiko, Clara Ahiel, William Stevens, Oliver Wellington, and Kisto.* University of Arizona Press, Tucson.

Bahr, Donald, Juan Smith, William Smith Allison, and Julian Hayden

1994 *The Short Swift Time of Gods on Earth: The Hohokam Chronicles.* University of California Press, Berkeley.

Baillie, Michael G. L.

1999 *Exodus to Arthur: Catastrophic Encounters with Comets.* Batsford, London.

In press Tree-Rings Focus Attention on Global Environmental Events That Could Be Caused by Extraterrestrial Forcing. In *Asteroid/Comet Impacts on Human Society*, edited by P. Bobrowsky and H. Rickman. Springer, Berlin.

Baker, V. R.

1987 Paleoflood Hydrology and Extraordinary Flood Events. *Journal of Hydrology* 96:79–99.

Bannister, Bryant, Elizabeth A. M. Gell, and John W. Hannah

1966 *Tree-Ring Dates from Arizona N–Q, Verde–Showlow–St. Johns Area.* Laboratory of Tree-Ring Research, University of Arizona, Tucson.

Barnes, C. W.

1987 Geology of the Gray Mountain Area, Arizona. In *Geological Society of America Centennial Field Guide—Rocky Mountain Section*, edited by S. S. Beus, pp. 379–384. Geological Society of America, Boulder.

Barrett, Elinore M.

2002 *Conquest and Catastrophe: Changing Rio Grande Pueblo Settlement Patterns in the Sixteenth and Seventeenth Centuries.* University of New Mexico Press, Albuquerque.

Bartlett, Katherine

1942 Notes upon the Routes of Espejo and Farfan to the Mines in the Sixteenth Century. *New Mexico Historical Review* 17:21–46.

Bayham, Frank E.

1980 Animal Exploitation and Seasonality: Inferences from the Faunal Data. In *Prehistory in Dead Valley, East-Central Arizona: The TG&E Springerville Project*, edited by D. E. Doyel and S. Debowski, pp. 389–399. Arizona State Museum Archaeological Series 144. University of Arizona, Tucson.

Becher, M. E.

1992 Archaeological Investigations at a Pueblo II Kayenta Anasazi Ruin Located in the Tusayan Ranger District, Kaibab National Forest, Arizona. Unpublished senior thesis, Department of Anthropology, University of Cincinnati, Cincinnati.

Beck, C., and G. Jones

1992 New Directions? Great Basin Archaeology in the 1990s. *Journal of California and Great Basin Anthropology* 14:22–36.

Beckett, Patrick H., and Terry L. Corbett

1992 Indian Cultural Diversity in Southern New Mexico, 1581–1988. In *Current Research on the Late Prehistory and Early History of New Mexico*, edited by Bradley J. Vierra, pp. 3–9. New Mexico Archaeological Council, Special Publication No. 1. Albuquerque.

Beeson, William L.

1966 Archaeological Survey near St. Johns, Arizona: A Methodological Study. Unpublished Ph.D. dissertation, Department of Anthropology, University of Arizona, Tucson.

Berlin, G. L., J. R. Ambler, R. H. Hevly, and G. G. Schaber

1977 Identification of a Sinagua Agricultural Field by Aerial Thermography, Soil Chemistry, Pollen/Plant Analysis, and Archaeology. *American Antiquity* 42:588–600.

Berlin, G. L., D. E. Salas, and P. R. Geib

1990 A Prehistoric Sinagua Agricultural Site in the Ashfall Zone of Sunset Crater, Arizona. *Journal of Field Archaeology* 17:1–16.

Bernard-Shaw, M.

1988 Hohokam Canal Systems and Late Archaic Wells: The Evidence from the Los Morteros Site. In *Recent Research on Tucson Basin Prehistory*, edited by W. Doelle and P. Fish, pp. 153–174. Anthropological Papers No. 10. Institute for American Research, Tucson.

Berry, C., and M. Berry

1986 Chronological and Conceptual Models in the Southwestern Archaic. In *Anthropology of the Desert West: Essays in Honor of Jesse D. Jennings*, edited by C. Condie and D. Fowler, pp. 253–327. University of Utah Anthropological Papers No. 110. University of Utah Press, Salt Lake City.

Berry, M.

1982 *Time, Space, and Transition in Anasazi Prehistory.* University of Utah Press, Salt Lake City.

Betancourt, Julio L., Elizabeth A. Pierson, Kate Aasen Rylander, James A. Fairchild-Parks, and Jeffrey S. Dean

1993 Influence of History and Climate on New Mexico Piñon-Juniper Woodlands. In *Managing Piñon-Juniper Ecosystems for Sustainability and Social Needs*, coordinated by Earl F. Aldon and Douglas W. Shaw, pp. 42–62. USDA Forest Service General Technical Report RM-236. Rocky Mountain Forest and Range Experiment Station, Fort Collins, Colorado.

Betancourt, J. L., T. R. Van Devender, and P. S. Martin

1990 *Packrat Middens: The Past 40,000 Years of Biotic Change.* University of Arizona Press, Tucson.

Bierhorst, John

1992 *History and Mythology of the Aztecs: The Codex Chimalpopoca.* University of Arizona Press, Tucson.

Biersack, A.

1999 Introduction: From the "New Ecology" to the New Ecologies. *American Anthropologist* 101:5–18.

Binford, Lewis R.

1972 *An Archaeological Perspective.* Seminar Press, New York.

1980 Willow Smoke and Dogs' Tails: Hunter-Gatherer Settlement Systems and Archaeological Site Formation. *American Antiquity* 45:4–20.

Biondi, F., and T. W. Swetnam

1987 Box-Jenkins Models of Forest Interior Tree-Ring Chronologies. *Tree-Ring Bulletin* 47:71–96.

Blake, Michael, Steven LeBlanc, and Paul Minnis

1986 Changing Settlement and Population in the Mimbres Valley, SW New Mexico. *Journal of Field Archaeology* 13:439–464.

Blong, R. J.

1982 *The Time of Darkness: Local Legends and Volcanic Reality in Papua New Guinea.* University of Washington Press, Seattle.

Bloom, A. L.

1978 *Geomorphology: A Systematic Analysis of Late Cenozoic Landforms.* Prentice-Hall, Englewood Cliffs, New Jersey.

Bode, M. F., and A. Evans

1989 *Classical Novae.* Wiley, New York.

Bohannon, P.

1965 Introduction. In *Houses and House-Life of the American Aborigines* [1881], by L. H. Morgan, pp. v–xxi. University of Chicago Press, Chicago.

Bohrer, V. L.

1975 The Prehistoric and Historic Role of the Cool-Season Grasses in the Southwest. *Economic Botany* 29:199–207.

1981 Methods of Recognizing Cultural Activity from Pollen in Archaeological Sites. *Kiva* 46:135–142.

1982 Plant Remains from Rooms at Grasshopper Pueblo. In *Multidisciplinary Research at Grasshopper Pueblo, Arizona*, edited by W. A. Longacre, S. J. Holbrook, and M. W. Graves, pp. 97–105. Anthropological Papers of the University of Arizona No. 40. University of Arizona Press, Tucson.

Bolton, Herbert E.

1964 *Coronado: Knight of Pueblos and Plains.* Reprinted. University of New Mexico Press, Albuquerque. Originally published 1949.

Bonnicksen, T. M.

2000 *America's Ancient Forests: From the Ice Age to the Age of Discovery.* Wiley, New York.

Boone, Elizabeth Hill

2000 *Stories in Red and Black: Pictorial Histories of the Aztecs and Mixtecs.* University of Texas Press, Austin.

Bostwick, Todd W., David H. Greenwald, and Mary-Ellen Walsh-Anduze

1996 The Hohokam Post-classic Period Occupation and an Early Piman Presence on the Salt River Floodplain. In *Life on the Floodplain: Further Investigations at Pueblo Salado for Phoenix Sky Harbor International Airport*, edited by D. H. Greenwald, J. H. Ballagh, D. Mitchell, and R. A. Anduze, pp. 461–490. Anthropological Papers No. 4, Vol. 2. Pueblo Grande Museum, Phoenix.

Bostwick, Todd W., and Stan Plum

2005 The Shaw Butte Hilltop Site: A Prehistoric Hohokam Observatory. In *Current Studies in Archaeoastronomy: Conversations Across Time and Space*, edited by J. W. Fountain and R. M. Sinclair, pp. 151–160. Carolina Academic Press, Durham, North Carolina.

Box, G. E. P., and G. M. Jenkins

1976 *Time Series Analysis: Forecasting and Control.* Holden Day, San Francisco.

Bozarth, S.

1992 Fossil Pollen and Phytolith Analysis. In *Archaeological Investigations at Lee Canyon: Kayenta Anasazi Farmsteads in the Upper Basin, Coconino County, Arizona*, edited by S. M. Whittlesey, pp. 135–144. Technical Series No. 38. Statistical Research, Tucson.

Bradfield, R. M.

1971 *The Changing Pattern of Hopi Agriculture.* Occasional Paper No. 30. Royal Anthropological Institute of Great Britain and Ireland, London.

1995 *An Interpretation of Hopi Culture.* Derby, U.K.

Brandt, C.

1991 Sites MU 38, MU 235, and MU 236 Archaeobotanical Analysis. Report on file, Zuni Archaeology Program, Zuni Pueblo, New Mexico.

1995 Traditional Agriculture on the Zuni Indian Reservation in the Recent Historic Period. In *Soil, Water, Biology, and Belief in Prehistoric and Traditional Southwestern Agriculture*, edited by H. Wolcott Toll, pp. 291–301. Special Publication No. 2. New Mexico Archaeological Council, Albuquerque.

Brewer, D. G., R. K. Jorgensen, L. P. Munk, W. A. Robbie, and J. L. Travis

1991 *Terrestrial Ecosystems Survey of the Kaibab National Forest.* USDA Forest Service, Albuquerque.

Bronitsky, G., and J. D. Merritt

1985 *The Archaeology of Southeast Arizona: A Class I Cultural Resource Inventory.* Cultural Resource Series No. 2. Bureau of Land Management, Arizona State Office, Phoenix.

Brown, D. E.
1981 The Effects of Drought on White-Tailed Deer Recruitment in the Arid Southwest. In *Arizona Big Game Investigations, 1980–1981*. Project W53R. Arizona Game and Fish Department, Phoenix.
Brown, G. B.
1996 Direct Crop Production Evidence from Prehistoric Agricultural Fields. Master's thesis, Department of Anthropology, Northern Arizona University, Flagstaff.
Brown, J. H., and G. B. West (editors)
2000 *Scaling in Biology*. Oxford University Press, New York.
Brown, Peter Lancaster
1973 *Comets, Meteorites, and Men*. Taplinger, New York.
Brown, R. B., and Rolf M. Sinclair
1998 A Triple Crown for Padre Newmann: The First Reported Earthquake, Comet, and Eclipse in Chihuahua. In *1998 Yearbook of the Conference of Latin Americanist Geographers*, edited by D. J. Keeling and J. Wiley, pp. 121–126. University of Texas Press, Austin.
Bryan, Kirk
1929 Floodwater Farming. *Geographical Review* 19:444–456.
Bull, William B.
1991 *Geomorphic Responses to Climatic Change*. Oxford University Press, Oxford.
1997 Discontinuous Ephemeral Streams. *Geomorphology* 19:227–276.
Burkham, D. E.
1972 *Channel Changes of the Gila River in the Safford Valley, Arizona, 1846–1970*. U.S. Geological Survey Professional Paper 655-Gila. U.S. Government Printing Office, Washington, D.C.
Burns, Barney Tillman
1983 *Simulated Anasazi Storage Behavior Using Crop Yields Reconstructed from Tree Rings, A.D. 652–1968*. Ph.D. dissertation, University of Arizona. University Microfilms, Ann Arbor.
Butzer, Karl W.
1982 *Archaeology as Human Ecology: Method and Theory for a Contextual Approach*. Cambridge University Press, Cambridge.
Bye, R. A., Jr.
1981 Quelites—Ethnoecology of Edible Plants—Past, Present, and Future. *Journal of Ethnobiology* 1:109–123.
Cable, John, and David E. Doyel
1987 Pioneer Period Village Structure and Settlement Pattern in the Phoenix Basin. In *The Hohokam Village: Site Structure and Organization*, edited by David E. Doyel, pp. 21–70. American Association for the Advancement of Science, Glenwood Springs, Colorado.
Cantley, Gary J.
1991 Archaeology of a Rockpile Field in the Santan Mountains, Arizona. Unpublished Master's thesis. Department of Anthropology, Arizona State University, Tempe.
Carrasco, Davíd
1990 *Religions of Mesoamerica*. Harper, San Francisco.
1992 *Quetzalcoatl and the Irony of Empire*. University of Chicago Press, Chicago.
Chamberlain, Von Del
1982 *When Stars Came Down to Earth: Cosmology of the Skidi Pawnee Indians of North America*. Ballena Press, Los Altos, California, and University of Maryland Center for Archaeoastronomy, College Park.
Chenault, M. L.
1996 The Hohokam Post-classic Polverón Phase. In *The Sky Harbor Project: Early Desert Farming and Irrigation Settlements: Archaeological Investigations in the Phoenix Sky*

Harbor Venter, Vol. 4: Special Studies, Synthesis, and Conclusions, edited by D. H. Greenwald and J. H. Ballagh, pp. 117–140. Anthropological Research Paper No. 4. SWCA, Flagstaff.

Chorley, R. J., S. A. Schumm, and D. E. Sugden

1984 *Geomorphology*. Methuen, New York.

Christenson, A. L.

1992 Ceramics. In *Investigations at Lee Canyon: Kayenta Anasazi Farmsteads in the Upper Basin, Coconino County, Arizona*, edited by S. M. Whittlesey, pp. 75–89. Technical Series No. 38. Statistical Research, Tucson.

2003 Archaeological Survey in Tsegi Canyon, 1988–1995. In *Climbing the Rocks: Papers in Honor of Helen and Jay Crotty*, edited by Regge N. Wiseman, Thomas C. O'Laughlin, and Cordelia T. Snow, pp. 69–82. Papers of the Archaeological Society of New Mexico No. 29. Albuquerque.

Chyba, C. F., P. Thomas, and K. Zahnle

1993 The 1908 Tunguska Explosion: Atmospheric Disruption of a Stony Meteorite. *Nature* 361:40–44.

Ciolek-Torrello, Richard

1995 The Houghton Road Site, the Agua Caliente Phase, and the Early Formative Period in the Tucson Basin. *Kiva* 60:531–574.

Ciolek-Torrello, R., S. M. Whittlesey, and J. R. Welch

1994 A Synthetic Model of Prehistoric Land Use. In *The Roosevelt Rural Sites Study: Changing Land Use in the Tonto Basin*, edited by R. Ciolek-Torrello and J. R. Welch, pp. 437–472. Technical Series No. 28, Vol. 3. Statistical Research, Tucson.

Clark, David H., and F. Richard Stephenson

1977 *The Historical Supernovae*. Pergamon Press, Oxford.

Clark, Jeffery J.

1995 The Role of Migration in Social Change. In *The Roosevelt Community Development Study: New Perspectives on Tonto Basin Prehistory*, edited by M. D. Elson, M. T. Stark, and D. A. Gregory, pp. 369–384. Anthropological Papers No. 15. Center for Desert Archaeology, Tucson.

Clark, J. J., A. P. Dutt, and J. M. Vint

1999 *U.S. 70 Safford to Thatcher Segment: Testing Results and a Plan for Data Recovery*. Technical Report No. 99-6. Center for Desert Archaeology, Tucson.

Clark, J. J., M. K. Woodson, and M. C. Slaughter

1999 Those Who Went to the Land of the Sun: Puebloan Migrations into Southeastern Arizona. In *Archaeology of a Land Between: Regional Dynamics in the Prehistory and History of Southeastern Arizona*, edited by H. D. Wallace. University of New Mexico Press, Albuquerque.

Clarke, David L.

1978 *Analytical Archaeology*. Rev. ed. Columbia University Press, New York.

Coder, C. M.

2000 *An Introduction to Grand Canyon Prehistory*. Grand Canyon Association, Grand Canyon, Arizona.

Cohodas, Marvin

1991 Ballgame Imagery in the Mayan Lowlands: History and Iconography. In *The Mesoamerican Ballgame*, edited by V. L. Scarborough and D. R. Wilcox, pp. 251–288. University of Arizona Press, Tucson.

Cole, K. L.

1990 Late Quaternary Vegetation Gradients Through the Grand Canyon. In *Packrat Middens: The Past 40,000 Years of Biotic Change*, edited by J. L. Betancourt, T. R. Van Devender, and P. S. Martin, pp. 240-258. University of Arizona Press, Tucson.

Colton, Harold Sellers

1939 *Prehistoric Culture Units and Their Relationships in Northern Arizona*. Museum of Northern Arizona Bulletin No. 17. Flagstaff.

Cook, Edward R., and Richard L. Holmes

1986 Users Manual for Program ARSTAN. In *Tree-Ring Chronologies of Western North*

America: California, Eastern Oregon, and Northern Great Basin, by Richard L. Holmes, Rex K. Adams, and Harold C. Fritts, pp. 50–65. Chronology Series VI. Laboratory of Tree-Ring Research, University of Arizona, Tucson.

Cook, E. R., and L. A. Kairiukstis (editors)

1990 *Methods of Dendrochronology: Applications in the Environmental Sciences.* Kluwer Academic, Boston.

Cook, Edward R., and Kenneth Peters

1981 The Smoothing Spline: A New Approach to Standardizing Forest Interior Tree-Ring Width Series for Dendroclimatic Studies. *Tree-Ring Bulletin* 41:45–53.

Cook, R. A.

1995 Long-Term Upland Wild-Resource Subsistence Technology: Evidence from Fire-Cracked Rock Piles in the Upper Basin, Kaibab National Forest, Northern Arizona. Master's thesis, Department of Anthropology, University of Cincinnati, Cincinnati.

Cooke, Ronald U., and Richard W. Reeves

1976 *Arroyos and Environmental Change in the American South-West.* Clarendon Press, Oxford.

Cooley, Maurice E.

1962 Late Pleistocene and Recent Erosion and Alluviation in Parts of the Colorado River System, Arizona and Utah. In *Geological Survey Research 1962: Short Papers in Geology, Hydrology, and Topography, Articles 1-59,* pp. 48–50. United States Geological Survey Professional Paper 450-B.

Cordell, L. S.

1984 Southwestern Archaeology. *Annual Review of Anthropology* 13:301–332.

1997 *Archaeology of the Southwest.* 2nd ed. Academic Press, New York.

Cordell, Linda S., David E. Doyel, and Keith W. Kintigh

1994 Processes of Aggregation in the Prehistoric Southwest. In *Themes in Southwest Prehistory,* edited by G. J. Gumerman, pp. 109–133. Advanced Seminar Series. School of American Research Press, Santa Fe.

Cordell, Linda S., and Fred Plog

1979 Escaping the Confines of Normative Thought: A Reevaluation of Puebloan Prehistory. *American Antiquity* 44:405–429.

Cosgrove, C. B., and H. S. Cosgrove

1932 *The Swarts Ruin, a Typical Mimbres Site in Southwestern New Mexico.* Papers of the Peabody Museum of American Archaeology and Ethnology, Vol. 15, No. 1. Harvard University, Cambridge.

Courlander, Harold

1971 *The Fourth World of the Hopis.* University of New Mexico Press, Albuquerque.

Coyne, G. V., and R. M. Sinclair (editors)

1995 *The Inspiration of Astronomical Phenomena. Vistas in Astronomy* 39:381–733.

Creel, Darrell

1997 Interpreting the End of the Mimbres Classic. In *Prehistory of the Borderlands,* edited by John Carpenter and Guadalupe Sanchez, pp. 25–32. Arizona State Museum Archaeological Series 186. University of Arizona, Tucson.

1999a The Black Mountain Phase in the Mimbres Area. In *The Casas Grandes World: A Diversity of Views,* edited by Curtis Schaafsma and Carroll Riley, pp. 107–120. University of Utah Press, Salt Lake City.

1999b *Status Report on Excavations at the Old Town Site (LA 1113), Luna County, New Mexico, Summer 1998.* Submitted to the U.S. Bureau of Land Management, New Mexico State Office, by the Department of Anthropology and the Texas Archeological Research Laboratory, University of Texas at Austin.

Crown, P. L.

1987 Classic Period Hohokam Settlement and Land Use in the Casa Grande Ruins Area. *Journal of Field Archaeology* 14:147–162.

1991 The Hohokam: Current Views of Prehistory and the Regional System. In *Chaco and Hohokam: Prehistoric Regional Systems in the American Southwest,* edited by P. L. Crown and W. J. Judge, pp. 135–158. School of American Research Press, Santa Fe.

1994 *Ceramics and Ideology: Salado Polychrome Pottery.* University of New Mexico Press, Albuquerque.

Crown, Patricia L., and W. James Judge (editors)

1991 *Chaco and Hohokam: Prehistoric Regional Systems in the American Southwest.* Advanced Seminar Series. School of American Research Press, Santa Fe.

Crown, P. L., and E. W. Sires, Jr.

1984 The Hohokam Chronology and Salt-Gila Aqueduct Project Research. In *Hohokam Archaeology along the Salt-Gila Aqueduct, Central Arizona Project,* edited by L. S. Teague and P. L. Crown, pp. 73–85. Arizona State Museum Archaeological Series 150, Vol. 9. University of Arizona, Tucson.

Crumley, C. L.

1994 Historical Ecology: A Multidimensional Ecological Orientation. In *Historical Ecology: Cultural Knowledge and Changing Landscapes,* edited by C. L. Crumley, pp. 1–16. School of American Research Press, Santa Fe.

Cummings, L. S.

1986 Analysis of Selected Macrofloral Samples from AZ I:1:17 (ASM). In *Prehistory of the Upper Basin, Coconino County, Arizona,* edited by A. P. Sullivan III, pp. 339–346. Arizona State Museum Archaeological Series 167. University of Arizona, Tucson.

1994 Anasazi Diet: Variety in the Hoy House and Lion House Coprolite Record and Nutritional Analysis. In *Paleonutrition: The Diet and Health of Prehistoric Americans,* edited by K. D. Sobolik, pp. 134–150. Southern Illinois University Press, Carbondale.

1995 Agriculture and the Mesa Verde Area Anasazi Diet: Description and Nutritional Analysis. In *Soil, Water, Biology, and Belief in Prehistoric and Traditional Southwestern Agriculture,* edited by H. W. Toll, pp. 335–352. New Mexico Archaeological Council, Albuquerque.

2001 *Pollen Analysis at Simkins Flat, Arizona.* Technical Report 01-40. Paleo Research Institute, Golden, Colorado.

Cummings, L. S., and K. Puseman

1995 *Pollen and Macrofloral Analysis at Sites MU 123.1, MU 125, and MU 235 in Kaibab National Forest, North-Central Arizona.* Technical Report 95-10. Paleo Research Labs, Denver.

1997 *Pollen and Macrofloral Analysis at Sites MU 125 and MU 125A, Kaibab National Forest, North-Central Arizona.* Technical Report 97-45. Paleo Research Labs, Denver.

Cutler, H. C.

1963 Plant Remains from a Grand Canyon Rock Shelter Granary. *American Antiquity* 28:301.

Cutler, H. C., and L. W. Blake

1980 Plant Materials from Grand Canyon Sites. In *Archaeology of the Grand Canyon: Unkar Delta,* edited by D. W. Schwartz, R. C. Chapman, and J. Kepp, pp. 209–212. School of American Research Press, Santa Fe.

Damp, J. E., S. A. Hall, and S. J. Smith

2002 Early Irrigation on the Colorado Plateau near Zuni Pueblo, New Mexico. *American Antiquity* 67:665–676.

Danson, Edward B.

1957 *An Archaeological Survey of West Central New Mexico and East Central Arizona.* Papers of the Peabody Museum of Archaeology and Ethnology, Vol. 44, No. 1. Harvard University, Cambridge.

Dart, Allen

1994 *Archaeological Studies of the Avra Valley, Arizona: Excavations in the Schuk Toak District.* Anthropological Papers No. 16. Center for Desert Archaeology, Tucson.

Davis, A. P.

1897 *Irrigation near Phoenix, Arizona.* U.S. Geological Survey Water-Supply Paper No. 2. U.S. Government Printing Office, Washington, D.C.

Davis, O. K.

1986 Pollen Analysis from AZ I:1:17 (ASM). In *Prehistory of the Upper Basin, Coconino*

County, Arizona, edited by A. P. Sullivan III, pp. 333–338. Arizona State Museum Archaeological Series 167. University of Arizona, Tucson.

Dean, Jeffrey S.

1969 *Chronological Analysis of Tsegi Phase Sites in Northeastern Arizona.* Papers of the Laboratory of Tree-Ring Research No. 3. University of Arizona Press, Tucson.

1970 Aspects of Tsegi Phase Social Organization: A Trial Reconstruction. In *Reconstructing Prehistoric Pueblo Societies*, edited by William A. Longacre, pp. 140–174. University of New Mexico Press, Albuquerque.

1988a Dendrochronology and Paleoenvironmental Reconstruction on the Colorado Plateaus. In *The Anasazi in a Changing Environment*, edited by George J. Gumerman, pp. 119–167. Cambridge University Press, Cambridge.

1988b A Model of Anasazi Behavioral Adaptation. In *The Anasazi in a Changing Environment*, edited by George J. Gumerman, pp. 25–44. Cambridge University Press, Cambridge.

1989 Temporal Aspects of Tsegi Phase Household Organization. In *Households and Communities: Proceedings of the 21st Annual Chacmool Conference*, edited by Scott MacEachern, David J. W. Archer, and Richard D. Garvin, pp. 196–200. Archaeological Association of the University of Calgary, Calgary.

1991 Some Thoughts on Hohokam Chronology. In *Exploring the Hohokam: Prehistoric Desert Peoples of the American Southwest*, edited by G. J. Gumerman, pp. 61–150. Amerind Foundation, Dragoon, Arizona, and University of New Mexico Press, Albuquerque.

1992 Environmental Factors in the Evolution of the Chacoan Sociopolitical System. In *Anasazi Regional Organization and the Chaco System*, edited by D. E. Doyel, pp. 35–44. Papers in Anthropology No. 5. Maxwell Museum of Anthropology, Albuquerque.

1994 The Medieval Warm Period on the Southern Colorado Plateau. *Climatic Change* 26:225–241.

1996a Demography, Environment, and Subsistence Stress. In *Evolving Complexity and Environmental Risk in the Prehistoric Southwest*, edited by Joseph A. Tainter and Bonnie Bagley Tainter, pp. 25–56. Santa Fe Institute Studies in the Sciences of Complexity, Proceedings Vol. 24. Addison-Wesley, Reading, Massachusetts.

1996b Kayenta Anasazi Settlement Transformations in Northeastern Arizona, A.D. 1150–1350. In *The Prehistoric Pueblo World, A.D. 1150–1350*, edited by Michael A. Adler, pp. 29–47. University of Arizona Press, Tucson.

2000 Complexity Theory and Sociocultural Change in the American Southwest. In *The Way the Wind Blows: Climate, History, and Human Action*, edited by Roderick J. McIntosh, Joseph A. Tainter, and Susan Keech McIntosh, pp. 89–118. Columbia University Press, New York.

2002 Late Pueblo II–Pueblo III in Kayenta-Branch Prehistory. In *Prehistoric Culture Change on the Colorado Plateau: Ten Thousand Years on Black Mesa*, edited by Shirley Powell and Francis E. Smiley, pp. 121–157. University of Arizona Press, Tucson.

Dean, Jeffrey S., William H. Doelle, and Janet D. Orcutt

1994 Adaptive Stress, Environment, and Demography. In *Themes in Southwest Prehistory*, edited by George J. Gumerman, pp. 53–86. School of American Research Press, Santa Fe.

Dean, Jeffrey S., Robert C. Euler, George J. Gumerman, Fred Plog, Richard H. Hevly, and Thor N. V. Karlstrom

1985 Human Behavior, Demography, and Paleoenvironment on the Colorado Plateau. *American Antiquity* 50:537–554.

Dean, Jeffrey S., and Gary S. Funkhouser

1995 Dendroclimatic Reconstructions for the Southern Colorado Plateau. In *Climate Change in the Four Corners and Adjacent Regions: Implications for Environmental Restoration and Land-Use Planning*, edited by W. J. Waugh, pp. 85–104. U.S. Department of Energy, Grand Junction Projects Office, Grand Junction, Colorado.

Dean, Jeffrey S., George J. Gumerman, Joshua M. Epstein, Robert L. Axtell, Alan C. Swedlund, Miles T. Parker, and Stephen McCarroll

2000 Understanding Anasazi Culture Change Through Agent-Based Modeling. In *Dynamics in Human and Primate Societies: Agent-Based Modeling of Social and Spatial Processes*, edited by Timothy A. Kohler and George J. Gumerman, pp. 179–205. Santa Fe Institute Studies in the Sciences of Complexity. Oxford University Press, New York.

Dean, Jeffrey S., Alexander J. Lindsay, Jr., and William J. Robinson

1978 Prehistoric Settlement in Long House Valley, Northeastern Arizona. In *Investigations of the Southwestern Anthropological Research Group: An Experiment in Archaeological Cooperation: The Proceedings of the 1976 Conference*, edited by Robert C. Euler and George J. Gumerman, pp. 25–44. Museum of Northern Arizona, Flagstaff.

Dean, Jeffrey S., and William J. Robinson

1977 *Dendroclimatic Variability in the American Southwest, A.D. 680 to 1970*. Laboratory of Tree-Ring Research, University of Arizona, Tucson.

1978 *Expanded Tree-Ring Chronologies for the Southwestern United States*. Chronology Series 3. Laboratory of Tree-Ring Research, University of Arizona, Tucson.

1982 Dendrochronology of Grasshopper Pueblo. *In Multidisciplinary Research at Grasshopper Pueblo, Arizona,* edited by W. A. Longacre, S. J. Holbrook, and M. W. Graves, pp. 46–60. Anthropological Papers of the University of Arizona No. 40. University of Arizona Press, Tucson.

Dean, Jeffrey S., Mark C. Slaughter, and Dennie O. Bowden III

1996 Desert Dendrochronology: Tree-Ring Dating Prehistoric Sites in the Tucson Basin. *Kiva* 62:7–26.

Dean, Jeffrey S., and Carla R. Van West

2002 Environment-Behavior Relationships in Southwestern Colorado. In *Seeking the Center Place: Archaeology and Ancient Communities in the Mesa Verde Region*, edited by Mark D. Varien and Richard H. Wilshusen, pp. 81–99. University of Utah Press, Salt Lake City.

Deaver, W. L.

1997 The Lower Verde Archaeological Project: Overview, Synthesis, and Conclusions. In *Vanishing River: Landscapes and Lives of the Lower Verde Valley*, edited by S. M. Whittlesey, R. Ciolek-Torrello, and J. H. Altschul, pp. 447–490. Statistical Research, Tucson.

Deaver, William L., and Richard Ciolek-Torrello

1995 Early Formative Chronology for the Tucson Basin. *Kiva* 60:481–530.

Debowski, Sharon S.

1980 Shell and Miscellaneous Artifacts. In *Prehistory in Dead Valley, East-Central Arizona: The TG&E Springerville Project*, edited by D. E. Doyel and S. Debowski, pp. 341–347. Arizona State Museum Archaeological Series 144. University of Arizona, Tucson.

Debowski, S. S., A. George, R. Goddard, and D. Mullon

1976 An Archaeological Survey of the Buttes Dam Reservoir. Arizona State Museum Archaeological Series 93. University of Arizona, Tucson.

DeGarmo, Glen

1975 Coyote Creek Pueblo. Unpublished Ph.D. dissertation, Department of Anthropology, University of California, Los Angeles.

deMonocal, P. B.

2001 Cultural Responses to Climate Change During the Late Holocene. *Science* 292:667–673.

Dennell, R.

1979 Prehistoric Diet and Nutrition: Some Food for Thought. *World Archaeology* 11:121–135.

Densmore, Frances

1929 *Papago Music*. Bureau of American Ethnology, Bulletin 90. Smithsonian Institution, Washington, D.C.

Dering, J. P.
1996 Dynamic Variation in Agricultural Practices During the Classic Period in the Tonto Basin, Arizona. Unpublished Ph.D. dissertation, Texas A&M University, College Station.
Diamond, J.
1997 *Guns, Germs, and Steel: The Fates of Human Societies.* W. W. Norton, New York.
Dick, H.
1965 *Bat Cave.* Monograph No. 27. School of American Research, Santa Fe.
Diehl, Michael
1996 The Intensity of Maize Processing and Production in Upland Mogollon Pithouse Villages, A.D. 200–1000. *American Antiquity* 61:102–115.
1997 Changes in Architecture and Land Use Strategies in the American Southwest: Upland Mogollon Pithouse Dwellers, A.C. 200–1000. *Journal of Field Archaeology* 24:179–194.
Diehl, M. (editor)
1997 *Archaeological Investigations of the Early Agricultural Period Settlement at the Base of A-Mountain, Tucson, Arizona.* Technical Report No. 97-5. Center for Desert Archaeology, Tucson.
Dincauze, D. F.
2000 *Environmental Archaeology: Principles and Practice.* Cambridge University Press, Cambridge.
DiPeso, C. C.
1958 *The Reeve Ruin of Southeastern Arizona.* Amerind Foundation Publication No. 8. Amerind Foundation, Dragoon, Arizona.
1974 *Casas Grandes: A Fallen Trade Center of the Gran Chicimeca. Volume 4, Architecture and Dating Methods.* Amerind Foundation Publication No. 9. Amerind Foundation, Dragoon, Arizona.
Dobyns, H.
1972 Altitude Sorting of Ethnic Groups in the Southwest. *Plateau* 47:42–48.
Doebley, J. F.
1984 "Seeds" of Wild Grasses: A Major Food of Southwestern Indians. *Economic Botany* 38:52–64.
Doelle, W. H.
1995 Tonto Basin Demography in a Regional Perspective. In *The Roosevelt Community Development Study: New Perspectives on Tonto Basin Prehistory,* edited by M. D. Elson, M. T. Stark, and D. A. Gregory, pp. 201–226. Anthropological Papers No. 15. Center for Desert Archaeology, Tucson.
Doelle, William H., Allen Dart, and Henry D. Wallace
1985 *The Southern Tucson Basin Survey: Intensive Survey Along the Santa Cruz River.* Institute for American Research Technical Report 85-3. Institute for American Research, Tucson.
Doelle, William H., David A. Gregory, and Henry D. Wallace
1995 Classic Period Platform Mound Systems in Southern Arizona. In *The Roosevelt Community Development Study: New Perspectives on Tonto Basin Prehistory,* edited by M. D. Elson, M. T. Stark, and D. A. Gregory, pp. 385–440. Anthropological Papers No. 15. Center for Desert Archaeology, Tucson.
Dolnick, E.
2001 *Down the Great Unknown: John Wesley Powell's 1869 Journey of Discovery and Tragedy Through the Grand Canyon.* HarperCollins, New York.
Donaldson, Bruce R.
1993 Late Mogollon Pueblo Prehistory: A View from the Rim. Paper presented at the 1993 Pecos Conference, Springerville, Arizona.
Doolittle, William E.
1984 Agricultural Change as an Incremental Process. *Annals of the Association of American Geographers* 74:124–137.

2000 *Cultivated Landscapes of North America.* Oxford University Press, Oxford.

Doolittle, W. E., and C. D. Frederick

1991 Phytoliths as Indicators of Prehistoric Maize (*Zea mays* subsp. *mays*, Poaceae) Cultivation. *Plant Systematics and Evolution* 177:175–184.

Downum, Christian E. (editor)

1993 *Between Desert and River: Hohokam Settlement and Land Use in the Los Robles Community.* Anthropological Papers of the University of Arizona No. 57. University of Arizona Press, Tucson.

Downum, Christian E., Paul R. Fish, and Suzanne K. Fish

1994 Refining the Role of *Cerros de Trincheras* in Southern Arizona Settlement. *Kiva* 59:271–296.

Doyel, David E.

1976 Classic Period Hohokam in the Gila River Basin. *The Kiva* 42:27–37.

1979 Sand Dune Farming in the Upper Little Colorado, Arizona. Paper presented at the 1978 Ethnobiology Conference, Flagstaff. Ms. on file, Arizona State Museum Library, Tucson.

1980a Archaeological Survey. In *Prehistory in Dead Valley, East-Central Arizona: The TG&E Springerville Project,* edited by D. E. Doyel and S. S. Debowski, pp. 401–410. Arizona State Museum Archaeological Series 144. University of Arizona, Tucson.

1980b Functional Analysis of Ceramic Assemblages from Dead Valley. In *Prehistory in Dead Valley, East-Central Arizona: The TG&E Springerville Project,* edited by D. E. Doyel and S. S. Debowski, pp. 205–221. Arizona State Museum Archaeological Series 144. University of Arizona, Tucson.

1980c Stylistic and Petrographic Analysis of Dead Valley Ceramic Assemblages. In *Prehistory in Dead Valley, East-Central Arizona: The TG&E Springerville Project,* edited by D. E. Doyel and S. S. Debowski, pp. 141–187. Arizona State Museum Archaeological Series 144. University of Arizona, Tucson.

1980d Synthesis. In *Prehistory in Dead Valley, East-Central Arizona: The TG&E Springerville Project,* edited by D. E. Doyel and S. S. Debowski, pp. 413–425. Arizona State Museum Archaeological Series 144. University of Arizona, Tucson.

1981 Environment, Subsistence, and Land Use in Dead Valley, East-Central Arizona. *The Kiva* 46:143–154.

1984 Stylistic and Petrographic Variability in Pueblo II Period Cibola Whiteware from the Upper Little Colorado. In *Regional Analysis of Prehistoric Ceramic Variation: Contemporary Studies of the Cibola Whitewares,* edited by Alan Sullivan and Jeffrey Hantman, pp. 4–16. Anthropological Research Papers No. 41. Arizona State University, Tempe.

1991a Hohokam Cultural Evolution in the Phoenix Basin. In *Exploring the Hohokam: Prehistoric Desert Peoples of the American Southwest,* edited by George J. Gumerman, pp. 231–278. Amerind Foundation, Dragoon, Arizona, and University of New Mexico Press, Albuquerque.

1991b Hohokam Exchange and Interaction. In *Chaco and Hohokam: Prehistoric Regional Systems in the American Southwest,* edited by P. L. Crown and W. J. Judge, pp. 225–252. School of American Research Press, Santa Fe.

1993a On Rivers and Boundaries in the Phoenix Basin, Arizona. *Kiva* 58:455–474.

1993b *Prehistoric Non-irrigated Agriculture in Arizona: A Context for Study.* Estrella Cultural Research Paper No. 3. Arizona State Parks, State Historic Preservation Office, Phoenix.

1994 Charles C. DiPeso: Expanding the Frontiers of New World Prehistory. *American Antiquity* 51:9–20.

1996a Resource Mobilization and Hohokam Society: Analysis of Obsidian Artifacts from the Gatlin Site, Arizona. *Kiva* 62:45–60.

1996b Settlement Organization at Gila Bend: A View from the West. Paper presented at the American Association for the Advancement of Science Symposium "New Perspectives on the Hohokam Village," June, Flagstaff, Arizona.

2000 The Santan Phase in the Phoenix Basin. In *The Hohokam Village Revisited*, edited by D. E. Doyel, S. K. Fish, and P. R. Fish, pp. 221–244. Southwestern and Rocky Mountain Division of the American Association for the Advancement of Science, Fort Collins, Colorado.

2004 Edge Work: The Late Prehistory of the Gila Bend Frontier. In *Fragile Patterns: Perspectives on Western Papaguería Archaeology*, edited by J. Altschul and A. Rankin. Statistical Research, Tucson. In preparation.

Doyel, David E., and Sharon S. Debowski (editors)

1980 *Prehistory in Dead Valley, East-Central Arizona: The TG&E Springerville Project.* Arizona State Museum Archaeological Series 144. University of Arizona, Tucson.

Doyel, David E., and Steven Lekson

1992 Regional Organization in the Southwest. *In Anasazi Regional Organization and the Chaco System*, edited by David E. Doyel, pp. 15–22. Papers in Anthropology No. 5. Maxwell Museum of Anthropology, Albuquerque. Reprinted, Maxwell Museum of Anthropology, 2001.

Dozier, Edward P.

1970 *The Pueblo Indians of North America.* Holt, Rinehart and Winston, New York.

Draper, N., and H. Smith

1981 *Applied Regression Analysis.* 2nd ed. Wiley, New York.

Duff, Andrew

1999 Regional Interaction and Transformation of Western Pueblo Identities, A.D. 1275–1400. Unpublished Ph.D. dissertation, Department of Anthropology, Arizona State University, Tempe.

2000 Scale, Interaction, and Regional Analysis in Late Pueblo Prehistory. In *The Archaeology of Regional Interaction*, edited by Michelle Hegmon, pp. 71–98. University Press of Colorado, Boulder.

2002 *Western Pueblo Identities: Regional Interaction, Migration, and Transformation.* University of Arizona Press, Tucson.

Dundes, Alan (editor)

1984 *Sacred Narrative: Selected Readings in the Theory of Myth.* University of California Press, Berkeley.

Dunmire, W. W., and G. D. Tierney

1997 *Wild Plants and Native Peoples of the Four Corners.* Museum of New Mexico Press, Santa Fe.

Durrenberger, R. W., and X. Murrieta

1978 *Clima del estado de Sonora, Mexico.* Climatological Publications, Mexico Climatology Series No. 3. Office of the State Climatologist, Arizona State University, Tempe.

EarthInfo, Inc.

1990 HYDRODATA Users Manual. EarthInfo, Boulder, Colorado.

Eddy, Frank W., and Maurice E. Cooley

1983 *Cultural and Environmental History of Cienega Valley, Southeastern Arizona.* Anthropological Papers of the University of Arizona No. 43. University of Arizona Press, Tucson.

Effland, R. W., Jr., A. T. Jones, and R. C. Euler

1981 *The Archaeology of Powell Plateau: Regional Interaction at Grand Canyon.* Monograph No. 3. Grand Canyon Natural History Association, Grand Canyon, Arizona.

Eggan, F.

1989 Great Basin Models for Hopi Institutions. In *Seasons of the Kachina*, edited by S. B. Vane, pp. 17–40. Ballena Press Anthropological Papers No. 34. California State University, Hayward.

Ellen, R.

1982 *Environment, Subsistence, and System: The Ecology of Small-Scale Social Formations.* Cambridge University Press, Cambridge.

Ellis, G. Lain

1995 An Interpretive Framework for Radiocarbon Dates from Soil Organic Matter from

Prehistoric Water Control Features. In *Soil, Water, Biology, and Belief in Prehistoric and Traditional Southwestern Agriculture*, edited by H. Wolcott Toll, pp. 155–186. Special Publication No. 2. New Mexico Archaeological Council, Albuquerque.
1998 Epistemology and the Evaluation of Archaeological Theories, with a Case Study from the Mimbres Region of Southwestern New Mexico. Unpublished Ph.D. dissertation, Texas A&M University, College Station.

Elson, Mark D.
1996 A Revised Chronology and Phase Sequence for the Lower Tonto Basin of Central Arizona. *Kiva* 62:117–147.

Elson, Mark D., and David A. Gregory
1995 Tonto Basin Chronology and Phase Sequence. In *The Roosevelt Community Development Study: New Perspectives on Tonto Basin Prehistory*, edited by M. D. Elson, M. T. Stark, and D. A. Gregory, pp. 61–77. Anthropological Papers No. 15. Center for Desert Archaeology, Tucson.

Elson, Mark D., Michael H. Ort, S. Jerome Hesse, and Wendell A. Duffield
2002 Lava, Corn, and Ritual in the Northern Southwest. *American Antiquity* 67:119–135.

Elson, M. D., M. T. Stark, and D. A. Gregory (editors)
1995 *The Roosevelt Community Development Study: New Perspectives on Tonto Basin Prehistory*. Anthropological Papers No. 15. Center for Desert Archaeology, Tucson.

Ely, L. L., and V. R. Baker
1985 Reconstructing Paleoflood Hydrology with Slackwater Deposits: Verde River, Arizona. *Physical Geography* 5(2):103–126.

Ely, L. L., Y. Enzel, V. R. Baker, and D. R. Cayan
1993 A 500-Year Record of Extreme Floods and Climate Change in the Southwestern United States. *Science* 262:410–412.

Emslie, S. D., R. C. Euler, and J. I. Mead
1987 A Desert Culture Shrine in Grand Canyon, Arizona, and the Role of Split-Twig Figurines. *National Geographic Research* 3:511–516.

Emslie, S. D., J. I. Mead, and L. Coats
1995 Split-Twig Figurines in Grand Canyon, Arizona: New Discoveries and Interpretations. *Kiva* 61:145–173.

Ensor, Bradley E., Marisa O. Ensor, and Gregory W. De Vries
2003 Hohokam Political Ecology and Vulnerability: Comments on Waters and Ravesloot. *American Antiquity* 68(1):169–181.

Euler, R. C.
1969 The Archaeology of the Canyon Country. In *John Wesley Powell and the Anthropology of the Canyon Country*, by D. D. Fowler, R. C. Euler, and C. S. Fowler, pp. 8–20. Geological Survey Professional Paper No. 670. U.S. Government Printing Office, Washington, D.C.
1983 The Pinto Basin Complex at Grand Canyon, Arizona. *The Kiva* 49:67–73.
1987 A Kayenta Anasazi Agricultural Community North of the Rio Colorado. Ms. on file, Department of Anthropology, Prescott College, Prescott, Arizona.
1988 Demography and Cultural Dynamics on the Colorado Plateaus. In *The Anasazi in a Changing Environment*, edited by G. J. Gumerman, pp. 192–229. Cambridge University Press, Cambridge.

Euler, Robert C., George J. Gumerman, Thor N. V. Karlstrom, Jeffrey S. Dean, and Richard H. Hevly
1979 The Colorado Plateaus: Cultural Dynamics and Paleoenvironment. *Science* 205: 1089–1101.

Ezzo, J. A.
1993 *Human Adaptation at Grasshopper Pueblo, Arizona: Social and Ecological Perspectives.* Archaeological Series No. 4. International Monographs in Prehistory, Ann Arbor.

Ezzo, J., and W. Deaver
1998 *Watering the Desert: Late Archaic Farming at the Costello-King Site.* Technical Series No. 68. Statistical Research, Tucson.

Ezzo, J. A., and Douglas Price
2002 Migration, Regional Reorganization, and Spatial Group Composition at Grasshopper Pueblo, Arizona. *Journal of Archaeological Science* 29:499–520.
Feldman, D. A.
1975 The History of the Relationship Between Environment and Culture in Ethnological Thought: An Overview. *Journal of the History of the Behavioral Sciences* 11:67–81.
Ferg, Alan, and Jim I. Mead
1993 *Red Cave: A Prehistoric Cave Shrine in Southeastern Arizona.* Arizona Archaeologist No. 26. Arizona Archaeological Society, Phoenix.
Fewkes, Jesse Walter
1997 *Casa Grande, Arizona.* Extract from the 28th Annual Report of the Bureau of American Ethnology. U.S. Government Printing Office, Washington, D.C., 1913. Image clarified and copyrighted by Arthur W. McGraw.
Fish, Paul R., Suzanne K. Fish, George J. Gumerman, and J. Jefferson Reid
1994 Toward an Explanation for Southwestern Abandonments. In *Themes in Southwest Prehistory,* edited by George J. Gumerman, pp. 148–163. School of American Research Press, Santa Fe.
Fish, Paul, S. Fish, A. Long, and C. Miksicek
1986 Early Corn Remains from Tumamoc Hill, Southern Arizona. *American Antiquity* 51:567–571.
Fish, Suzanne K.
1994 Archaeological Palynology of Gardens and Fields. In *The Archaeology of Garden and Field,* edited by N. F. Miller and K. L. Gleason, pp. 44–69. University of Pennsylvania Press, Philadelphia.
1995 Mixed Agricultural Technologies in Southern Arizona and Their Implications. In *Soil, Water, Biology, and Belief in Prehistoric and Traditional Southwestern Agriculture,* edited by H. W. Toll, pp. 101–116. Special Publication No. 2. New Mexico Archaeological Council, Albuquerque.
2000 Hohokam Impacts on Sonoran Desert Environments. In *Imperfect Balance: Precolumbian Landscape Transformation in the New World,* edited by David Lentz, pp. 251–280. Columbia University Press, New York.
Fish, Suzanne K., and Marcia Donaldson
1991 Production and Consumption in the Archaeological Record: A Hohokam Example. *Kiva* 56:255–275.
Fish, Suzanne K., and Paul R. Fish
1991 Comparative Aspects of Paradigms for the Neolithic Transition in the Levant and the American Southwest. In *Perspectives on the Past: Theoretical Biases in Mediterranean Hunter-Gatherer Research,* edited by G. Clark, pp. 396–410. University of Pennsylvania Press, Philadelphia.
1994 Prehistoric Desert Farmers of the Southwest. *Annual Review of Anthropology* 23:83–108.
1997 Cross-cultural Perspectives on Prehistoric Agricultural Potential in the Horseshoe Basin. In *Vanishing River, Vol. 2: Agriculture, Subsistence, and Environmental Studies,* assembled by Jeffrey A. Homberg, pp. 219–249. Statistical Research, Tucson.
Fish, Suzanne, and Paul Fish (editors)
1984 *Prehistoric Agricultural Strategies in the Southwest.* Anthropological Research Papers No. 25. Arizona State University, Tempe.
Fish, S. K., P. R. Fish, and J. H. Madsen
1992a An Introduction to Time, Place, and Research. In *The Marana Community in the Hohokam World,* edited by S. K. Fish, P. R. Fish, and J. H. Madsen, pp. 1–10. Anthropological Papers of the University of Arizona No. 56. University of Arizona Press, Tucson.
1992b Early Sedentism and Agriculture in the Northern Tucson Basin. In *The Marana Community in the Hohokam World,* edited by S. K. Fish, P. R. Fish, and J. H. Madsen, pp. 11–19. Anthropological Papers of the University of Arizona No. 56. University of Arizona Press, Tucson.

Fish, Suzanne K., Paul R. Fish, and John H. Madsen (editors)

1992 *The Marana Community in the Hohokam World.* Anthropological Papers of the University of Arizona No. 56. University of Arizona Press, Tucson.

Flannery, K.

1973 The Origins of Agriculture. *Annual Review of Anthropology* 2:271–310.

Forbes, Jack D.

1960 *Apache, Navaho, and Spaniard.* University of Oklahoma Press, Norman.

Ford, Richard

1984 Ecological Consequences of Early Agriculture in the Southwest. In *Selected Papers on the Archaeology of Black Mesa,* vol. 2, edited by S. Plog and S. Powell, pp. 127–138. Southern Illinois University Press, Carbondale.

1988 Commentary: Little Things Mean a Lot—Quantification and Qualification in Paleoethnobotany. In *Current Paleoethnobotany: Analytical Methods and Cultural Interpretations of Archaeological Plant Remains,* edited by C. A. Hastorf and V. S. Popper, pp. 215–222. University of Chicago Press, Chicago.

Forde, C. Daryll

1931 Hopi Agriculture and Land Ownership. *Journal of the Royal Anthropological Institute of Great Britain and Ireland* 61:357–407.

Fountain, John

2005 A Database of Rock Art Solar Markers. In *Current Studies in Archaeoastronomy: Conversations Across Time and Space,* edited by J. W. Fountain and R. M. Sinclair, pp. 101–108. Carolina Academic Press, Durham, North Carolina.

Fountain, John W., and Rolf M. Sinclair (editors)

2005 *Current Studies in Archaeoastronomy: Conversations Across Time and Space.* Carolina Academic Press, Durham, North Carolina.

Freeman, A.

1998 *Archaeological Investigations at the Wetlands Site, AZ AA:12–90 (ASM).* Technical Report No. 97-5. Center for Desert Archaeology, Tucson.

2000 Application of High-Resolution Alluvial Stratigraphy in Assessing the Hunter-Gatherer/Agricultural Transition in the Santa Cruz River Valley, Southeastern Arizona. *Geoarchaeology* 15:559–589.

Fritts, H. C.

1976 *Tree Rings and Climate.* Academic Press, New York.

1991 *Reconstructing Large-Scale Climatic Patterns from Tree-Ring Data: A Diagnostic Analysis.* University of Arizona Press, Tucson.

Gaines, S. W., and W. M. Gaines

2000 Impact of Small-Group Decision Making in Reducing Stress Conditions. *Journal of Anthropological Archaeology* 19:103–130.

Gasser, R. E.

1982 Anasazi Diet. In *The Coronado Project Archaeological Investigations: The Specialists' Volume; Biocultural Analyses,* compiled by R. E. Gasser, pp. 8–94. Research Paper No. 23. Museum of Northern Arizona, Flagstaff.

Gebaeur, A., and T. Price

1992 Foragers to Farmers: An Introduction. In *Transitions to Agriculture in Prehistory,* edited by A. Gebauer and T. Price, pp. 1–10. Monographs in World Archaeology No. 4. Prehistory Press, Madison, Wisconsin.

Geib, P. R.

1996 Competing Models of Pueblo II Anasazi Settlement-Subsistence Strategies in Glen Canyon. In *Glen Canyon Revisited,* edited by P. R. Geib, pp. 180–184. Anthropological Papers No. 119. University of Utah Press, Salt Lake City.

Gillespie, Susan D.

1991 Ballgames and Boundaries. In *The Mesoamerican Ballgame,* edited by V. L. Scarborough and D. R. Wilcox, pp. 317–345. University of Arizona Press, Tucson.

Gilman, Patricia A.

1987 Architecture as Artifact: Pit Structures and Pueblos in the American Southwest. *American Antiquity* 52:538–564.

Gilpin, D.

1994 Lukachukai and Salina Springs: Late Archaic/Early Basketmaker Habitation Sites in the Chinle Valley, Northeastern Arizona. *Kiva* 60:203–218.

Gladwin, Harold S.

1957 *History of the Ancient Southwest.* Bond-Wheelwight, Portland, Maine.

Gladwin, Harold S., Emil W. Haury, E. B. Sayles, and Nora Gladwin

1965 *Excavations at Snaketown: Material Culture.* Medallion Papers No. 25. Reprinted. University of Arizona Press, Tucson. Originally published 1938.

Graf, W. L.

1979 Catastrophe Theory as a Model for Change in Fluvial Systems. In *Adjustments of the Fluvial System*, edited by D. D. Rhodes and G. P. Williams, pp. 13–32. Kendall/Hunt, Dubuque, Iowa.

1981 Channel Instability in a Sand-Bed River. *Water Resources Research* 17:1087–1094.

1983 Flood-Related Channel Change in an Arid-Region River. *Earth Surface Processes and Landforms* 8:125–139.

1988a Definition of Flood Plains along Arid-Region Rivers. In *Flood Geomorphology*, edited by V. R. Baker, R. C. Kochel, and P. C. Patton, pp. 231–242. Wiley, New York.

1988b *Fluvial Processes in Dryland Rivers.* Springer, New York.

Graves, M. W.

1982 Anomalous Tree-Ring Dates and the Sequence of Room Construction at Canyon Creek Ruin, East-Central Arizona. *The Kiva* 47:107–131.

1983 Growth and Aggregation at Canyon Creek Ruin: Implications for Evolutionary Change in East-Central Arizona. *American Antiquity* 48:290–315.

1991 Estimating Ring Loss on Tree-Ring Specimens from East-Central Arizona: Implications for Prehistoric Pueblo Growth at Grasshopper Ruin. *Journal of Quantitative Anthropology* 3:83–115.

Graybill, D. A.

1979 Revised Computer Programs for Tree-Ring Research. *Tree-Ring Bulletin* 39:77–82.

1982 Chronology Development and Analysis. In *Climate from Tree Rings*, edited by M. K. Hughes, P. M. Kelly, J. R. Pilcher, and V. C. Lamarche, pp. 21–30. Cambridge University Press, London.

1985 Paleoclimate of the Hohokam Area: Problems and Prospects. In *Proceedings of the 1983 Hohokam Symposium, Part I*, edited by A. E. Dittert, Jr., and D. E. Dove, pp. 29–44. Occasional Paper No. 2. Arizona Archaeological Society, Phoenix.

1987 A Network of High-Elevation Conifers in the Western United States for Detection of Tree-Ring Growth Response to Increasing Carbon Dioxide. In *Proceedings of the International Symposium on Ecological Aspects of Tree-Ring Analysis*, compiled by G. C. Jacoby and J. W. Hornbeck, pp. 463–474. U.S. Department of Commerce, Springfield, Virginia.

1989 The Reconstruction of Prehistoric Salt River Stream-Flow. In *The 1982–1984 Excavations at Las Colinas: Environment and Subsistence*, by D. A. Graybill, D. A. Gregory, F. L. Nials, S. K. Fish, R. E. Gasser, C. H. Miksicek, and C. R. Szuter, pp. 25–38. Arizona State Museum Archaeological Series 162, Vol. 5. University of Arizona, Tucson.

Graybill, Donald A., and David E. Doyel

1980 Computer Seriation of Sites in Dead Valley. In *Prehistory in Dead Valley, East-Central Arizona: The TG&E Springerville Project*, edited by D. E. Doyel and S. S. Debowski, pp. 235–304. Arizona State Museum Archaeological Series 144. University of Arizona, Tucson.

Graybill, D. A., and D. A. Gregory

1989 Introduction. In *The 1982–1984 Excavations at Las Colinas: Environment and*

Subsistence, by D. A. Graybill, D. A. Gregory, F. L. Nials, S. K. Fish, R. E. Gasser, C. H. Miksicek, and C. R. Szuter, pp. 1–4. Arizona State Museum Archaeological Series 162, Vol. 5. University of Arizona, Tucson.

Graybill, D. A., D. A. Gregory, F. L. Nials, S. K. Fish, R. E. Gasser, C. H. Miksicek, and C. R. Szuter

1989 *The 1982–1984 Excavations at Las Colinas: Environment and Subsistence.* Arizona State Museum Archaeological Series 162, Vol. 5. University of Arizona, Tucson.

Graybill, D. A., and D. Meko

1995 Tree-Ring Reconstruction of Upper Gila River Discharge. *Water Resources Bulletin* 31(4):605–616.

Graybill D. A., and F. L. Nials

1989 Aspects of Climate, Streamflow, and Geomorphology Affecting Irrigation Systems in the Salt River Valley. In *The 1982–1984 Excavations at Las Colinas: Environment and Subsistence*, by D. A. Graybill, D. A. Gregory, F. L. Nials, S. K. Fish, R. E. Gasser, C. H. Miksicek, and C. R. Szuter, pp. 5–24. Arizona State Museum Archaeological Series 162, Vol. 5. University of Arizona, Tucson.

Graybill, D. A., and M. R. Rose

1992 Coniferous Forests of Arizona and New Mexico. In *Effects of Air Pollution on Coniferous Forests of the Western United States,* edited by R. K. Olson, D. Binkley, and M. Bohm, pp. 403–431. Springer, New York.

Green, C. R., and W. D. Sellers

1964 *Arizona Climate.* University of Arizona Press, Tucson.

Greenleaf, J. Cameron

1975 The Fortified Hill Site near Gila Bend, Arizona. *The Kiva* 40:213–282.

Greenwald, D. H., and K. C. Anderson

1994 The Phoenix Sky Harbor Center Canals. In *The Sky Harbor Project: Early Desert Farming and Irrigation Settlements; Archaeological Investigations in the Phoenix Sky Harbor Center, Vol. 4: Special Studies, Synthesis, and Conclusions*, edited by D. H. Greenwald and J. H. Ballagh, pp. 77–106. Anthropological Research Paper No. 4. SWCA, Flagstaff.

Gregory, David A.

1975 Defining Variability in Prehistoric Settlement Morphology. *Fieldiana: Anthropology* 65:40–46.

1991 Form and Variation in Hohokam Settlement Patterns. In *Chaco and Hohokam: Prehistoric Regional Systems in the American Southwest*, edited by P. L. Crown and W. J. Judge, pp. 159–194. School of American Research Press, Santa Fe.

1995a Prehistoric Settlement Patterns in the Eastern Tonto Basin. In *The Roosevelt Community Development Study: New Perspectives on Tonto Basin Prehistory*, edited by M. D. Elson, M. T. Stark, and D. A. Gregory, pp. 127–184. Anthropological Papers No. 15. Center for Desert Archaeology, Tucson.

1995b Prehistoric Site Variability: Typology, Chronology, and Location. In *The San Carlos Reservoir Cultural Resources Survey*, edited by A. T. Black and M. Green, pp. 51–121. Cultural Resources Report No. 87. Archaeological Consulting Services, Tempe.

Gregory, David (editor)

2001 *The Early Agricultural Occupation at Los Pozos (AZAA:12:91, ASM).* Anthropological Papers No. 21. Center for Desert Archaeology, Tucson.

Gregory, D. A., and M. D. Elson

1995 Tonto Basin Chronology. In *The Roosevelt Community Development Study: New Perspectives on Tonto Basin Prehistory*, edited by M. D. Elson, M. T. Stark, and D. A. Gregory, pp. 61-78. Anthropological Papers No. 15. Center for Desert Archaeology, Tucson.

Gregory, D. A., and G. Huckleberry

1994 *An Archaeological Survey in the Blackwater Area, Vol. 1: A History of Human Settlement in the Blackwater Area.* Cultural Resources Report No. 86. Archaeological Consulting Services, Tempe.

Gregory, D. A., and F. L. Nials

1985 Observations Concerning the Distribution of Classic Period Hohokam Platform Mounds. In *Proceedings of the 1983 Hohokam Symposium, Part I*, edited by A. E. Dittert, Jr., and D. E. Dove, pp. 373–388. Occasional Paper No. 2. Arizona Archaeological Society, Phoenix.

Griffin-Pierce, Trudy

1992 *Earth Is My Mother, Sky Is My Father: Space, Time, and Astronomy in Navajo Sandpainting.* University of New Mexico Press, Albuquerque.

Grissino-Mayer, Henri D.

1996 A 2,129-Year Annual Reconstruction of Precipitation for Northwestern New Mexico, USA. In *Tree-Rings, Environment, and Humanity: Proceedings of the International Conference, Tucson, Arizona, 17–21 May 1994*, edited by Jeffrey S. Dean, David M. Meko, and Thomas W. Swetnam, pp. 191–204. Radiocarbon, Tucson.

Grissino-Mayer, Henri D., Christopher H. Baisan, and Thomas W. Swetnam

1997 *A 1,373-Year Reconstruction of Annual Precipitation for the Southern Rio Grande Basin.* Final report submitted to the Directorate of Environment, Natural Resources Division, Fort Bliss, Texas, for the Legacy Program. Laboratory of Tree-Ring Research, University of Arizona, Tucson.

Gumerman, G. J.

1984 *A View from Black Mesa: The Changing Face of Archaeology.* University of Arizona Press, Tucson.

1988 A Historical Perspective on Environment and Culture in Anasazi Country. In *The Anasazi in a Changing Environment*, edited by G. J. Gumerman, pp. 1–24. Cambridge University Press, Cambridge.

2001 Southwestern Foodways: Beyond Nutrition. In *Examining the Course of Southwest Archaeology: The Durango Conference, September 1995*, edited by D. A. Phillips, Jr., and Lynne Sebastian, pp. 79–94. Special Publication No. 3. New Mexico Archaeological Council, Albuquerque.

Gumerman, G. J. (editor)

1988 *The Anasazi in a Changing Environment.* Cambridge University Press, Cambridge.

1991 *Exploring the Hohokam: Prehistoric Desert Peoples of the American Southwest.* Amerind Foundation, Dragoon, Arizona, and University of New Mexico Press, Albuquerque.

Gumerman, G. J., and J. S. Dean

1989 Prehistoric Cooperation and Competition in the Western Anasazi Area. In *Dynamics of Southwest Prehistory*, edited by L. S. Cordell and G. J. Gumerman, pp. 99–148. School of American Research, Santa Fe.

Gumerman, G. J., and M. Gell-Mann (editors)

1994 *Understanding Complexity in the Prehistoric Southwest.* Addison-Wesley, Reading, Massachusetts.

Gumerman, George J., Alan C. Swedlund, Jeffrey S. Dean, and Joshua M. Epstein

2003 The Evolution of Social Behavior in the Prehistoric American Southwest. *Artificial Life* 9:435–444.

Gunn, Joel D.

2000 *The Years Without Summer: Tracing A.D. 536 and Its Aftermath.* BAR International Series 872. Archaeopress, Oxford.

Haas, Jonathan, and Winifred Creamer

1993 *Stress and Warfare Among the Kayenta Anasazi of the Thirteenth Century A.D.* Fieldiana Anthropology, New Series, No. 21, Publication 1450. Field Museum of Natural History, Chicago.

Hack, John T.

1942 *The Changing Physical Environment of the Hopi Indians of Arizona.* Papers of the Peabody Museum of American Archaeology and Ethnology, Vol. 35, No. 1. Harvard University, Cambridge.

1945 Recent Geology of the Tsegi Canyon. In *Archaeological Studies in Northeast Arizona*, edited by Ralph L. Beals, George W. Brainerd, and Watson Smith, pp. 151–158.

University of California Publications in American Archaeology and Ethnology, Vol. 44, No. 1. University of California Press, Berkeley.

Hackbarth, M. R.

1992 Archaeological Investigations at La Escuela Cuba, AZ U:6:213 (ASM). *In Prehistoric and Historic Occupation of the Lower Verde Valley: The State Route 87 Verde Bridge Project*, edited by M. R. Hackbarth, pp. 489–501. Northland Research, Flagstaff.

Hackett, Charles W. (editor)

1937 *Historical Documents Relating to New Mexico, Nueva Vizcaya, and Approaches Thereto, to 1733*. Vol. III. Carnegie Institution of Washington, Publication No. 330. Washington, D.C.

Halbirt, C. D., and S. G. Dosh

1986 Archaeological Investigations: White Mountain Apache Indian Tribe, Tribal Development Office, Canyon Day Wastewater Treatment Facility, Fort Apache Indian Reservation, Gila and Navajo Counties, Arizona. Ms. on file, Museum of Northern Arizona, Flagstaff.

Hall, E. T.

1942 *Archaeological Survey of Walhalla Glades*. Bulletin No. 20. Museum of Northern Arizona, Flagstaff.

Hanson, R. T., J. S. McLean, and R. S. Miller

1994 *Hydrogeologic Framework and Preliminary Simulation of Ground-Water Flow in the Mimbres Basin, Southwestern New Mexico*. U.S. Geological Survey, Water-Resources Investigations Report No. 94-4011. Albuquerque.

Hantman, Jeffrey

1989 Surplus Production and Complexity in the Upper Little Colorado Province, East-Central Arizona. In *The Socio-Political Structure of Prehistoric Southwestern Societies*, edited by Steadman Upham, Kent Lightfoot, and Roberta Jewett, pp. 419–445. Westview Press, Boulder, Colorado.

Hard, R. J., R. P. Mauldin, and G. R. Raymond

1996 Mano Size, Stable Carbon Isotope Ratios, and Macrobotanical Remains as Multiple Lines of Evidence of Maize Dependence in the American Southwest. *Journal of Archaeological Method and Theory* 3:253–318.

Hard, R. J., and J. R. Roney

1998 A Massive Terraced Village Complex in Chihuahua, Mexico, 3000 Years Before Present. *Science* 279:1661–1664.

1999 Cerro Juanaqueña. *Archaeology Southwest* 13(1):4–5.

Harrington, John Peabody

1916 *The Ethnogeography of the Tewa Indians*. 29th Annual Report of the Bureau of American Ethnology, 1907–1908. Washington, D.C.

Harris, M.

1979 *Cultural Materialism: The Struggle for a Science of Culture*. Random House, New York.

Hartmann, William K.

1991 *Astronomy: The Cosmic Journey*. Wadsworth, Belmont, California.

Hassan, Fekri A.

1978 Demographic Archaeology. In *Advances in Archaeological Method and Theory*, Vol. 1, edited by Michael B. Schiffer, pp. 49–103. Academic Press, New York.

Hastings, J., and R. Humphrey

1969 *Climatological Data and Statistics for Sonora and Northern Sinaloa*. Technical Reports on the Meteorology and Climatology of Arid Regions No. 19. University of Arizona Institute of Atmospheric Physics, Tucson.

Hastorf, C. A.

1988 The Use of Paleoethnobotanical Data in Prehistoric Studies of Crop Production, Processing, and Consumption. In *Current Paleoethnobotany: Analytical Methods and Cultural Interpretations of Archaeological Plant Remains*, edited by C. A. Hastorf and V. S. Popper, pp. 119–144. University of Chicago Press, Chicago.

Haug, G. H., D. Gunther, L. C. Peterson, D. M. Sigman, D. A. Hughen, and B. Aeschlimann
2003 Climate and the Collapse of Maya Civilization. *Science* 299:1731–1735.
Haury, Emil W.
1931 *Kivas of the Tusayan Ruin, Grand Canyon, Arizona.* Medallion Papers No. 9. Gila Pueblo, Globe, Arizona.
1932 *Roosevelt:9:6, a Hohokam Site of the Colonial Period.* Medallion Papers No. 11. Gila Pueblo, Globe, Arizona.
1934 *The Canyon Creek Ruin and the Cliff Dwellings of the Sierra Ancha.* Medallion Papers No. 14. Gila Pueblo, Globe, Arizona.
1936 *The Mogollon Culture of Southwestern New Mexico.* Medallion Papers No. 20, Gila Pueblo, Globe, Arizona.
1945 *The Excavation of Los Muertos and Neighboring Ruins in the Salt River Valley, Southern Arizona.* Papers of the Peabody Museum of American Archaeology and Ethnology, Vol. 24, No. 1. Harvard University, Cambridge.
1962 The Greater American Southwest. In *Courses Toward Urban Life,* edited by R. J. Braidwood and G. R. Willey, pp. 106–131. Aldine, Chicago.
1967 The Hohokam, First Masters of the American Desert. *National Geographic* 131(5):670–695.
1976 *The Hohokam: Desert Farmers and Craftsmen.* University of Arizona Press, Tucson.
Haynes, C. Vance, Jr., and Bruce B. Huckell
1986 *Sedimentary Successions of the Prehistoric Santa Cruz River, Tucson, Arizona.* Open-File Report. Arizona Bureau of Mines and Geology, Tucson.
Hays-Gilpin, Kelley, and Jane H. Hill
1999 The Flower World in Material Culture: An Iconographic Complex in the Southwest and Mesoamerica. *Journal of Anthropological Research* 55:1–37.
Hegmon, Michelle, Margaret C. Nelson, Roger Anyon, Darrell Creel, Steven A. LeBlanc, and Harry J. Shafer
1999 Systematics of Late and Post Mimbres Occupations in the American Southwest. *Kiva* 65:143–166.
Hegmon, Michelle, Margaret Nelson, and Susan Ruth
1998 Abandonment, Reorganization, and Social Change: Analysis of Pottery and Architecture from the Mimbres Region of the American Southwest. *American Anthropologist* 100:148–162.
Henderson, T. Kathleen
1989a The Chronology of the Las Acequias Canals. In *Prehistoric Agricultural Activities on the Lehi-Mesa Terrace: Perspectives on Hohokam Irrigation Cycles,* edited by N. W. Ackerly and T. K. Henderson. Northland Research, Flagstaff.
1989b Dating the Tempe Outer Loop Features. In *Prehistoric Agricultural Activities on the Lehi-Mesa Terrace: Excavations at La Cuenca del Sedimento,* edited by N. W. Ackerly and T. K. Henderson. Northland Research, Flagstaff.
Hendricks, D. M.
1985 *Arizona Soils.* University of Arizona, Tucson.
Herrington, LaVerne
1979 Settlement Patterns and Water Control Systems of the Mimbres Classic Phase, Grant County, New Mexico. Unpublished Ph.D. dissertation, University of Texas, Austin.
1982 Water-Control Systems of the Mimbres Classic Phase. In *Mogollon Archaeology: Proceedings of the 1980 Mogollon Conference,* edited by Patrick Beckett, pp. 75–90. Acoma Books, Ramona, California.
Hetherington, Barry
1996 *A Chronicle of Pre-telescopic Astronomy.* Wiley, New York.
Hewett, Edgar L., and Bertha P. Dutton
1945 *The Pueblo Indian World.* University of New Mexico Press, Albuquerque.
Hickerson, Nancy Parrott
1994 *The Jumanos: Hunters and Traders of the South Plains.* University of Texas Press, Austin.

Hill, Jane H.

1992 The Flower World of the Old Uto-Aztecan. *Journal of Anthropological Research* 48:117–144.

Hinkes, M. J.

1983 Skeletal Evidence of Stress in Subadults: Trying to Come of Age at Grasshopper Pueblo. Unpublished Ph.D. dissertation, University of Arizona, Tucson.

Hirschboeck, K. K.

1985 Hydroclimatology of Flow Events in the Gila River Basin, Central and Southern Arizona. Unpublished Ph.D. dissertation, Department of Geosciences, University of Arizona, Tucson.

Ho Peng-Yoke

1962 Ancient and Mediaeval Observations of Comets and Novae in Chinese Sources. *Vistas in Astronomy* 5:127–225.

Ho Peng-Yoke and Ang Tian-Se

1970 Chinese Astronomical Records on Comets and "Guest Stars." *Oriens Extremus* 1/2: 63–99.

Hodge, Frederick Webb, George P. Hammond, and Agapito Rey

1945 *Fray Alonso de Benavides' Revised Memorial of 1634.* University of New Mexico Press, Albuquerque.

Holbrook, S. J., and M. W. Graves

1982 Modern Environment of the Grasshopper Region. In *Multidisciplinary Research at Grasshopper Pueblo, Arizona,* edited by W. A. Longacre, S. J. Holbrook, and M. W. Graves, pp. 5–11. Anthropological Papers of the University of Arizona No. 40. University of Arizona Press, Tucson.

Holmes, Richard L.

1983 Computer-Assisted Quality Control in Tree-Ring Dating and Measurement. *Tree-Ring Bulletin* 43:69–78.

Homburg, J.

1992 Soil Fertility Study. In *Archaeological Investigations at Lee Canyon: Kayenta Anasazi Farmsteads in the Upper Basin, Coconino County, Arizona,* edited by S. Whittlesey, pp. 145–161. Technical Series No. 38. Statistical Research, Tucson.

Homburg, Jeffrey A., and Richard Ciolek-Torrello (editors)

1997 *Vanishing River: Landscapes and Lives of the Lower Verde Valley, Vol. 2: Agricultural, Subsistence, and Environmental Studies.* Statistical Research, Tucson.

Hopkins, R. L.

1990 Kaibab Formation. In *Grand Canyon Geology,* edited by S. S. Beus and M. Morales, pp. 225–245. Oxford University Press, Oxford.

Hoskinson, Tom

1992 Saguaro Wine, Ground Figures, and Power Mountains: Investigations at Sears Point, Arizona. In *Earth and Sky: Visions of the Cosmos in Native American Folklore,* edited by R. A. Williamson and C. Farrer, pp. 131–161. University of New Mexico Press, Albuquerque.

2005 Calendric Investigations of a Complex Petroglyph Panel at a Gila River Archaeological Site in Arizona. In *Current Studies in Archaeoastronomy: Conversations Across Time and Space,* edited by J. W. Fountain and R. M. Sinclair, pp. 169–179. Carolina Academic Press, Durham, North Carolina.

Hough, W.

1930 Ancient Pueblo Subsistence. In *Proceedings of the 23rd International Congress of Americanists,* pp. 67–69. New York.

Howard, J. B.

1993 A Paleohydraulic Approach to Examining Agricultural Intensification in Hohokam Irrigation Systems. *Research in Economic Anthropology Supplement* 7:261–322.

Howard, J. B., and G. Huckleberry

1991 *The Operation and Evolution of an Irrigation System: The East Papago Canal Study.* Publications in Archaeology No. 18. Soil Systems, Phoenix.

Huckell, B. B.
1990 Late Preceramic Farmer-Foragers in Southeastern Arizona: A Cultural and Ecological Consideration of the Spread of Agriculture in the Arid Southwestern United States. Unpublished Ph.D. dissertation, University of Arizona, Tucson.
1995 *Of Marshes and Maize: Preceramic Agricultural Settlements in the Cienega Valley, Southeastern Arizona.* Anthropological Papers of the University of Arizona No. 59. University of Arizona Press, Tucson.
Huckell, B. B., and L. W. Huckell
1984 Excavations at Milagro, a Late Archaic Site in the Eastern Tucson Basin. Ms. on file, Arizona State Museum, University of Arizona, Tucson.
1999 McEuen Cave. *Archaeology Southwest* 15:12.
Huckell, L. W.
1992 Plant Remains. In *Archaeological Investigations at Lee Canyon: Kayenta Anasazi Farmsteads in the Upper Basin, Coconino County, Arizona*, edited by S. Whittlesey, pp. 119–133. Technical Series No. 38. Statistical Research, Tucson.
Huckleberry, G. A.
1993 Late-Holocene Stream Dynamics on the Middle Gila River, Pinal County, Arizona. Unpublished Ph.D. dissertation, Department of Geosciences, University of Arizona, Tucson.
1994 Archaeological Implications of Late Holocene Channel Changes on the Middle Gila River, Arizona. *Geoarchaeology* 10:159–182.
Huckleberry, G. A., and B. R. Billman
1998 Floodwater Farming, Discontinuous Ephemeral Streams, and Puebloan Abandonment in Southwestern Colorado. *American Antiquity* 63:595–616.
Hudson, Dee T.
1973 A Meteorite Discovery with Archaeological Importance from the Camp Verde Area, Central Arizona. *Plateau* 45:41–43.
Hudson, Travis, and Ernest Underhay
1978 *Crystals in the Sky: An Intellectual Odyssey Involving Chumash Astronomy, Cosmology, and Rock Art.* Anthropological Papers No. 10. Ballena Press, Santa Barbara.
Hughes, M. K., P. M. Kelly, J. R. Pilcher, and V. C. Lamarche (editors)
1982 *Climate from Tree Rings.* Cambridge University Press, London.
Huntington, Ellsworth
1914 *The Climatic Factor in Arid America.* Carnegie Institution of Washington, Publication No. 192. Washington, D.C.
Huntoon, P. W.
1990 Planerozoic Structural Geology of the Grand Canyon. In *Grand Canyon Geology*, edited by S. S. Beus and M. Morales, pp. 261–309. Oxford University Press, Oxford.
Hutira, J.
1986 Analysis of Plants, Pollen, and Coprolites. In *A Cross Section of Grand Canyon Archaeology: Excavations at Five Sites Along the Colorado River*, edited by A. T. Jones, pp. 269–332. Publications in Anthropology No. 28. Western Archeological and Conservation Center, Tucson.
Imoto, Susumu, and Ichiro Hasegawa
1958 Historical Records of Meteor Showers in China, Korea, and Japan. *Smithsonian Contributions to Astrophysics* 2(6):131–143.
Irwin-Williams, C.
1967 Picosa: The Elemental Southwestern Culture. *American Antiquity* 32:441–455.
1973 *The Oshara Tradition: Origins of Anasazi Culture.* Contributions in Anthropology Vol. 5, No. 1. Eastern New Mexico University, Portales.
1979 Post-Pleistocene Archaeology, 7000-2000 B.C. In *Southwest*, edited by A. Ortiz, pp. 31–42. Handbook of North American Indians, Vol. 9, William C. Sturtevant, general editor. Smithsonian Institution Press, Washington, D.C.
Jeter, Marvin D.
1980 Analysis of Dead Valley Lithic Assemblages. In *Prehistory in Dead Valley, East-Central*

Arizona: The TG&E Springerville Project, edited by D. E. Doyel and S. S. Debowski, pp. 235–304. Arizona State Museum Archaeological Series 144. University of Arizona, Tucson.

John, Elizabeth A. H.

1981 *Storms Brewed in Other Men's Worlds: The Confrontation of Indians, Spanish, and French in the Southwest, 1540–1795*. Reprinted. University of Nebraska Press, Lincoln. Originally published 1975, Texas A&M University Press, College Station.

Jones, A. T.

1986 Spatial and Temporal Variations in Grand Canyon Subsistence and Technology. *Western Anasazi Reports* 3:260-272.

Jorde, Lynn B.

1977 Precipitation Cycles and Cultural Buffering in the Prehistoric Southwest. In *For Theory Building in Archaeology*, edited by Lewis R. Binford, pp. 385–396. Academic Press, New York.

Judd, N. M.

1926 *Archeological Investigations North of the Rio Colorado*. Bureau of American Ethnology Bulletin No. 82. Smithsonian Institution, Washington, D.C.

Judge, W. J.

1982 The Paleo-Indian and Basketmaker Periods: An Overview and Some Research Problems. In *The San Juan Tomorrow*, edited by F. Plog and W. Wait, pp. 5–57. National Park Service, Southwest Region, and School of American Research, Santa Fe.

Karlstrom, Eric T.

1983 Soils and Geomorphology of Northern Black Mesa. *In Excavations on Black Mesa, 1981: A Descriptive Report*, edited by F. E. Smiley, Deborah L. Nichols, and Peter P. Andrews, pp. 317–342. Center for Archaeological Investigations Research Paper No. 36. Southern Illinois University, Carbondale.

Karlstrom, Eric T., and Thor N. V. Karlstrom

1986 Late Quaternary Alluvial Stratigraphy and Soils of the Black Mesa–Little Colorado River Areas, Northern Arizona. In *Geology of Central and Northern Arizona, Geological Society of America Rocky Mountain Section Guidebook*, edited by J. D. Nations, C. M. Conway, and G. A. Swann, pp. 71–92. Flagstaff.

Karlstrom, Thor N. V.

1988 Alluvial Chronology and Hydrologic Change of Black Mesa and Nearby Regions. In *The Anasazi in a Changing Environment*, edited by George J. Gumerman, pp. 45–91. Cambridge University Press, Cambridge.

Kauffman, William J., III

1993 *Discovering the Universe*. 3rd ed. W. H. Freeman, New York.

Keber, Eloise Quiñones

1995 *Codex Telleriano-Remensis: Ritual, Divination, and History in a Pictorial Aztec Manuscript*. University of Texas Press, Austin.

Kehoe, A.

1981 Revisionist Anthropology: Aboriginal North America. *Current Anthropology* 22:503–517.

Kelso, Gerald K.

1980 Palynology and Human Paleoecology in Dead Valley. In *Prehistory in Dead Valley, East-Central Arizona: The TG&E Springerville Project*, edited by D. E. Doyel and S. S. Debowski, pp. 349–370. Arizona State Museum Archaeological Series 144. University of Arizona, Tucson.

Kessell, John L.

1979 *Kiva, Cross, and Crown: The Pecos Indians and New Mexico, 1540–1840*. National Park Service, U.S. Department of the Interior, Washington, D.C.

King, F. B.

1994 Interpreting Wild Plant Foods in the Archaeological Record. In *Eating on the Wild Side*, edited by N. L. Etkin, pp. 185–209. University of Arizona Press, Tucson.

Kintigh, Keith

1996 The Cibola Region in the Post-Chaco Era. In *The Prehistoric Pueblo World, A.D. 1150–1350*, edited by Michael A. Adler, pp. 131–144. University of Arizona Press, Tucson.

Kirch, Patrick V.

1980 The Archaeological Study of Adaptation: Theoretical and Methodological Issues. In *Advances in Archaeological Method and Theory*, Vol. 3, edited by Michael B. Schiffer, pp. 101–156. Academic Press, New York.

Kirch, P., and T. L. Hunt (editors)

1997 *Historical Ecology in the Pacific Islands: Prehistoric Environmental and Landscape Changes.* Yale University Press, New Haven.

Kohl, P. L.

1981 Materialist Approaches in Prehistory. *Annual Review of Anthropology* 10:89–118.

Kohler, M. A.

1949 On the Use of Double-Mass Analysis for Testing the Consistency of Meteorological Records and for Making Required Adjustments. *Bulletin of the American Meteorological Society* 30:188–189.

Kohler, Timothy A.

1993 News from the North American Southwest: Prehistory on the Edge of Chaos. *Journal of Archaeological Research* 1:267–321.

2000 Putting Social Sciences Together Again: An Introduction to the Volume. In *Dynamics in Human and Primate Societies: Agent-Based Modeling of Social and Spatial Processes*, edited by T. A. Kohler and G. J. Gumerman, pp. 1–18. Santa Fe Institute Studies in the Sciences of Complexity. Oxford University Press, New York.

Kohler, T. A., J. Kresl, C. Van West, E. Carr, and R. H. Wilshusen

2000 Be There Then: A Modeling Approach to Settlement Determinants and Spatial Efficiency Among Late Ancestral Pueblo Populations of the Mesa Verde Region, U.S. Southwest. In *Dynamics in Human and Primate Societies: Agent-Based Modeling of Social and Spatial Processes*, edited by T. A. Kohler and G. J. Gumerman, pp. 145–178. Santa Fe Institute Studies in the Sciences of Complexity. Oxford University Press, New York.

Kronk, Gary W.

1999 *Cometography: A Catalog of Comets, Vol. 1: Ancient–1799.* Cambridge University Press, Cambridge.

Krupp, E. C.

1997 *Skywatchers, Shamans, and Kings: Astronomy and the Archaeology of Power.* Wiley, New York.

Lange, Richard C.

1996 The Little Colorado River, Farming, and Prehistory in the Homol'ovi Area. In *River of Change: Prehistory of the Middle Little Colorado River Valley, Arizona*, edited by E. Charles Adams, pp. 239–257. Arizona State Museum Archaeological Series 185. University of Arizona, Tucson.

Lanner, R. M.

1981 *The Piñon Pine.* University of Nevada Press, Reno.

Lansing, J. S.

1991 *Priests and Programmers: Technologies of Power in the Engineered Landscape of Bali.* Princeton University Press, Princeton.

LeBlanc, Steven A.

1983 *The Mimbres People: Ancient Pueblo Painters of the American Southwest.* Thames and Hudson, London.

1989 Cultural Dynamics in the Southern Mogollon Area. In *Dynamics of Southwestern Prehistory*, edited by Linda S. Cordell and George J. Gumerman, pp. 179–207. School of American Research, Santa Fe.

1999 *Prehistoric Warfare in the American Southwest.* University of Utah Press, Salt Lake City.

Lekson, Stephen

1992 The Surface Archaeology of Southwestern New Mexico. *The Artifact* 30(3):1–36.

1999 *The Chaco Meridian: Centers of Political Power in the Ancient Southwest.* AltaMira Press, Walnut Creek, California.

León-Portilla, Miguel

1963 *Aztec Thought and Culture.* University of Oklahoma Press, Norman.

1992 *The Aztec Image of Self and Society: An Introduction to Nahua Culture.* University of Utah Press, Salt Lake City.

1995 On the Meanings of the Celestial Bodies in Pre-Hispanic Mexico. In *The Inspiration of Astronomical Phenomena,* edited by G. V. Coyne, S.J., and R. M. Sinclair. *Vistas in Astronomy* 39:451–461.

Leonard, R. D., and H. E. Reed

1993 Population Aggregation in the Prehistoric American Southwest: A Selectionist Model. *American Antiquity* 58:648–661.

Leopold, L. B., and M. G. Wolman

1957 *River Channel Patterns: Braided, Meandering, and Straight.* U.S. Geological Survey Professional Paper No. 282-B, 39-85. Washington, D.C.

Lesser, Alexander

1933 *The Pawnee Ghost Dance Hand Game.* Contributions to Anthropology 16. Columbia University, New York.

Levy, Jerrold

1992 *Orayvi Revisited: Social Stratification in an "Egalitarian" Society.* School of American Research Press, Santa Fe.

Leyenaar, Ted J. J.

1992 Ulama, the Survival of the Mesoamerican Ballgame, *Ullamalitzli. Kiva* 58:115–153.

Lightfoot, Kent

1978 Population Movements and Social Interaction in the Prehistoric Southwest: An Experiment from Springerville, Arizona. In *An Analytical Approach to Cultural Resource Management: The Little Colorado Planning Unit,* edited by Fred Plog, pp. 188–200. Anthropological Research Papers No. 13. Arizona State University, Tempe.

Lightfoot, Kent, and Fred Plog

1984 Intensification Along the North Side of the Mogollon Rim. In *Prehistoric Agricultural Strategies in the Southwest,* edited by Suzanne Fish and Paul Fish, pp. 179–196. Anthropological Research Papers No. 25. Arizona State University, Tempe.

Lindauer, Owen

1984 Sherd Areas as Loci of Subsistence in the Middle Gila. In *Prehistoric Agricultural Strategies in the Southwest,* edited by Paul R. Fish and Suzanne K. Fish, pp. 73–84. Anthropological Research Papers No. 25. Arizona State University, Tempe.

1992 Centralized Storage: Evidence from a Salado Platform Mound. In *Developing Perspective on Tonto Basin Prehistory,* edited by C. L. Redman, G. E. Rice, and K. E. Pedrick, pp. 33–44. Anthropological Field Studies No. 26. Office of Cultural Resource Management, Department of Anthropology, Arizona State University, Tempe.

1995 *Where the Rivers Converge: Roosevelt Platform Mound Study; Report on the Rock Island Complex.* Roosevelt Monograph Series No. 4, Anthropological Field Studies No. 33. Office of Cultural Resource Management, Department of Anthropology, Arizona State University, Tempe.

1996 *The Place of the Storehouses: Roosevelt Platform Mound Study; Report on Schoolhouse Point Mound, Pinto Creek Complex.* Roosevelt Monograph Series No. 6, Anthropological Field Studies No. 35. Office of Cultural Resource Management, Department of Anthropology, Arizona State University, Tempe.

Lindsay, Alexander J., Jr.

1969 *The Tsegi Phase of the Kayenta Cultural Tradition in Northeastern Arizona.* Ph.D. dissertation, University of Arizona. University Microfilms International, Ann Arbor.

1987 Anasazi Population Movements to Southern Arizona. *American Archaeology* 6(3): 190–198.

Lipe, W. D., and R. G. Matson

1975 Archaeology and Alluvium in the Grand Gulch–Cedar Mesa Area, Southeastern Utah. In *Canyonlands Country*, edited by James E. Fassett, pp. 67–71. Four Corners Geological Society, Durango, Colorado.

Littmann, Mark

1998 *Heavens on Fire: The Great Leonid Meteor Storms.* Cambridge University Press, Cambridge.

Littmann, Mark, Ken Willcox, and Fred Espenak

1999 *Totality: Eclipses of the Sun.* 2nd ed. Oxford University Press, New York.

Lloyd, J. William

1911 *Aw-aw-tam Indian Nights, Being the Myths and Legends of the Pimas of Arizona.* Lloyd Group, Westfield, New Jersey.

Longacre, William A.

1964 A Synthesis of Upper Little Colorado Prehistory. *Fieldiana: Anthropology* 55:201–216. Chicago.

1975 Population Dynamics at the Grasshopper Pueblo, Arizona. *American Antiquity* 40:71–74.

1976 Population Dynamics at the Grasshopper Pueblo, Arizona. In *Demographic Anthropology: Quantitative Approaches,* edited by E. B. W. Zubrow, pp. 169–184. University of New Mexico Press, Albuquerque.

López Austin, Alfredo

1996 *The Rabbit on the Face of the Moon: Mythology in the Mesoamerican Tradition,* translated by B. R. Ortiz de Montellano and T. Ortiz de Montellano. University of Utah Press, Salt Lake City.

Lorentzen, L. H.

1993 From Atlatl to Bow: The Impact of Improved Weapons on Wildlife in the Grasshopper Region. Master's paper, University of Arizona, Tucson.

Lucas, Jason

1996 Three Circle Phase Architecture at Old Town, a Prehistoric Mimbres Site in Luna County, Southwestern New Mexico. Master's thesis, University of Texas, Austin.

Lyneis, M. M.

1996 Pueblo II–Pueblo III Change in Southwestern Utah, the Arizona Strip, and Southern Nevada. In *The Prehistoric Pueblo World, A.D. 1150–1350,* edited by M. A. Adler, pp. 11–28. University of Arizona Press, Tucson.

Mabry, J.

1990 *A Late Archaic Occupation at AZ AA:12:105(ASM).* Technical Report No. 90-6. Center for Desert Archaeology, Tucson.

1997 Chapter 1—Introduction. In *Archaeological Investigations of Early Village Sites in the Middle Santa Cruz Valley—Descriptions of the Santa Cruz Bend, Square Hearth, Stone Pipe, and Canal Sites,* by Jonathan Mabry, D. Swartz, H. Wöcherl, J. Clark, G. Archer, and M. Lindeman, pp. 1–6. Anthropological Papers No. 18. Center for Desert Archaeology, Tucson.

1999 Las Capas and Early Irrigation Farming. *Archaeology Southwest* 13(1):14.

2000a The Red Mountain Phase and the Origins of Hohokam Villages. In *The Hohokam Village Revisited,* edited by D. E. Doyel, S. K. Fish, and P. R. Fish, pp. 37–63. Southwestern and Rocky Mountain Division of the American Association for the Advancement of Science, Fort Collins, Colorado.

2000b The Role of Irrigation in the Transition to Agriculture and Sedentism in the Southwest. Paper presented at the Southwest Symposium 2000, January 13–15, Santa Fe, New Mexico.

Mabry, J. (editor)

1997a *Archaeological Investigations of Early Village Sites in the Middle Santa Cruz Valley—Analyses and Synthesis, Part I.* Anthropological Papers No. 19. Center for Desert Archaeology, Tucson.

1997b *Archaeological Investigations of Early Village Sites in the Middle Santa Cruz Valley—*

Analyses and Synthesis, Part II. Anthropological Papers No. 19. Center for Desert Archaeology, Tucson.

Mabry, J., D. Swartz, H. Wöcherl, J. Clark, G. Archer, and M. Lindeman

1997 *Archaeological Investigations of Early Village Sites in the Middle Santa Cruz Valley—Descriptions of the Santa Cruz Bend, Square Hearth, Stone Pipe, and Canal Sites.* Anthropological Papers No. 18. Center for Desert Archaeology, Tucson.

Mackin, J. H.

1948 Concept of the Graded River. *Bulletin of the Geological Society of America* 59:463–512.

MacNeish, R., and M. Eubanks

2000 Comparative Analysis of the Río Balsas and Tehuacán Models for the Origin of Maize. *Latin American Antiquity* 11:3–20.

Madsen, D. B.

1986 Great Basin Nuts: A Short Treatise on the Distribution, Productivity, and Prehistoric Use of Pinyon. In *Anthropology of the Desert West: Essays in Honor of Jesse D. Jennings,* edited by C. J. Condie and D. D. Fowler, pp. 21–41. Anthropological Papers No. 110. University of Utah Press, Salt Lake City.

Madsen, David, and J. O'Connell (editors)

1982 *Man and Environment in the Great Basin.* Society for American Archaeology Papers No. 2. Washington, D. C.

Mails, Thomas E., and Dan Evehema

1995 *Hotevilla: Hopi Shrine of the Covenant, Microcosm of the World.* Marlowe, New York.

Marshall, Michael P., and Henry J. Walt

1984 *Rio Abajo: Prehistory and History of a Rio Grande Province.* New Mexico Historic Preservation Program, Historic Preservation Division, Santa Fe.

Martin, Paul S.

1970 Explanation as an Afterthought and as a Goal. *In Reconstructing Prehistoric Pueblo Societies,* edited by William A. Longacre, pp. 194–201. University of New Mexico Press, Albuquerque.

Martin, Paul S., and Fred Plog

1973 *The Archaeology of Arizona.* Doubleday Natural History Press, New York.

Martin, Paul S., and John B. Rinaldo

1950 *Sites of the Reserve Phase.* Fieldiana: Anthropology 38(3). Chicago.

1960 *Excavations in the Upper Little Colorado, Eastern Arizona.* Fieldiana: Anthropology 51(1). Chicago.

Martin, Paul S., John B. Rinaldo, and William A. Longacre

1961 *Mineral Creek Site and Hooper Ranch Pueblo, Eastern Arizona.* Fieldiana: Anthropology 52. Chicago.

Martin, Paul S., John B. Rinaldo, W. A. Longacre, C. Cronin, L. G. Freeman Jr., and J. Schoenwetter

1962 *Chapters in the Prehistory of Eastern Arizona I.* Fieldiana: Anthropology 53. Chicago.

Mason, J. Alden

1921 The Papago Migration Legend. *Journal of American Folk-Lore* 34:254–268.

Masse, W. Bruce

1981 A Reappraisal of the Protohistoric Sobaipuri Indians of Southeastern Arizona. In *The Protohistoric Period in the North American Southwest, A.D. 1450–1700,* edited by D. R. Wilcox and W. B. Masse, pp. 28–56. Anthropological Research Papers No. 24. Arizona State University, Tempe.

1982 Hohokam Ceramic Art: Regionalism and the Imprint of Societal Change. In *Southwestern Ceramics: A Comparative View,* edited by A. H. Schroeder. *Arizona Archaeologist* 15:71–105.

1991 The Quest for Subsistence Sufficiency and Civilization in the Sonoran Desert. In *Chaco and Hohokam: Prehistoric Regional Systems in the American Southwest,* edited by P. L. Crown and W. J. Judge, pp. 195–223. Advanced Seminar Series. School of American Research Press, Santa Fe.

1995 The Celestial Basis of Civilization. In *The Inspiration of Astronomical Phenomena*, edited by G. V. Coyne, S.J., and R. M. Sinclair. *Vistas in Astronomy* 39:463–477.

1997 Errata for "The Celestial Basis of Civilization" (1995). *Vistas in Astronomy* 40:441.

1998 Earth, Air, Fire, and Water: The Archaeology of Bronze Age Cosmic Catastrophes. In *Natural Catastrophes During Bronze Age Civilisations: Archaeological, Geological, Astronomical, and Cultural Perspectives*, edited by B. J. Peiser, T. Palmer, and M. E. Bailey, pp. 53–92. BAR International Series 728. Archaeopress, Oxford.

2003a Elder Brother's Creations. In *Fragile Patterns: Archaeology and History in the Arizona Papaguería*, edited by J. H. Altschul and A. G. Rankin. Statistical Research, Tucson. In press.

2003b Transient Celestial Events in Traditional Hawai'i. In *Islands in the Sky: Astronomy and the Role of Celestial Phenomena in Hawaiian Myth, Language, Religion, and Chiefly Power*, by W. B. Masse, R. K. Johnson, and H. D. Tuggle. University of Hawaii Press, Honolulu. In preparation.

In press The Archaeology and Anthropology of Quaternary Period Cosmic Impact. In *Asteroid/Comet Impacts on Human Society*, edited by P. Bobrowsky and H. Rickman. Springer, Berlin.

Masse, W. Bruce, Laura A. Carter, and Gary F. Somers

1991 Waha'ula *Heiau*: The Regional and Symbolic Context of Hawai'i Island's "Red Mouth" Temple. *Asian Perspectives* 30(1):19–56.

Masse, W. Bruce, and Robert Soklow

2005 Black Suns and Dark Times: Cultural Responses to Solar Eclipses in the Ancient Puebloan Southwest. In *Current Studies in Archaeoastronomy: Conversations Across Time and Space*, edited by J. W. Fountain and R. M. Sinclair, pp. 47–67. Carolina Academic Press, Durham, North Carolina.

Masse, W. Bruce, and H. David Tuggle

1998 The Date of Hawaiian Colonization. In *Easter Island in Pacific Context: South Seas Symposium*, edited by C. M. Stevenson, G. Lee, and F. J. Morin, pp. 229–235. Easter Island Foundation Occasional Paper No. 4. Bearsville and Cloud Mountain Presses, Los Ocos, California.

Matson, R. G.

1991 *The Origins of Southwestern Agriculture*. University of Arizona Press, Tucson.

1999 The Spread of Maize to the Colorado Plateau. *Archaeology Southwest* 13(1):10–11.

May, R.

1973 *Stability and Complexity in Model Ecosystems*. Monographs in Ecological Studies No. 6. Princeton University Press, Princeton.

McDonald, J. E.

1957 A Note on the Precision of Estimation of Missing Precipitation Data. *Transactions of the American Geophysical Union* 38:657–661.

McGregor, John C.

1941 *Southwestern Archaeology*. Wiley, London.

McGuire, Randall H., and María Elisa Villalpando C.

1993 *An Archaeological Survey of the Altar Valley, Sonora, Mexico*. Arizona State Museum Archaeological Series 184. University of Arizona, Tucson.

McKusick, C. R.

1982 Avifauna from Grasshopper Pueblo. In *Multidisciplinary Research at Grasshopper Pueblo, Arizona*, edited by W. A. Longacre, S. J. Holbrook, and M. W. Graves, pp. 87–96. Anthropological Papers of the University of Arizona No. 40. University of Arizona Press, Tucson.

McNutt, C. H., and R. C. Euler

1966 The Red Butte Lithic Sites near Grand Canyon, Arizona. *American Antiquity* 31:410–419.

McPherson, Gale, and David E. Doyel

1980 Ground Stone Assemblages from Sites in Dead Valley. In *Prehistory in Dead Valley, East-Central Arizona: The TG&E Springerville Project*, edited by D. E. Doyel and

S. Debowski, pp. 305–340. Arizona State Museum Archaeological Series 144. University of Arizona, Tucson.

Meeus, Jean

1982 The Frequency of Total and Annular Solar Eclipses at a Given Place. *Journal of the British Astronomical Association* 92:124–126.

Meko, D., and D. A. Graybill

1993 *Gila River Streamflow Reconstruction.* Report submitted to the U.S. Bureau of Land Management. Laboratory of Tree-Ring Research, University of Arizona, Tucson.

Metzger, D. G.

1961 *Geology in Relation to Availability of Water Along the South Rim, Grand Canyon National Park, Arizona.* Geological Survey Water Supply Paper No. 1475-C. U.S. Government Printing Office, Washington, D.C.

Midvale, Frank

1946 Prehistoric Irrigation in the Lower Verde Valley. Privately printed map, Phoenix.

1965 Prehistoric Irrigation in the Casa Grande Ruins Area. *The Kiva* 30:82–86.

1968 Prehistoric Irrigation in the Salt River Valley. *The Kiva* 34:28–34.

Miksicek, C. H.

1983 Archaeobotanical Aspects of Las Fosas: A Statistical Approach to Prehistoric Plant Remains. In *Hohokam Archaeology Along the Salt-Gila Aqueduct,* edited by L. S. Teague and P. L. Crown, pp. 671–700. Arizona State Museum Archaeological Series 150, Vol. VI. University of Arizona, Tucson.

1987 Formation Processes of the Archaeobotanical record. In *Advances in Archaeological Method and Theory,* Vol. 10, edited by M. B. Schiffer, pp. 211–247. Academic Press, New York.

Milbrath, Susan

1995 Eclipse Imagery in Mexica Sculpture of Central Mexico. In *The Inspiration of Astronomical Phenomena,* edited by G. V. Coyne, S.J., and R. M. Sinclair. *Vistas in Astronomy* 39:479–502.

2000 *Star Gods of the Maya: Astronomy in Art, Folklore, and Calendars.* University Press of Texas, Austin.

Miller, Mary, and Karl Taube

1993 *The Gods and Symbols of Ancient Mexico and the Maya.* Thames and Hudson, London.

Miller, N. F.

1988 Ratios in Paleoethnobotanical Analysis. In *Current Paleoethnobotany: Analytical Methods and Cultural Implications of Archaeological Plant Remains,* edited by C. A. Hastorf and V. S. Popper, pp. 72–85. University of Chicago Press, Chicago.

Mindeleff, Cosmos

1896 *Aboriginal Remains in the Verde Valley, Arizona.* 13th Annual Report, Bureau of Ethnology. Smithsonian Institution, Washington, D.C.

Mindeleff, Victor

1989 *A Study of Pueblo Architecture: Tusayan and Cibola.* Reprinted. Smithsonian Institution, Washington, D.C. Originally published 1891.

Minnis, Paul E.

1980 Prehistoric Puebloan Food and Fuel: The Macroplant Evidence from Dead Valley, Arizona. In *Prehistory in Dead Valley, East-Central Arizona: The TG&E Springerville Report,* edited by D. E. Doyel and S. S. Debowski, pp. 371–388. Arizona State Museum Archaeological Series 144. University of Arizona, Tucson.

1981 Seeds in Archaeological Sites: Sources and Some Interpretive Problems. *American Antiquity* 46:143–152.

1985a Domesticating Plants and People in the Greater American Southwest. In *Prehistoric Food Production in North America,* edited by R. Ford, pp. 309–340. Anthropological Papers No. 75. Museum of Anthropology, University of Michigan, Ann Arbor.

1985b *Social Adaptation to Food Stress: A Prehistoric Southwestern Example.* University of Chicago Press, Chicago.

1991 Famine Foods of the Northern American Desert Borderlands in Historical Context. *Journal of Ethnobiology* 11:231–257.
1992 Earliest Plant Cultivation in the Desert Borderlands of North America. In *The Origins of Agriculture: An International Perspective*, edited by C. Wesley Cowan and P. Watson, pp. 121–141. Smithsonian Institution Press, Washington, D.C.
1996 Notes on Economic Uncertainty and Human Behavior in the Prehistoric North American Southwest. In *Evolving Complexity and Environmental Risk in the Prehistoric Southwest*, edited by J. A. Tainter and B. B. Tainter, pp. 57–78. Addison-Wesley, Reading, Massachusetts.

Mitchell, Douglas R.
1995 Chronological Considerations: Area 6, 15, and 16 of the Pueblo Salado Community. In *Life on the Floodplain: Further Investigations at Pueblo Salado for Phoenix Sky Harbor International Airport*, edited by D. H. Greenwald, J. H. Ballagh, D. Mitchell, and R. A. Anduze, pp. 317–338. Anthropological Papers No. 4. Pueblo Grande Museum, Phoenix.

Mixon, Benjamin, and Raymond E. White
1991 Skywatchers of the Salt River Valley Hohokam. *Astronomy Quarterly* 8:245–259.

Monserud, R. A.
1986 Time Series Analyses of Tree-Ring Chronologies. *Forest Science* 32(2):349–372.

Montgomery, Douglas C., and Elizabeth A. Peck
1992 *Introduction to Linear Regression Analysis*. Wiley, New York.

Mooney, James
1973 *The Ghost-Dance Religion and Wounded Knee*. Reprinted. Dover, New York. Originally published 1896.

Moore, J. and J. Winters (editors)
1980 *Human Adaptations in a Marginal Environment: The UII Mitigation Project*. Office of Contract Archaeology, University of New Mexico, Albuquerque.

Moran, E. F.
1990 Ecosystem Ecology in Biology and Anthropology: A Critical Assessment. In *The Ecosystem Approach in Anthropology*, edited by E. F. Moran, pp. 3–40. University of Michigan Press, Ann Arbor.

Morgan, L. H.
1881 *Houses and House-Life of the American Aborigines*. Contributions to North American Ethnology Vol. VI. U.S. Government Printing Office, Washington, D.C.

Morris, Clyde P.
1971 A Brief Economic History of the Camp and Middle Verde Reservations. *Plateau* 44:43–52.

Morris, D. H.
1970 Walnut Creek Village: A Ninth-Century Hohokam-Anasazi Settlement in the Mountains of Central Arizona. *American Antiquity* 35:49–61.

Morrison, David
1997 The Contemporary Hazard of Cometary Impacts. In *Comets and the Origin and Evolution of Life*, edited by P. J. Thomas, C. F. Chyba, and C. P. McKay, pp. 243–258. Springer, New York.

Mucke, H., and J. Meeus
1983 *Canon of Solar Eclipses: -2003 to +2526*. Astronomisches Büro, Vienna.

Muenchrath, Deborah A.
1995 Productivity, Morphology, Phenology, and Physiology of a Desert-Adapted Native American Maize (*Zea mays* L.) Cultivar. Unpublished Ph.D. dissertation, Department of Agronomy, Iowa State University, Ames.

Muenchrath, Deborah A., Maya Kuratomi, Jonathan Sandor, and Jeffrey Homberg
2002 Observational Study of Maize Production Systems of Zuni Farmers in Semiarid New Mexico. *Journal of Ethnobiology* 22:1–34.

Muenchrath, Deborah A., and Ricardo J. Salvador
1995 Maize Productivity and Agroecology: Effects of Environment and Agricultural

Practices on the Biology of Maize. In *Soil, Water, Biology, and Belief in Prehistoric and Traditional Southwestern Agriculture*, edited by H. W. Toll, pp. 303–333. Special Publication No. 2. New Mexico Archaeological Council, Albuquerque.

Murdin, Paul

2001 Effects of the 2001 Total Solar Eclipse on African Wildlife. *Astronomy and Geophysics* 42(4):4–8.

Murie, James R.

1981 *Ceremonies of the Pawnee: Part I, the Skiri, and Part II, the South Bands.* Edited by D. R. Parks. Smithsonian Contributions to Anthropology No. 27. Washington, D.C.

Muro, M.

1998a New Finds Explode Old Views of the American Southwest. *Science* 279:653–654.

1998b Not Just Another Roadside Attraction. *American Archaeology* 2(4):10-16.

Nabhan, Gary P.

1984 Soil Fertility Renewal and Water Harvesting in Sonoran Desert Agriculture: The Papago Example. *Arid Lands Newsletter* 20:21–28.

1985 *Gathering the Desert.* University of Arizona Press, Tucson.

1986 Ak-chin "Arroyo Mouth" and the Environmental Setting of the Papago Indian Fields in the Sonoran Desert. *Applied Geography* 6:61–75.

Neely, J. A.

1997 Foothill Irrigation and Domestic Water Systems of the Safford Valley, Southeastern Arizona. Paper presented at the 62nd Annual Meeting of the Society for American Archaeology, April 2–6, Nashville.

Neely, J. A., and J. S. Crary

1998 The Marijilda Canyon Canal: A Complex Irrigation and Domestic Water System in the Safford Valley, Southeastern Arizona. Paper presented at the 63rd Annual Meeting of the Society for American Archaeology, March 25–29, Seattle.

Neely, J. A., J. S. Crary, and M. K. Woodson

1997 An Archaeological Overview of Upper (Northern) Southeastern Arizona: The Formative and Classic Periods. In *Archaeology of a Land Between: Regional Dynamics in the Prehistory and History of Southeastern Arizona*, edited by H. D. Wallace. University of New Mexico Press, Albuquerque. In preparation.

Neitzel, Jill

1991 Hohokam Material Culture and Behavior: The Dimensions of Organizational Change. In *Exploring the Hohokam: Prehistoric Desert Peoples of the American Southwest*, edited by G. J. Gumerman, pp. 177–230. Amerind Foundation, Dragoon, Arizona, and University of New Mexico Press, Albuquerque.

Nelson, Ben A.

1990 Comments: Southwestern Sedentism Reconsidered. In *Perspectives on Southwestern Prehistory*, edited by P. Minnis and C. Redman, pp. 157–166. Westview Press, Boulder, Colorado.

Nelson, Ben A., and Roger Anyon

1996 Fallow Valleys: Asynchronous Occupations in the Mimbres Region. *Kiva* 61:275–294.

Nelson, Margaret C.

1999 *Mimbres During the Twelfth Century: Abandonment, Continuity, and Reorganization.* University of Arizona Press, Tucson.

Nelson, M. C., and M. W. Diehl

1999 Foraging and Farming. In *Mimbres During the Twelfth Century*, by M. C. Nelson, pp. 142–166. University of Arizona Press, Tucson.

Nequatewa, Edmund

1967 *Truth of a Hopi.* Reprinted. Northland Press, Flagstaff. Originally published 1936.

Netting, Robert McC.

1975 *Cultural Ecology.* Cummings, Menlo Park, California.

1993 *Smallholders, Householders.* Stanford University Press, Palo Alto.

Nials, F. L., and D. A. Gregory

1989 Irrigation Systems in the Lower Salt River Valley. In *The 1982–1984 Excavations at Las Colinas: Environment and Subsistence*, by D. A. Graybill, D. A. Gregory, F. L. Nials,

S. K. Fish, R. E. Gasser, C. H. Miksicek, and C. R. Szuter, pp. 39–58. Arizona State Museum Archaeological Series 162, Vol. 5. University of Arizona, Tucson.

Nials, Fred L., David A. Gregory, and Donald A. Graybill

1989 River Streamflow and Hohokam Irrigation Systems. In *The 1982–1984 Excavations at Las Colinas: Environment and Subsistence*, by D. A. Graybill, D. A. Gregory, F. L. Nials, S. K. Fish, R. E. Gasser, C. H. Miksicek, and C. R. Szuter, pp. 59–76. Arizona State Museum Archaeological Series 162, Vol. 5. University of Arizona, Tucson.

Norr, J. A.

1997 Archaeological Investigation of a Prehistoric Water Collection Feature in the Upper Basin, Kaibab National Forest, Northern Arizona. Master's thesis, Department of Anthropology, University of Cincinnati.

Olsen, J. W.

1982 Prehistoric Environmental Reconstruction by Vertebrate Faunal Analysis, Grasshopper Pueblo. In *Multidisciplinary Research at Grasshopper Pueblo, Arizona,* edited by W. A. Longacre, S. J. Holbrook, and M. W. Graves, pp. 63–72. Anthropological Papers of the University of Arizona No. 40. University of Arizona Press, Tucson.

1990 *Vertebrate Faunal Remains from Grasshopper Pueblo, Arizona.* Anthropological Papers No. 83. Museum of Anthropology, University of Michigan, Ann Arbor.

Olsen, S. J.

1982 Water Resources and Aquatic Fauna at Grasshopper Pueblo. In *Multidisciplinary Research at Grasshopper Pueblo, Arizona,* edited by W. A. Longacre, S. J. Holbrook, and M. W. Graves, pp. 61–62. Anthropological Papers of the University of Arizona No. 40. University of Arizona Press, Tucson.

Ortiz, Alfonso

1969 *The Tewa World: Space, Time, Being, and Becoming in a Pueblo Society.* University of Chicago Press, Chicago.

Palmer, W. C.

1965 *Meteorological Drought.* U.S. Weather Bureau Research Paper No. 45. U.S. Department of Commerce, Washington, D.C.

Parsons, Elsie Clews

1996 *Pueblo Indian Religion, Vols. 1 and 2.* Reprinted. University of Nebraska Press, Lincoln. Originally published 1939.

Pearsall, D. M.

1988 Interpreting the Meaning of Macro-remain Abundance: The Impact of Source and Context. In *Current Paleoethnobotany: Analytical Methods and Cultural Interpretations of Archaeological Plant Remains,* edited by C. A. Hastorf and V. S. Popper, pp. 97–118. University of Chicago Press, Chicago.

Phillips, David A., and Lynne Sebastian

2001 *Examining the Course of Southwestern Archaeology: The Durango Conference, September 1995.* Special Publication No. 3. New Mexico Archaeological Council, Albuquerque.

Pierson, William M.

1874 Correspondence Relative to the Discovery of a Large Meteorite in Mexico. In *Smithsonian Institution Annual Report, 1873*, pp. 419–422. Washington, D.C.

Piperno, D., and K. Flannery

2001 The Earliest Archaeological Maize (*Zea mays* L.) from Highland Mexico: New Accelerator Mass Spectrometry Dates and Their Implications. *Proceedings of the National Academy of Science U.S.A.* 98:2101–2103.

Plog, Fred T.

1974 *The Study of Prehistoric Change.* Academic Press, New York.

Plog, Fred T. (editor)

1978 *An Analytical Approach to Cultural Resource Management: The Little Colorado Planning Unit.* Anthropological Research Papers No. 13. Arizona State University, Tempe.

Plog, Fred, Richard Effland, Jeffrey S. Dean, and Sylvia W. Gaines

1978 SARG: Future Research Directions. In *Investigations of the Southwestern Anthropological Research Group: An Experiment in Archaeological Cooperation; The Proceedings of the 1976 Conference*, edited by Robert C. Euler and George J. Gumerman, pp. 177–186. Museum of Northern Arizona, Flagstaff.

Plog, Fred, and Cheryl Garrett

1972 Explaining Variability in Prehistoric Water Control Systems. In *Contemporary Archaeology: A Guide to Theory and Contributions*, edited by Mark Leone, pp. 280-289. Southern Illinois University Press, Carbondale.

Plog, Fred, George J. Gumerman, Robert C. Euler, Jeffrey S. Dean, Thor N. V. Karlstrom, and Richard H. Hevly

1988 Anasazi Adaptive Strategies: The Model, Predictions, and Results. In *The Anasazi in a Changing Environment*, edited by George J. Gumerman, pp. 230-276. Cambridge University Press, Cambridge.

Pope, K., M. Pohl, J. Jones, D. Lentz, C. von Nagy, F. Vega, and R. Quitmyer

2001 Origin and Environmental Setting of Ancient Agriculture in the Lowlands of Mesoamerica. *Science* 292:1370-1373.

Popper, V. S.

1988 Selecting Quantitative Measurements in Paleoethnobotany. In *Current Paleoethnobotany: Analytical Methods and Cultural Interpretations of Archaeological Plant Remains*, edited by C. A. Hastorf and V. S. Popper, pp. 53–71. University of Chicago Press, Chicago.

Powell, J. W.

1875 The Cañons of the Colorado. *Scribner's Monthly* 9:523–537.

Powell , Shirley

1983 *Mobility and Adaptation: The Anasazi of Black Mesa, Arizona.* Southern Illinois University Press, Carbondale.

2002 The Puebloan Florescence and Dispersion: Dinnebito and Beyond, A.D. 800-1150. In *Prehistoric Culture Change on the Colorado Plateau*, edited by S. Powell and F. E. Smiley, pp. 79–117. University of Arizona Press, Tucson.

Powell, S., and G. J. Gumerman

1987 *People of the Mesa: The Archaeology of Black Mesa, Arizona.* Southern Illinois University Press, Carbondale.

Preston, Robert A., and Ann L. Preston

2005 Consistent Forms of Solstice Sunlight Interaction with Petroglyphs Throughout the Prehistoric American Southwest. In *Current Studies in Archaeoastronomy: Conversations Across Time and Space*, edited by J. W. Fountain and R. M. Sinclair, pp. 109–119. Carolina Academic Press, Durham, North Carolina.

Pyne, S. J.

1998 *How the Canyon Became Grand: A Short History.* Viking, New York.

Rada, W. S., and F. R. Stephenson

1992 A Catalogue of Meteor Showers in Medieval Arab Chronicles. *Quarterly Journal of the Royal Astronomical Society* 33:5–16.

Rand, P. J.

1965 Factors Related to the Distribution of Ponderosa and Pinyon Pines at Grand Canyon, Arizona. Unpublished Ph.D. dissertation, Department of Botany, Duke University, Chapel Hill.

Rappaport, R. A.

1971 The Flow of Energy in an Agricultural Society. *Scientific American* 224:116–133.

Rautman, Alison E.

1994 Regional Climate Records and Local Experience: "Drought" and the Decline of Dry Farming in Central New Mexico. *C & A: Climate and Agriculture* 49:12–15.

Ravesloot, John C.

1979 The Animas Phase: The Post Classic Mimbres Occupation of the Mimbres Valley, New Mexico. Unpublished Master's thesis, Southern Illinois University, Carbondale.

Ravesloot, John C. (editor)

1987 *The Archaeology of the San Xavier Bridge Site (AZ BB:13–14), Tucson Basin, Southern*

Arizona. Archaeological Series 171, 3 parts. Cultural Resource Management Division, Arizona State Museum, University of Arizona, Tucson.

Ravesloot, J. R., and M. H. Regan

1995 Demographic, Health, Genetic, and Mortuary Characteristics of Late Prehistoric Central Arizona Populations. In *Salado*, edited by Jeffrey S. Dean, pp. 57–80. Amerind Foundation, Dragoon, Arizona, and University of New Mexico Press, Albuquerque.

Redman, Charles L.

1999 *Human Impact on Ancient Environments.* University of Arizona Press, Tucson.

Redman, Charles L., Steven R. James, Paul R. Fish, and J. Daniel Rogers (editors)

2004 *The Archaeology of Global Change: The Impact of Humans on Their Environment.* Smithsonian Institution Press, Washington, D.C.

Reed, Erik K.

1956 Types of Village Plan Layouts in the Southwest. In *Prehistoric Settlement Patterns in the New World*, edited by G. R. Willey, pp. 11–17. Viking Fund Publications in Anthropology No. 23. Wenner-Gren Foundation for Anthropological Research, New York.

Reed, P. F.

2000 Fundamental Issues in Basketmaker Archaeology. In *Foundations of Anasazi Culture: The Basketmaker-Pueblo Transition*, edited by P. F. Reed, pp. 3–16. University of Utah Press, Salt Lake City.

Regan, Marcia, Joel D. Irish, and Christy G. Turner II

1994 Skeletal and Dental Indicators of Health Among the Salado from Tonto Basin, Arizona. *Abstracts of the 59th Annual Meeting*, p. 118. Society for American Archaeology, Anaheim, California.

Reichardt, Karen, and Gary Nabhan

1980 Ecology and Ethnobotany of Dead Valley. In *Prehistory in Dead Valley, East-Central Arizona: The TG&E Springerville Project*, edited by D. E. Doyel and S. S. Debowski, pp. 19–33. Arizona State Museum Archaeological Series 144. University of Arizona, Tucson.

Reid, J. Jefferson

1973 Growth and Response to Stress at Grasshopper Pueblo, Arizona. Unpublished Ph.D. dissertation, University of Arizona, Tucson.

1978 Response to Stress at Grasshopper Pueblo, Arizona. In *Discovering Past Behavior: Experiments in the Archaeology of the American Southwest*, edited by P. F. Grebinger, pp. 195–213. Gordon and Breach, London.

1989 A Grasshopper Perspective on the Mogollon of the Arizona Mountains. In *Dynamics of Southwest Prehistory*, edited by L. Cordell and G. Gumerman, pp. 65–97. Smithsonian Institution Press, Washington, D.C.

Reid, J. Jefferson, and Jeffrey S. Dean

1996 Donald Alan Graybill (1942–1993). *Kiva* 61:309–315.

Reid, J. J., and I. Shimada

1982 Pueblo Growth at Grasshopper: Methods and Models. In *Multidisciplinary Research at Grasshopper Pueblo, Arizona*, edited by W. A. Longacre, S. J. Holbrook, and M. W. Graves, pp. 12–18. Anthropological Papers of the University of Arizona No. 40. University of Arizona Press, Tucson.

Reid, J. J., J. R. Welch, B. K. Montgomery, and M. N. Zedeño

1996 A Demographic Overview of the Late Pueblo III Period in the Mountains of East-Central Arizona. In *The Prehistoric Pueblo World, A.D. 1150–1350*, edited by M. A. Adler, pp. 73–85. University of Arizona Press, Tucson.

Reid, J. J., and S. M. Whittlesey

1997 *The Archaeology of Ancient Arizona.* University of Arizona Press, Tucson.

1999 *Grasshopper Pueblo: A Story of Archaeology and Ancient Life.* University of Arizona Press, Tucson.

Rice, G. E.

1992 Modeling the Development of Complexity in the Sonoran Desert of Arizona. In

Developing Perspectives on Tonto Basin Prehistory, edited by C. L. Redman, G. E. Rice, and K. E. Pedrick, pp. 11–26. Anthropological Field Studies No. 26. Arizona State University, Tempe.

Rice, G., R. Effland, and L. Blank-Roper

1980 *A Sample Survey of Tusayan Planning Unit 1, Kaibab National Forest, Arizona.* Cultural Resources Report No. 33. USDA Forest Service, Albuquerque.

Richards, K.

1982 *Rivers, Form, and Process in Alluvial Channels.* Methuen, London.

Riggs, C. R.

2001 *The Architecture of Grasshopper Pueblo.* University of Utah Press, Salt Lake City.

Rinaldo, John B., and Elaine A. Bluhm

1956 *Late Mogollon Pottery Types of the Reserve Area.* Fieldiana: Anthropology 36(7). Chicago.

Ritter, D. F.

1978 *Process Geomorphology.* W. C. Brown, Dubuque, Iowa.

Ritter, Dale F., R. Craig Kochel, and Jerry R. Miller

1995 *Process Geomorphology.* 3rd ed. W. C. Brown, Dubuque, Iowa.

Robinson, W. J., and C. M. Cameron

1991 *A Directory of Tree-Ring Dated Prehistoric Sites in the American Southwest.* Laboratory of Tree-Ring Research, University of Arizona, Tucson.

Robinson, William J., and Robert Evans

1980 A Microcomputer-Based Tree-Ring Measurement System. *Tree-Ring Bulletin* 40: 59–63.

Robinson, W. J., B. W. Harrill, and R. L. Warren

1975 *Tree-Ring Dates from Arizona N–O, Flagstaff Area.* Laboratory of Tree-Ring Research, University of Arizona, Tucson.

Rocek, T. R.

1995 Sedentarization and Agricultural Dependence: Perspectives from the Pithouse-to-Pueblo Transition in the American Southwest. *American Antiquity* 60:218–239.

Rose, M. R.

1989 Present and Past Environmental Conditions. In *Man, Models, and Management: An Overview of the Arizona Strip and the Management of Its Cultural Resources*, edited by J. H. Altschul and H. C. Fairley, pp. 7–52. Statistical Research, Tucson.

Rose, Martin R., Jeffrey S. Dean, and William J. Robinson

1981 *The Past Climate of Arroyo Hondo, New Mexico, Reconstructed from Tree-Rings.* Arroyo Hondo Archaeological Series Vol. IV. School of American Research Press, Santa Fe.

Roth, B. J.

1988 Recent Research on the Late Archaic Occupation of the Northern Tucson Basin. In *Recent Research on Tucson Basin Prehistory*, edited by W. Doelle and P. Fish, pp. 81–85. Anthropological Papers No. 10. Institute for American Research, Tucson.

1989 Late Archaic Settlement and Subsistence in the Tucson Basin. Unpublished Ph.D. dissertation, University of Arizona, Tucson.

1992 Sedentary Agriculturalists or Mobile Hunter-Gatherers? Recent Evidence of the Late Archaic Occupation of the Northern Tucson Basin. *Kiva* 57:291–314.

1993 Changing Perceptions of the Late Archaic: An Example from the Southern Southwest. *North American Archaeologist* 14:123–137.

Roth, B., and K. Wellman

2001 New Insights into the Early Agricultural Period in the Tucson Basin: Excavations at the Valley Farms Site (AZ AA:12:736). *Kiva* 67:59–79.

Rugge, Dale

1980 Geophysical Characteristics and Resources of Dead Valley. In *Prehistory in Dead Valley, East-Central Arizona: The TG&E Springerville Project*, edited by D. E. Doyel and S. S. Debowski, pp. 35–41. Arizona State Museum Archaeological Series 144. University of Arizona, Tucson.

Rugge, Dale, and David E. Doyel
1980 Petrographic Analysis of Ceramics from Dead Valley. In *Prehistory in Dead Valley, East-Central Arizona: The TG&E Springerville Project,* edited by D. E. Doyel and S. S. Debowski, pp. 189–203. Arizona State Museum Archaeological Series 144. University of Arizona, Tucson.

Russell, Frank
1975 *The Pima Indians.* Reprinted. University of Arizona Press, Tucson. Originally published 1908.

Sahagún, Fray Bernardino de
1953 *Florentine Codex: General History of the Things of New Spain. Book 7—The Sun. Moon, and Stars, and the Binding of the Years.* Monographs 14(7). School of American Research, Santa Fe.

Saxton, Dean, and Lucille Saxton
1973 *O'othham Hoho'ok A'agitha: Legends and Lore of the Papago and Pima Indians.* University of Arizona Press, Tucson.

Scarborough, Vernon L., and David R. Wilcox (editors)
1991 *The Mesoamerican Ballgame.* University of Arizona Press, Tucson.

Schaafsma, Curtis F., and Carroll L. Riley (editors)
1999 *The Casas Grandes World.* University of Utah Press, Salt Lake City.

Schaefer, Bradley E.
1996 Peak Brightness of Historical Supernovae and the Hubble Constant. *Astrophysical Journal* 459:438–454.
2005 Astronomical Omens. In *Current Studies in Archaeoastronomy: Conversations Across Time and Space,* edited by J. W. Fountain and R. M. Sinclair, pp. 37–45. Carolina Academic Press, Durham, North Carolina.

Schele, Linda, and David A. Freidel
1990 *A Forest of Kings: The Untold Story of the Ancient Maya.* Morrow, New York.
1991 The Courts of Creation: Ballcourts, Ballgames, and Portals to the Maya Otherworld. In *The Mesoamerican Ballgame,* edited by V. L. Scarborough and D. R. Wilcox, pp. 289–315. University of Arizona Press, Tucson.

Schoenwetter, J., and V. DaCosta
1976 Pollen Studies in the Marble Canyon Area, Arizona. Ms. on file, Department of Anthropology, Arizona State University, Tempe.

Schoenwetter, J., and A. E. Dittert, Jr.
1968 An Ecological Interpretation of Anasazi Settlement Patterns. In *Anthropological Archaeology in the Americas,* edited by Betty J. Meggers, pp. 41–66. Anthropological Society of Washington, Washington, D.C.

Scholes, France V.
1937 Troublous Times in New Mexico—1659–1670. *New Mexico Historical Review* XII(2, 4). Continued in XIII(1), 1938; XV(3, 4), 1940; XVI(1, 2, 3), 1941. Albuquerque.

Schove, D. Justin, and Alan Fletcher
1984 *Chronology of Eclipses and Comets, A.D. 1–1000.* Boydell Press, Wolfeboro, New Hampshire.

Schreiber, Katarina, and Alan P. Sullivan
1984 *The Prehistoric Occupation of Voigt Mesa, Arizona: The 1983 TEP Springerville Project.* Arizona State Museum Archaeological Series 166. University of Arizona, Tucson.

Schroeder, Albert H.
1979 Pueblos Abandoned in Historic Times. In *Southwest,* edited by Alfonso Ortiz, pp. 236–254. Handbook of North American Indians, Vol. 9, William C. Sturtevant, general editor. Smithsonian Institution, Washington, D.C.

Schumm, S. A.
1968 River Adjustment to Altered Hydrologic Regimen—Murrumbidgee River and Paleochannels, Australia. *U.S. Geological Survey Professional Paper* 598:1–65.
1969 River Metamorphosis. *Journal of the Hydraulic Division, American Society of Civil Engineers* 95:255–273.

1977 *The Fluvial System.* Wiley, New York.

Schumm, S. A., and G. R. Brackenridge

1987 River Response. In *North America and Adjacent Oceans During the Last Deglaciation.* Geological Society of America, *The Geology of North America* K-3:221–240.

Schumm, S. A., and R. W. Lichty

1963 Channel Widening and Flood-Plain Construction Along Cimarron River in Southwestern Kansas. *U.S. Geological Survey Professional Paper* 352-D:71–88.

Schwartz, D. W.

1957 Climate Change and Culture History in the Grand Canyon. *American Antiquity* 22:372–377.

1963 An Archaeological Survey of Nankoweap Canyon, Grand Canyon National Park. *American Antiquity* 28:289–302.

1965 Nankoweap to Unkar: An Archaeological Survey of the Upper Grand Canyon. *American Antiquity* 30:278–296.

1966 A Historical Analysis and Synthesis of Grand Canyon Archaeology. *American Antiquity* 31:469–484.

1990 *On the Edge of Splendor: Exploring Grand Canyon's Human Past.* School of American Research Press, Santa Fe.

Schwartz, D. W., R. C. Chapman, and J. Kepp

1980 *Archaeology of the Grand Canyon: Unkar Delta.* School of American Research Press, Santa Fe.

Schwartz, D. W., J. Kepp, and R. C. Chapman

1981 *Archaeology of the Grand Canyon: The Walhalla Plateau.* School of American Research Press, Santa Fe.

Schwartz, D. W., M. P. Marshall, and J. Kepp

1979 *Archaeology of the Grand Canyon: The Bright Angel Site.* School of American Research Press, Santa Fe.

Scoones, I.

1999 New Ecology and the Social Sciences: What Prospects for a Fruitful Engagement? *Annual Review of Anthropology* 28:479–507.

Sellers, William D., and Richard H. Hill (editors)

1974 *Arizona Climate, 1931–1972.* 2nd ed. University of Arizona Press, Tucson.

Seymour, Deni (editor)

1997 *Excavations at Lru-Kish Kachreau and Other Sites at the Sandoval County Landfill.* Report No. 15. Lone Mountain Archaeological Services, Albuquerque.

Shackley, M. Stephen

1995 Sources of Archaeological Obsidian in the Greater American Southwest: An Update and Quantitative Analysis. *American Antiquity* 60:531–551.

Shafer, Harry J.

1995 Architecture and Symbolism in Transitional Pueblo Development in the Mimbres Valley, SW New Mexico. *Journal of Field Archaeology* 22:23–47.

Shaul, David Leedom, and Jane H. Hill

1998 Tepimans, Yumans, and Other Hohokam. *American Antiquity* 63:375–396.

Shaw, Anna

1968 *Pima Indian Legends.* University of Arizona Press, Tucson.

Sheridan, Thomas E.

1988 *Where the Dove Calls: The Political Ecology of a Peasant Corporate Community in Northwestern Mexico.* University of Arizona Press, Tucson.

Sheridan, Thomas E., and Gary P. Nabhan

1978 Living with a River: Traditional Farmers of the Rio San Miguel. *Journal of Arizona History* 19:1-17.

Shoemaker, E. M., D. J. Roddy, C. B. Moore, R. Pfeilsticker, C. L. Curley, T. Dunkelman, K. Kuerzel, M. Taylor, C. Shoemaker, and P. Donnelly

1995 Impact Crater Identified on the Navajo Nation near Chinle, Arizona. *Meteoritics* 30(5):578.

Short, H. L.
1979 *Deer in Arizona and New Mexico: Their Ecology and a Theory Explaining Recent Population Decreases.* General Technical Report RM-70. Rocky Mountain Forest and Range Experimental Station, U.S. Department of Agriculture, Fort Collins, Colorado.

Simmons, Alan H.
1981 The "Other" Archaeology of Northwestern New Mexico—Perspectives on the Aceramic Occupation of the San Juan Basin. *Contract Abstracts and CRM Archaeology* 2(2):12-20.
1986 New Evidence for the Early Use of Cultigens in the American Southwest. *American Antiquity* 51(1):73-89.
1987 Paleoeconomy and Paleoenvironments in the Chaco Canyon Region During the Archaic: A Preliminary Report on the Chaco Shelters Project. *American Archaeology* 6:166-169.
1989 The Unknown Archaeology of the Southwest: The Archaic. In *Human Adaptations and Cultural Change in the Greater Southwest: An Overview of Archaeological Resources in the Basin and Range Province,* by A. Simmons, A. Wiener Stodder, D. Dykeman, and P. Hicks. Arkansas Archaeological Survey, Research Series No. 32. University of Arkansas, Fayetteville.
1998 Surface Fragments, Buried Hippos—Potential Problems with Over-reliance on Surface Data as Mirrors. In *Surface Archaeology,* edited by A. Sullivan, pp. 159–167. University of New Mexico Press, Albuquerque.

Simmons, Alan H. (editor)
1982 *Prehistoric Adaptive Strategies in the Chaco Canyon Region, Northwestern New Mexico.* 3 vols. Navajo Nation Papers in Anthropology No. 9. Window Rock, Arizona.
1984 *Archaic Prehistory and Paleoenvironments in the San Juan Basin, New Mexico: The Chaco Shelters Project.* University of Kansas Museum of Anthropology, Project Report Series, No. 53. Lawrence.

Sires, Earl W.
1984 Excavations at El Polvoron (AZ U:15:59). In *Hohokam Archaeology Along the Salt-Gila Aqueduct, Central Arizona Project: Village Sites on Queen Creek and Siphon Draw,* edited by L. S. Teague, and P. L. Crown, pp. 221–326. Arizona State Museum Archaeological Series 150(4). University of Arizona, Tucson.

Smiley, F. E.
1994 The Agricultural Transition in the Northern Southwest: Patterns in the Current Chronometric Data. *Kiva* 60:165–189.

Smith, Bruce
1995 *The Emergence of Agriculture.* Scientific American Library, New York.
1998 Between Foraging and Farming. *Science* 279:1651–1652.

Smith, L. P.
1981 *Long-Term Streamflow Histories of the Salt and Verde Rivers, Arizona, as Reconstructed from Tree Rings.* Contract DACW-09-80-C-0071. U.S. Army Corps of Engineers, Los Angeles District Office. Los Angeles.

Smith, S. J.
2003 *Simkins Flat Pollen Analysis.* Report on file, Laboratory of Paleoecology, Northern Arizona University, Flagstaff.

Smith, Watson
1950 Preliminary Report on the Peabody Museum Upper Gila Expedition. *El Palacio* 57:393–399.

Snow, D. H.
1990 Tener comal y metate: Protohistoric Rio Grande Maize Use and Diet. In *Perspectives on Southwestern Prehistory,* edited by P. E. Minnis and C. L. Redman, pp. 289–300. Westview Press, Boulder, Colorado.

Southworth, C. H.
1919 History of Irrigation on the Gila River. In *Hearings Before the Committee on Indian*

Affairs, House of Representatives, 66th Congress, 1st Session, 2:103–223. Washington, D.C.

Spaulding, W.

1991 A Middle Holocene Vegetation Record from the Mojave Desert of North America and Its Paleoclimatic Significance. *Quaternary Research* 35:427–437.

Spielmann, Katherine A.

1991 Coercion or Cooperation? Plains-Pueblo Interaction in the Protohistoric Period. In *Farmers, Hunters, and Colonists: Interaction Between the Southwest and the Southern Plains,* edited by Katherine A. Spielmann, pp. 36–50. University of Arizona Press, Tucson.

Spier, Leslie

1978 *Yuman Tribes of the Gila River.* Reprinted. Dover, New York. Originally published 1933.

Stacy, Valeria Kay Pheriba

1977 Activity Patterning at Cerros de Trincheras in South Central Arizona. *The Kiva* 43:11–18.

Stafford, C. R.

1978 *Archaeological Investigations at Seneca Lake, San Carlos Indian Reservation, Arizona.* Anthropological Research Papers No. 14. Arizona State University, Tempe.

Stark, M. T., J. J. Clark, and M. D. Elson

1995 Social Boundaries and Cultural Identity in the Tonto Basin. In *The Roosevelt Community Development Study: New Perspectives on Tonto Basin Prehistory,* edited by M.D. Elson, M. T. Stark, and D. A. Gregory, pp. 343–368. Anthropological Papers No. 15. Center for Desert Archaeology, Tucson.

Stegner, W.

1954 *Beyond the Hundredth Meridian: John Wesley Powell and the Second Opening of the West.* Penguin Books, New York.

Stein, G. J.

1999 *Rethinking World-Systems: Diasporas, Colonies, and Interaction in Uruk Mesopotamia.* University of Arizona Press, Tucson.

Stevens, M. A., D. B. Simmons, and E. V. Richardson

1975 Non-equilibrium River Form. *Journal of the Hydraulic Division, American Society of Civil Engineers* 101:557–566.

Steward, Julian H.

1955 *Theory of Culture Change.* University of Illinois Press, Urbana.

Stewart, G. R., and M. Donnelly

1943 Soil and Water Economy in the Pueblo Southwest. *Scientific Monthly* LVI:31–44, 134–144.

Stockton, C. W.

1975 *Long-Term Streamflow Records Reconstructed from Tree Rings.* Papers of the Laboratory of Tree-Ring Research, No. 5. University of Arizona Press, Tucson.

Stockton, C. W., and W. R. Boggess

1980 Augmentation of Hydrologic Records Using Tree Rings. In *Improved Hydrologic Forecasting—Why and How Engineering Foundation Conference,* pp. 239–265. American Society of Civil Engineers, Pacific Grove, California.

Stockton, C. W., and G. C. Jacoby

1976 *Long-Term Surface Water Supply and Streamflow Trends in the Upper Colorado River Basin Based on Tree-Ring Analyses.* Lake Powell Research Project Bulletin No. 18. Institute of Geophysics and Planetary Physics, University of California, Los Angeles.

Stokes, M. A., and T. L. Smiley

1968 *An Introduction to Tree-Ring Dating.* University of Chicago Press, Chicago.

Strahler, A. N.

1944 A Guide to the East Kaibab Monocline in the Grand Canyon Region. *Plateau* 17:1–13.

Sullivan, Alan P.

1982 Mogollon Agriculture: An Appraisal and a New Model. *The Kiva* 48:1–16.

1987 Seeds of Discontent: Implications of a "Pompeii" Archaeobotanical Assemblage for Grand Canyon Anasazi Subsistence Models. *Journal of Ethnobiology* 7:137–153.

1992 Pinyon Nuts and Other Wild Resources in Western Anasazi Subsistence Economies. In *Long-Term Subsistence Change in Prehistoric North America*, edited by D. L. Croes, R. A. Hawkins, and B. L. Isaac, pp. 195–239. JAI Press, Greenwich, Connecticut.

1995 Artifact Scatters and Subsistence Organization. *Journal of Field Archaeology* 22:49–64.

1996 Risk, Anthropogenic Environments, and Western Anasazi Subsistence. In *Evolving Complexity and Environmental Risk in the Prehistoric Southwest*, edited by J. A. Tainter and B. B. Tainter, pp. 145–167. Addison-Wesley, Reading, Massachusetts.

1997 Theoretical Implications of Regional Variation in Prehistoric Puebloan Subsistence and Settlement at Grand Canyon. Paper presented at the 62nd Annual Meeting of the Society for American Archaeology, April 2–6, Nashville.

2000 Effects of Small-Scale Prehistoric Runoff Agriculture on Soil Fertility: The Developing Picture from Upland Terraces in the American Southwest. *Geoarchaeology: An International Journal* 15:291–313.

Sullivan, Alan P. (editor)

1986 *Prehistory of the Upper Basin, Coconino County, Arizona.* Arizona State Museum Archaeological Series 167. University of Arizona, Tucson.

1998 *Surface Archaeology.* University of New Mexico Press, Albuquerque.

Sullivan, A. P., III, M. E. Becher, and C. E. Downum

1995 Tusayan White Ware Chronology: New Dendrochronological and Archaeological Evidence. *Kiva* 61:175–188.

Sullivan, A. P., III, R. A. Cook, M. P. Purtill, and P. M. Uphus

2001 Economic and Land-Use Implications of Prehistoric Fire-Cracked Rock Piles, Northern Arizona. *Journal of Field Archaeology* 28(3-4):367–382.

Sullivan, A. P., P. B. Mink, and P. M. Uphus

2000 From John W. Powell to Robert C. Euler: Testing Models of Grand Canyon's Prehistoric Puebloan Settlement History. In *Culture and Environment in the American Southwest: Essays in Honor of Robert C. Euler*, edited by D. A. Phillips, Jr., and J. A. Ware, pp. 49–68. Anthropological Papers No. 8 SWCA, Phoenix.

Sullivan, A. P., J. A. Tainter, and D. L. Hardesty

1999 Historical Science, Heritage Resources, and Ecosystem Management. In *Ecological Stewardship: A Common Reference for Ecosystem Management*, Vol. 2, edited by N. C. Johnson, A. J. Malk, R. Szaro, and W. T. Sexton, pp. 493–515. Elsevier Science, Oxford.

Swetnam, Thomas W., and Julio L. Betancourt

1998 Mesoscale Disturbance and Ecological Response to Decadal-Scale Climate Variability in the American Southwest. *Journal of Climate* 11:3128–3147.

Tagg, M.

1999 *Fresnal Shelter.* Archaeology Southwest 13:7.

Tainter, J. A.

1984 Foragers in the Southwestern Forests. In *Stage I Locational Modeling in the Southwestern Region*, edited by L. S. Cordell and D. F. Green, pp. 21–28. Cultural Resources Management Report No. 5. USDA Forest Service, Albuquerque.

Taube, Karl

2000 The Turquoise Hearth: Fire, Self-Sacrifice, and the Central Mexican Cult of War. In *Mesoamerica's Classic Heritage: From Teotihuacan to the Aztecs*, edited by C. Carrasco, L. Jones, and S. Sessions, pp. 269–340. University Press of Colorado, Boulder.

Taylor, W. W.

1958 A Brief Survey Through the Grand Canyon of the Colorado River. In *Two Archaeological Studies in Northern Arizona*, by W. W. Taylor, pp. 18–30. Bulletin No. 30. Museum of Northern Arizona, Flagstaff.

Teague, Lynn S.

1989 The Post-Classic and the Fate of the Hohokam. In *The 1982–1984 Excavations at Las Colinas: Synthesis and Conclusions*, edited by L. S. Teague and William L. Deaver, pp.

145–167. Arizona State Museum Archaeological Series 162, Vol. 6. University of Arizona, Tucson.

1993 Prehistory and the Traditions of the O'odham and Hopi. *Kiva* 58:435–454.

Thomas, D. H.

1983 *The Archaeology of Monitor Valley 1: Epistemology.* Anthropological Papers No. 58. American Museum of Natural History, New York.

Toll, M. S.

1988 Flotation Sampling: Problems and Some Solutions, with Examples from the American Southwest. In *Current Paleoethnobotany: Analytical Methods and Cultural Interpretations of Archaeological Plant Remains,* edited by C. A. Hastorf and V. S. Popper, pp. 36–52. University of Chicago Press, Chicago.

Tramer, Elliot J.

2000 Bird Behavior During a Total Solar Eclipse. *Wilson Bulletin* 112:431–432.

Travis, Scott E.

1990 The Prehistoric Agricultural Landscape of Wupatki National Monument. In *The Wupatki Archaeological Inventory Survey Project: Final Report,* edited by B. Anderson, pp. 4-1–4-54. Southwest Cultural Resources Center, Professional Paper No. 35. National Park Service, Santa Fe.

Tuggle, H. D., and J. J. Reid

2001 Conflict and Defense in the Grasshopper Region of East-Central Arizona. In *Deadly Landscapes: Case Studies in Prehistoric Southwestern Warfare,* edited by G. E. Rice and S. A. LeBlanc, pp. 86–107. University of Utah Press, Salt Lake City.

Tuggle, H. D., J. J. Reid, and R. C. Cole, Jr.

1984 Fourteenth-Century Mogollon Agriculture in the Grasshopper Region of Arizona. In *Prehistoric Agriculture Strategies in the Southwest,* edited by S. F. Fish and P. R. Fish, pp. 101–110. Anthropological Research Papers No. 33. Arizona State University, Tempe.

Turner, C. G., M. H. Regan, and J. D. Irish

1994 Physical Anthropology and Human Taphonomy. In *The Roosevelt Rural Sites Study: Changing Land Use in the Tonto Basin,* edited by R. Ciolek-Torrello and J. R. Welch, pp. 559–589. Technical Series No. 28, Vol. 2. Statistical Research, Tucson.

Turney, O. A.

1929 Prehistoric Irrigation. *Arizona Historical Review* 2(5).

Underhill, Ruth

1946 *Papago Indian Religion.* Columbia University Press, New York.

Underhill, Ruth M., Donald M. Bahr, Baptisto Lopez, Jose Pancho, and David Lopez

1997 *Rainhouse and Ocean: Speeches for the Papago Year.* University of Arizona Press, Tucson.

Upham, S., R. MacNeish, W. Galinant, and C. Stevenson

1987 Evidence Concerning the Origin of Maiz de Ocho. *American Anthropologist* 89:410-419.

Upham, S., R. MacNeish, and C. Stevenson

1988 The Age and Evolutionary Significance of Southwestern Maiz de Ocho. *American Anthropologist* 90:683–684.

U.S. Geological Survey

1960 *Compilation of Records of Surface Waters of the United States Through September 1950, Part 8, Western Gulf of Mexico Basins.* USGS Water-Supply Paper No. 1312. U.S. Government Printing Office, Washington, D.C.

1964 *Compilation of Records of Surface Waters of the United States, October 1950 to September 1960, Part 8, Western Gulf of Mexico Basins.* USGS Water-Supply Paper No. 1732. U.S. Government Printing Office, Washington, D.C.

Van Devender, T. R., and W. G. Spaulding

1979 Development of Vegetation and Climate in the Western United States. *Science* 204:701–710.

Vanin, Gabriele

1999 *Cosmic Phenomena.* Firefly Books, Buffalo, New York.

Vansina, Jan

1985 *Oral Tradition as History.* University of Wisconsin Press, Madison.

Van West, Carla R.

1994a *Modeling Prehistoric Agricultural Productivity in Southwestern Colorado: A GIS Approach.* Reports of Investigations No. 67. Department of Anthropology, Washington State University, Pullman, and Crow Canyon Archaeological Center, Cortez, Colorado.

1994b *River, Rain, or Ruin: Intermittent Prehistoric Land Use Along the Middle Little Colorado River.* Technical Series No. 53. Statistical Research, Tucson.

1996 Agricultural Potential and Carrying Capacity in Southwestern Colorado, A.D. 901–1300. In *The Prehistoric Pueblo World, A.D. 1150-1350*, edited by M. A. Adler, pp. 214–227. University of Arizona Press, Tucson.

Van West, C. R., and J. H. Altschul

1994 Agricultural Productivity and Carrying Capacity in the Tonto Basin. In *The Roosevelt Rural Sites Study: Changing Land Use in the Tonto Basin*, edited by R. Ciolek-Torrello and J. R. Welch, pp. 361–436. Technical Series No. 28, Vol. 2. Statistical Research, Tucson.

Van West, Carla R., and Jeffrey S. Dean

2000 Environmental Characteristics of the A.D. 900-1300 Period in the Central Mesa Verde Region. *Kiva* 66:19–44.

Van West, Carla R., and E. K. Huber

1995 *Data Recovery Plan for Archaeological Investigations in the Fence Lake Transportation Corridor, Arizona.* Technical Report. Statistical Research, Tucson.

Vierra, Brad (editor)

1994 *Archaic Hunter-Gatherer Archaeology in the American Southwest.* Contributions in Anthropology 13, No. 1. Eastern New Mexico University, Portales.

1999 Current Research on the Late Archaic Along the U.S./Mexican Borderlands. Symposium at the 64th Annual Meeting of the Society for American Archaeology, March 26, Chicago.

Vivian, Gordon

1964 *Excavations in a Seventeenth-Century Jumano Pueblo, Gran Quivira, New Mexico.* Archeological Research Series Vol. 8. National Park Service, U.S. Department of the Interior, Washington, D.C.

Vivian, Patricia

1994 Anthropomorphic Figures in Pottery Mound Murals. In *Kachinas in the Pueblo World*, edited by P. Schaafsma, pp. 81–91. University of New Mexico Press, Albuquerque.

Vivian, R. Gwinn

1974 Conservation and Diversion: Water Control Systems in the Anasazi Southwest. In *Irrigation's Impact on Society*, edited by T. Downing and Mc. Gibson. Anthropological Papers of the University of Arizona No. 25. University of Arizona Press, Tucson.

1992 Chacoan Water Management. In *Anasazi Regional Organization and the Chaco System*, edited by David E. Doyel, pp. 45–58. Papers in Anthropology No. 5. Maxwell Museum of Anthropology, Albuquerque.

Wade, Cameron

1997 Total Eclipses of the Sun in Anasazi Country, A.D. 700 to 1700. In *Layers of Time: Papers in Honor of Robert H. Weber*, edited by M. S. Duran and D. T. Kirkpatrick, pp. 99–114. Papers of the Archaeological Society of New Mexico No. 23. Albuquerque.

Wagner, G. E.

1988 Comparability Among Recovery Techniques. In *Current Paleoethnobotany: Analytical Methods and Cultural Interpretations of Archaeological Plant Remains*, edited by C. A. Hastorf and V. S. Popper, pp. 17–35. University of Chicago Press, Chicago.

Wallace, Henry D., James M. Heidke, and William H. Doelle
1995 Hohokam Origins. *Kiva* 60:575–618.
Wallace, Henry D., and James P. Holmlund
1986 *Petroglyphs of the Picacho Mountains, Southcentral Arizona.* Anthropological Papers No. 6. Institute for American Research, Tucson.
Wallace, Henry D., and Michael Lindeman
2000 Foundations of Hohokam Political Organization. Paper presented at the 65th Annual Meeting of the Society for American Archaeology, Philadelphia.
Ward, Albert (editor)
1978 *Limited Occupation and Activity Sites: A Collection of Conference Papers.* Contributions to Anthropological Studies No. 1. Center for Anthropological Studies, Albuquerque.
Waters, Frank
1963 *Book of the Hopi.* Penguin Books, New York.
Waters, Michael R.
1988a Holocene Alluvial Geology and Geoarchaeology of the San Xavier Reach of the Santa Cruz River, Arizona. *Geological Society of America Bulletin* 100:479–491.
1988b The Impact of Fluvial Processes and Landscape Evolution on Archaeological Sites and Settlement Patterns Along the San Xavier Reach of the Santa Cruz River, Arizona. *Geoarchaeology* 3:205–219.
1992 *Principles of Geoarchaeology: A North American Perspective.* University of Arizona Press, Tucson.
1998 The Effect of Landscape and Hydrologic Variables on the Prehistoric Salado: Geoarchaeological Investigations in the Tonto Basin, Arizona. *Geoarchaeology* 13:105–160.
Waters, Michael R., and C. Vance Haynes
2001 Late Quaternary Arroyo Formation and Climate Change in the American Southwest. *Geology* 29(5):399–402.
Waters, Michael R., and John C. Ravesloot
2000 Late Quaternary Geology of the Middle Gila River, Gila River Indian Reservation, Arizona. *Quaternary Research* (54):49–57.
2001 Landscape Change and Cultural Evolution of the Hohokam Along the Middle Gila River and Other River Valleys in South-Central Arizona. *American Antiquity* 66: 285–299.
Watson, R. A., and P. J. Watson
1969 *Man and Nature: An Anthropological Essay in Human Ecology.* Harcourt, New York.
Weiss, H., and H. S. Bradley
2001 Archaeology: What Drives Societal Collapse. *Science* 291:609–610.
Welch, John R.
1991 From Horticulture to Agriculture in the Late Prehistory of the Grasshopper Region, Arizona. In *Mogollon V,* edited by P. Beckett, pp. 75–92. COAS, Las Cruces.
1992 Irrigation Agriculture in the Tonto Basin. In *Proceedings of the Second Salado Conference, Globe, Arizona,* edited by R. C. Lange and S. Germick, pp. 161–171. Occasional Paper 1992. Arizona Archaeological Society, Phoenix.
1994 Archaeological Studies of Agricultural Contexts. In *The Roosevelt Rural Sites Study: Changing Land Use in the Tonto Basin,* edited by R. Ciolek-Torrello and J. R. Welch, pp. 223–252. Technical Series No. 28. Statistical Research, Tucson.
1996 The Archaeological Measures and Social Implications of Agricultural Commitment. Unpublished Ph.D. dissertation, University of Arizona, Tucson.
Welch, John, and Richard Ciolek-Torrello
1994 Analytic Approaches to Ancient Agroecology: Goals and Methods of the Agricultural Field Study. In *The Roosevelt Rural Sites Study: Changing Land Use in the Tonto*

Basin, edited by R. Ciolek-Torrello and J. R. Welch, pp. 41–56. Technical Series No. 28. Statistical Research, Tucson.

Weltfish, Gene

1977 *The Lost Universe: Pawnee Life and Culture*. Reprinted. University of Nebraska Press, Lincoln. Originally published 1965.

Westfall, Deborah A.

1981 *Prehistory of the St. Johns Area, East-Central Arizona: The TEP St. Johns Project*. Arizona State Museum Archaeological Series 153. University of Arizona, Tucson.

Whittaker, J. C.

1984 Arrowheads and Artisans: Stone Tool Manufacture and Individual Variation at Grasshopper Pueblo. Unpublished Ph.D. dissertation, University of Arizona, Tucson.

Whittlesey, S. M.

1993 Subsistence Strategies in Transitional Southwestern Environments. In *Culture and Environment: A Fragile Existence*, edited by R. W. Jamieson, S. Abonyi, and N. A. Mirau, pp. 261–271. University of Calgary Archaeological Association, Calgary.

1994 Ceramics. In *The Roosevelt Rural Sites Study: Changing Land Use in the Tonto Basin*, edited by R. Ciolek-Torrello and J. R. Welch, pp. 377–468. Technical Series No. 28, Vol. 2. Statistical Research, Tucson.

1997 An Overview of Research History and Archaeology of Central Arizona. In *The Vanishing River: Landscapes and Lives of the Lower Verde Valley*, Vol. 1, edited by Stephanie Whittlesey, Richard Ciolek-Torrello, and Jeffrey H. Altschul, pp. 59–142. Statistical Research, Tucson.

Whittlesey, S. M. (editor)

1992 *Investigations at Lee Canyon: Kayenta Anasazi Farmsteads in the Upper Basin, Coconino County, Arizona*. Technical Series No. 38. Statistical Research, Tucson.

Whittlesey, Stephanie, and R. Ciolek-Torrello

1996 The Archaic-Formative Transition in the Tucson Basin. In *Early Formative Adaptations in the Southern Southwest*, edited by B. Roth, pp. 49–64. Monographs in World Archaeology No. 25. Prehistory Press, Madison, Wisconsin.

Wigley, T. M. L., K. R. Briffa, and P. D. Jones

1984 On the Average Value of Correlated Time Series, with Applications in Dendroclimatology and Hydrometeorology. *Journal of Climate and Applied Meteorology* 23:201–213.

Wilcox, David R.

1979 Warfare Implications of Dry-Laid Masonry Walls on Tumamoc Hill. *The Kiva* 45:15–38.

1987 The Evolution of Hohokam Ceremonial Systems. In *Astronomy and Ceremony in the Prehistoric Southwest*, edited by J. Carlson and W. J. Judge, pp. 149–168. Papers in Anthropology No. 2. Maxwell Museum of Anthropology, Albuquerque.

1991a Hohokam Social Complexity. In *Chaco and Hohokam: Prehistoric Regional Systems in the American Southwest*, edited by P. L. Crown and W. J. Judge, pp. 253–275. School of American Research Press, Santa Fe.

1991b The Mesoamerican Ballgame in the American Southwest. In *The Mesoamerican Ballgame*, edited by V. L. Scarborough and D. R. Wilcox, pp. 101–125. University of Arizona Press, Tucson.

Wilcox, David R., and Jonathan Haas

1994 Scream of the Butterfly: Competition and Conflict in the Prehistoric Southwest. In *Themes in Southwest Prehistory*, edited by G. J. Gumerman, pp. 211–238. Advanced Seminar Series. School of American Research Press, Santa Fe.

Wilcox, D. R., T. R. McGuire, and C. Sternberg

1981 *Snaketown Revisited*. Arizona State Museum Archaeological Series 155. University of Arizona, Tucson.

Wilcox, David R., and Lynette O. Shenk

1977 *The Architecture of Casa Grande and Its Interpretation.* Arizona State Museum Archaeological Series 115. University of Arizona, Tucson.

Wilcox, David R., and Charles Sternberg

1981 Additional Studies of the Architecture of the Casa Grande and Its Interpretation. Arizona State Museum Archaeological Series 146. University of Arizona, Tucson.

1983 *Hohokam Ballcourts and Their Interpretation.* Arizona State Museum Archaeological Series 160. University of Arizona, Tucson.

Wilken, Gene C.

1987 *Good Farmers: Traditional Agricultural Resource Management in Mexico and Central America.* University of California Press, Berkeley.

Wills, W.

1985 Early Agriculture in the Mogollon Highland of New Mexico. Unpublished Ph.D. dissertation, University of Michigan, Ann Arbor.

1988a Early Agriculture and Sedentism in the American Southwest: Evidence and Interpretations. *Journal of World Prehistory* 2:445–488.

1988b *Early Prehistoric Agriculture in the American Southwest.* School of American Research Press, Santa Fe.

1992 Plant Cultivation and the Evolution of Risk-Prone Economies in the Prehistoric American Southwest. In *Transitions to Agriculture in Prehistory,* edited by A. B. Gebauer and T. D. Price, pp. 153–176. Monographs in World Archaeology No. 4. Prehistory Press, Madison, Wisconsin.

1995 Archaic Foraging and the Beginning of Food Production in the American Southwest. In *Last Hunters, First Farmers,* edited by T. Price and A. Gebauer, pp. 215–242. School of American Research Press, Santa Fe.

1999 Bat Cave. *Archaeology Southwest* 13:13.

Wills, W., R. Ford, and J. Speth

1984 Preliminary Report on the University of Michigan Investigations at Bat Cave (LA 4935). Ms. on file, Bureau of Land Management, Socorro, New Mexico.

Wills, W. H., and B. B. Huckell

1994 Economic Implications of Changing Land-Use Patterns in the Late Archaic. In *Themes in Southwest Prehistory,* edited by G. J. Gumerman, pp. 33–52. School of American Research Press, Santa Fe.

Wilson, Jack

1976 Proposed Tucson Gas and Electric Company Power Plant near Springerville, Arizona: A Study of Cultural Resources in the Area. Arizona State Museum Archive, University of Arizona, Tucson.

Winter, J. C., and P. F. Hogan

1986 Plant Husbandry in the Great Basin and Adjacent Northern Colorado Plateau. In *Anthropology of the Desert West: Essays in Honor of Jesse D. Jennings,* edited by C. J. Condie and D. D. Fowler, pp. 117–144. Anthropological Papers No. 110. University of Utah Press, Salt Lake City.

Wolf, E.

1982 *Europe and the People Without History.* University of California Press, Berkeley.

Wolman, M. G., and R. Gerson

1978 Relative Scales of Time and Effectiveness of Climate in Watershed Geomorphology. *Earth Surface Processes and Landforms* 3:189–208.

Wonnacott, R. J., and T. H. Wonnacott

1981 *Regression: A Second Course in Statistics.* Wiley, New York.

Wood, J. Scott

1979 Site Morphology and Function on the Planning Unit. In *An Analytical Approach to Cultural Resource Management: The Little Colorado Planning Unit,* edited by Fred Plog, pp. 64–82. Anthropological Research Papers No. 13. Arizona State University, Tempe.

Woodbury, R. B.
1961 *Prehistoric Agriculture at Point of Pines, Arizona.* Memoirs of the Society for American Archaeology No. 17. University of Utah Press, Salt Lake City.
Woodson, Kyle
1999 Migrations in Late Anasazi Prehistory: The Evidence from the Goat Hill Site. *Kiva* 65(1):63–84.
Woosley, Anne I.
1980 Agricultural Diversity in the Prehistoric Southwest. *The Kiva* 45(4):317–337.
Woosley, Anne, and Allan McIntyre
1996 *Mimbres Mogollon Archaeology: Charles C. DiPeso's Excavations at Wind Mountain.* Amerind Foundation Archaeology Series No. 10. University of New Mexico Press, Albuquerque.
Wright, Harold Bell
1929 *Long Ago Told (Huh-kew Ah-kah): Legends of the Papago Indians.* D. Appleton, New York.
Xu, Zhentao, David W. Pankenier, and Yaotiao Jiang
2000 *East Asian Archaeoastronomy: Historical Records of Astronomical Observations of China, Japan, and Korea.* Gordon and Breach, Amsterdam.
Yeomans, Donald K.
1991 *Comets: A Chronological History of Observation, Science, Myth, and Folklore.* Wiley, New York.
Zárate Salmerón, Jerónimo de
1966 *Relaciones, an Account of Things Seen and Learned by Father Jerónimo de Zárate Salmerón from the Year 1538 to Year 1626.* Translated by A. R. Milich. Horn and Wallace, Albuquerque.
Zarbin, Earl
1980 Salt River Valley Canals, 1867–1875. Ms. on file, Salt River Project, Phoenix.
Zirker, J. B.
1995 *Total Eclipses of the Sun.* Princeton University Press, Princeton.
Zubrow, Ezra B. W.
1974 *Population, Contact, and Climate in the New Mexican Pueblos.* Anthropological Papers of the University of Arizona No. 24. University of Arizona Press, Tucson.
Zyniecki, M.
1996 The Chronology of the Polverón Phase. In *The Sky Harbor Project: Early Desert Farming and Irrigation Settlements; Archaeological Investigations in the Phoenix Sky Harbor Center, Vol. 4: Special Studies, Synthesis, and Conclusions,* edited by D. H. Greenwald and J. H. Ballagh, pp. 141–148. Anthropological Research Paper No. 4. SWCA, Flagstaff.

Contributors

Julio L. Betancourt, United States Geological Survey, Tucson, AZ

Darrell Creel, Texas Archaeological Research Laboratory, University of Texas, Austin, TX

Jeffrey S. Dean, Laboratory of Tree-Ring Research, University of Arizona, Tucson, AZ

David E. Doyel, Estrella Cultural Research, Phoenix, AZ

Fred Espenak, Laboratory for Extraterrestrial Physics, NASA Goddard Space Flight Center, HI

Paul R. Fish, Arizona State Museum, University of Arizona, Tucson, AZ

Suzanne K. Fish, Arizona State Museum, University of Arizona, Tucson, AZ

Gary S. Funkhouser, Laboratory of Tree-Ring Research, University of Arizona, Tucson, AZ

Donald A. Graybill, Laboratory of Tree-Ring Research, University of Arizona, Tucson, AZ

David A. Gregory, Desert Archaeology, Inc., Tucson, AZ

George J. Gumerman, School of American Research and Santa Fe Institute, Santa Fe, NM

W. Bruce Masse, Los Alamos National Laboratory, Los Alamos, NM

Fred L. Nials, Desert Archaeology, Inc., Tucson, AZ

James A. Parks, Laboratory of Tree-Ring Research, University of Arizona, Tucson, AZ

J. Jefferson Reid, Department of Anthropology, University of Arizona, Tucson, AZ

Anthony H. Ruter, Department of Anthropology and Center for Climatic Research, University of Wisconsin, Madison, WI

Ann Clair Seiferle-Valencia, Department of Anthropology, Harvard University, Cambridge, MA

Alan H. Simmons, Department of Anthropology, University of Nevada, Las Vegas, NV

Alan P. Sullivan III, Department of Anthropology, University of Cincinnati, OH

Michael R. Waters, Department of Anthropology, Texas A&M University, College Station, TX

Index